Collins

CONCISE REVISION COURSE
CSEC®
Integrated Science

Anne Tindale and Peter DeFreitas
Reviewers: Naresh Birju and Shaun deSouza

William Collins' dream of knowledge for all began with the publication of his first book in 1819. A self-educated mill worker, he not only enriched millions of lives, but also founded a flourishing publishing house. Today, staying true to this spirit, Collins books are packed with inspiration, innovation and practical expertise. They place you at the centre of a world of possibility and give you exactly what you need to explore it.

Published by Collins
An imprint of HarperCollins*Publishers*
The News Building
1 London Bridge Street
London
SE1 9GF
UK

HarperCollins*Publishers*
Macken House
39/40 Mayor Street Upper
Dublin 1
D01 C9W8
Ireland

Browse the complete Collins Caribbean catalogue at
collins.co.uk/caribbeanschools

© HarperCollins*Publishers* Limited 2026

10 9 8 7 6 5 4 3 2 1

ISBN 978-0-00-876844-7

Collins CSEC® Integrated Science is an independent publication and has not been authorised, sponsored or otherwise approved by **CXC**®. CSEC® is a registered trademark of the ***Caribbean Examinations Council (CXC®)***.

All rights reserved. No part of this book may be reproduced, stored in a retrieval system, or transmitted in any form or by any means, electronic, mechanical, photocopying, recording or otherwise, without the prior permission in writing of the Publisher. This book is sold subject to the conditions that it shall not, by way of trade or otherwise, be lent, re-sold, hired out or otherwise circulated without the Publisher's prior consent in any form of binding or cover other than that in which it is published and without a similar condition including this condition being imposed on the subsequent purchaser.

Without limiting the exclusive rights of any author, contributor or the publisher of this publication, any unauthorised use of this publication to train generative artificial intelligence (AI) technologies is expressly prohibited. HarperCollins also exercise their rights under Article 4(3) of the Digital Single Market Directive 2019/790 and expressly reserve this publication from the text and data mining exception.

Entered words that we have reason to believe constitute trademarks have been designated as such. However, neither the presence nor absence of such designation should be regarded as affecting the legal status of any trademark.

The contents of this publication are believed correct at the time of printing. Nevertheless, the Publisher can accept no responsibility for errors or omissions, changes in the detail given or for any expense or loss thereby caused. HarperCollins does not warrant that any website mentioned in this title will be provided uninterrupted, that any website will be error-free, that defects will be corrected, or that the website or the server that makes it available are free of viruses or bugs. For full terms and conditions please refer to the site terms provided on the website.

British Library Cataloguing in Publication Data.

A CIP record of this book is available from the British Library.

Authors: Anne Tindale and Peter DeFreitas
Reviewers: Naresh Birju and Shaun deSouza
Publisher: Catherine Martin
Product manager: Saaleh Patel
Project manager: Julianna Dunn
Copy editor: Aidan Gill
Proofreader: Mitch Fitton
Illustrator: Ann Paganuzzi
Production controller: Alhady Ali
Typesetter: Six Red Marbles
Cover designer: Gordon MacGilp
Cover image: Gordon MacGilp
Printed and bound by Martins the Printers in the UK.

p65 Alfred Pasieka/Science Photo Library, p107 Phanie – Sipa Press/Alamy, p205tc G.I. Dobner/Alamy, p230 Science History Images/Alamy, p234c and r Anne Tindale, p235r Anne Tindale.

All other photos © Shutterstock.com

Contents

The pathway to success — v

Module 1 – Organisms and life processes

1 Units of life — 1
Cells — 1
Movement of substances into and out of cells — 3

2 Reproduction and growth in plants — 7
Asexual and sexual reproduction compared — 7
Asexual reproduction in plants — 9
Sexual reproduction in flowering plants — 11
Growth in plants and crop production — 14
Soil — 17

3 Reproduction and growth in animals — 23
Asexual reproduction in animals — 23
Sexual reproduction in humans — 24
Growth in animals — 33

4 Transport systems — 37
Transport systems in multicellular organisms — 37
The circulatory system in humans — 38
Transport systems in plants — 43

5 Excretion — 46
Excretion and egestion in living organisms — 46
The kidneys — 47
The skin — 49
Excretion in flowering plants — 51

6 Sense organs and coordination — 53
Sense organs in humans — 53
The eye — 53
The ear — 57
The nervous system — 59
The endocrine (hormonal) system — 62

7 Health — 65
Microbes — 65
Communicable or infectious diseases — 66
The immune system and immunity — 68
Non-communicable or non-infectious diseases — 70
Hygiene — 73
Drug use and abuse — 75
Pests and pest control — 76
Food contaminants and the growth of microorganisms — 80
Food preservation — 81

Exam-style questions — 84

Module 2 – Energy

8 Conservation of energy — 89
Energy, work and change — 89
Photosynthesis and energy transfer in the environment — 98

9 Energy in life processes — 103
The human diet — 103
Digestion in humans — 107
Respiration — 113
Breathing — 115
Gaseous exchange and smoking — 118

10 Fossil fuels and alternative sources of energy — 122
Fossil fuels — 122
Non-renewable, renewable and alternative sources of energy — 127

11 Electricity and lighting — 134
Electrical components and circuits — 134
Electricity in the home — 138
Energy conservation measures — 143
Electrical and fire hazards — 146

12 Temperature control and ventilation — 153
Processes of heat transfer — 153
Temperature and thermometers — 159
Temperature regulation in humans — 161
Ventilation — 161

Exam-style questions — 165

Module 3 – Our planet

13 The universe and our solar system 170
- Celestial bodies 170
- Human exploration of the universe 174

14 The terrestrial environment 180
- The Earth's atmosphere and weather 180
- Tides and their effects 185
- Volcanoes and earthquakes 187

15 Water and the aquatic environment 192
- Properties of water 192
- Hard and soft water 193
- Uses of water 195
- Fishing methods used in the Caribbean 196
- Water pollution 197
- Purifying water for domestic use 198
- Conditions for flotation 200
- Navigational devices used at sea 203
- Water safety devices 204
- Effects of diving on the human body 205

16 Forces 208
- Types of forces and their effects 208
- Linear momentum 214
- Understanding machines 215
- The human skeleton 220

17 Metals and non-metals 225
- Properties and uses of metals 225
- Properties and uses of non-metallic materials 226
- The reactivity of metals 228
- Aluminium cooking and canning utensils 229
- Alloys in the home and workplace 229
- Tarnishing and rusting 230

18 Household chemicals 233
- The uses of common household chemicals 233
- Properties of acids, bases and salts 236
- The effects of cleaning agents on household appliances 237
- Soapy and soapless detergents 239
- The states of matter 241
- The properties of mixtures 243
- Separation techniques 245

19 Pollutants and the environment 248
- Air pollution and its effects 248
- Community hygiene 249
- Plastics 252

Exam-style questions 255

Index 261

The pathway to success

About this book

This book has been written primarily as a **revision course** for students studying for the CXC® CSEC® Integrated Science examination. The facts are presented **concisely** using a variety of formats which makes them **easy to understand** and **learn**.

Key terms are highlighted in **bold** type and important **definitions** which must be learnt are written in *italics* and highlighted in colour. **Annotated diagrams** and **tables** have been used wherever possible, **worked examples** have been given where appropriate and the relationship between **structure** and **function** is continually emphasised. **Diagrams** marked with a **star** (★) are specifically identified in the syllabus as ones that should be known, and the important labels are highlighted in **bold** type. **Questions** to help test knowledge and understanding, and to provide practice for the actual examination, are included throughout the book.

The following sections provide **valuable information** on the format of the CSEC® examination, how to revise successfully, successful examination technique, key instruction words used on examination papers, drawing graphs and the School-Based Assessment.

The CSEC® Integrated Science syllabus and this book

The **CSEC® Integrated Science syllabus** is available online at **http://cxc-store.com**. The syllabus is divided into **three modules** and each module is divided into several distinct **topics**. Each chapter in **this book** covers a particular topic in the syllabus. You are strongly advised to read through the syllabus carefully since it provides detailed information on the specific objectives of each topic and the format of the CSEC® examination.

- **Chapters 1 to 7** cover topics in **Module 1 – Organisms and life processes**.
- **Chapters 8 to 12** cover topics in **Module 2 – Energy**.
- **Chapters 13 to 19** cover topics in **Module 3 – Our planet**.

At the end of each chapter, or section within a chapter, you will find a selection of **revision questions**. These questions test your **knowledge** and **understanding** of the topic covered in the chapter or section. At the end of Chapters 7, 12 and 19, you will find a selection of **exam-style questions** which also test how you **apply** the knowledge you have gained and help prepare you to answer the style of questions that you will encounter in your CSEC® examination. You will find the answers to all these questions online at **www.collins.co.uk/caribbeanschools**.

The format of the CSEC® Integrated Science examination

The examination consists of **two written papers**: **Paper 01**, an objective-type paper, and **Paper 02**, a structured-type paper, and the **School-Based Assessment (SBA)** (Paper 031 – see page x) or its **alternative**, a **practical examination** for private candidates (Paper 032). Your performance is evaluated using the following **three** profiles:

- **Knowledge and Comprehension**
- **Use of Knowledge**
- **Experimental Skills**.

You have **three options** for sitting **Paper 01** and **Paper 02**.

Option A – Regular Sitting

- **Paper 01** (1 hour 15 minutes) consists of **60 multiple choice questions**, 20 from **each module**.
- **Paper 02** (2 hours 30 minutes) consists of **six compulsory structured type questions**, two from **each module**.

Option B – Modular Sitting (One Module)

- **Paper 01 (25 minutes)** consists of **20 multiple choice questions** from **one selected module**.
- **Paper 02 (50 minutes)** consists of **two compulsory structured-type questions** from the **same selected module**.

Option C – Modular Sitting (Two Modules)

- **Paper 01 (50 minutes)** consists of **40 multiple choice questions** from **two selected modules**, 20 from **each module**.
- **Paper 02 (1 hour 40 minutes)** consists of **four compulsory structured-type questions** from the **same two selected modules**, **two** from **each module**.

You will be required to submit your **SBA** or complete its **alternative**, Paper 032, at your **first sitting** regardless of the option chosen.

If you chose to use the **modular approach** by attempting **one module** at each sitting (Option B), or **one module** at one sitting (Option B) and **two modules** at another sitting (Option C), you must successfully complete **all modules** within a **four-year period** to be awarded the **full CSEC® award**. You may **reuse** the **score** that you obtained from your moderated **SBA** or its **alternative** at your first sitting when attempting the other module or modules. Alternatively, you may choose to **redo** your SBA or **resit** its alternative if you wish to improve your score, since that score accounts for **21%** of your final mark for **each option**.

Paper 01

Paper 01 is worth **29%** of your final examination mark for each option. It consists of a **maximum of 60 multiple choice questions** drawn from **all** areas of the syllabus and organised under the **three modules**, with **20 questions** from **each module**. Each question is worth **1 mark**. Four **choices** of answer are provided for each question, of which one is correct.

- Make sure you read each question **thoroughly**; some questions may ask which answer is **incorrect**.
- Some questions may give two or more correct answers and ask which one is the **best** or **most suitable**; you must consider each answer very carefully before making your choice.
- If you do not know the answer, try to work it out by **eliminating** the incorrect answers. Never leave a question unanswered.

Paper 02

Paper 02 is worth **50%** of your final examination mark for each option. It consists of a **maximum** of **six compulsory structured-type questions** which test information drawn from **all** areas of the syllabus and it is organised into **three sections**, with each section containing **two** questions from a **single module**. A question may require knowledge of **several topics** within the module. Each question is divided into several parts and the answers should be written in **spaces** provided in the answer booklet. These spaces indicate the length of answer required and answers should be restricted to them. Take time to **read the entire paper** before beginning to answer any of the questions.

- The **first question** from **each module** has a **practical/investigative component** and is worth **20 marks**. You can expect to be provided with some form of **data** or **information** that you are required to analyse and answer questions about. This might be in the form of a table or a graph. If you are given a table, you may be asked to draw a **graph** using the data and then may be asked questions about the graph. Make sure you know how to draw graphs (see pages ix to x). You may also be asked to perform calculations, identify patterns, trends or relationships, make predictions and draw conclusions.
- The **second question** from **each module** is worth **15 marks**, and it may contain some kind of **stimulus material**, such as diagrams, which you will be asked questions about.

The marks allocated for the different parts of each question are clearly given. A total of **35 marks** is available for **each module** and the **time** allowed for each is **50 minutes**. For each module, you should allow about **25 minutes** for the first question and no more than **20 minutes** for the second question. This will allow you time to read the paper fully before you begin and time to check over your answers when you have finished.

Successful revision

The following should provide a guide for **successful revision**.

- **Begin your revision early**. You should start your revision at least two months before the examination and should plan a **revision timetable** to cover this period. Plan to revise in the evenings when you do not have much homework, at weekends, during the Easter vacation and during study leave.
- When you have a **full day** available for revision, consider the day as three sessions of about three to four hours each, **morning**, **afternoon** and **evening**. Study during two of these sessions only and do something non-academic and relaxing during the third.
- **Read through the topic** you plan to learn to make sure you **understand** it before starting to learn it; understanding is a lot safer than thoughtless learning.
- Try to understand and learn **one topic** in each revision session, more if topics are short and fewer if topics are long.
- **Revise every topic** in each module. Do not pick and choose topics since **all questions** on your examination paper are **compulsory**.
- **Learn the topics in order**. When you have learnt **all** the topics you have to learn **once**, go back to the first topic and begin again. Try to cover each topic **several times**.
- **Revise in a quiet location** without any form of distraction.
- **Sit up to revise**, preferably at a table. Do not sit in a comfy chair or lie on a bed where you can easily fall asleep.
- Obtain copies of **past CSEC® Integrated Science examination papers** and use them to practise answering exam-style questions, starting with the most recent papers. These can be purchased online from the CXC® Store.
- You can use a variety of different **methods** to **learn** your work. Chose which ones work best for you.
    - **Read the topic several times**, then close the book and try to write down the **main points**. Do not try to memorise your work 'word for word' since work learnt by heart is not usually understood, and questions test your **understanding** as well as your ability to repeat facts.
    - **Summarise** the **main points** of each topic on **flash cards** and use these to help you study.
    - **Draw simple diagrams** with **annotations**, **spider diagrams** and **flow charts** to summarise topics in visual ways that are easy to learn.
    - **Practise labelling diagrams** that you have been given. You may be asked to do this in your exam.
    - **Use memory aids** such as:
        - **acronyms**, e.g. **R**oy **G** **B**iv for the seven colours of the visible spectrum of light; **r**ed, **o**range, **y**ellow, **g**reen, **b**lue, **i**ndigo, **v**iolet.
        - **mnemonic phrases**, e.g. '**m**y **v**ery **e**ducated **m**other **j**ust **s**erved **u**s **n**achos' for the order of the planets in our solar system moving outwards from the Sun: Mercury, Venus, Earth, Mars, Jupiter, Saturn, Uranus, Neptune.
        - **associations between words**, e.g. t**ri**cuspid – **ri**ght (therefore the bicuspid valve must be on the left side of the heart), **a**rteries – **a**way (therefore veins must take blood towards the heart).
    - **Test yourself** using the questions throughout this book and others from past CSEC® examination papers.

Successful examination technique

- **Read the instructions** at the start of each paper very carefully and do **precisely** what they require.
- **Read through the entire paper** before you begin to answer any of the questions.
- **Read each question at least twice** before beginning your answer to ensure you clearly **understand** what is being asked.
- **Study diagrams**, **graphs** and **tables** in detail and make sure that you **understand** the information they are giving before answering the questions that follow.
- **Underline the important words** in each question to help you answer precisely what the question is asking.
- **Reread** the question when you are **part way through** your answer to check that you are answering what is being asked.
- Look at the **number of marks** allocated for each part of a question and make sure you include at least as many **points** in your answer as there are **marks**.
- **Give precise** and **factual answers**. You will not get marks for information which is 'padded out' or irrelevant. The number of marks awarded for each answer indicates how long and detailed it should be.
- **Use correct scientific terminology** throughout your answers.
- Give any **numerical answer** the appropriate **unit** using the proper abbreviation/symbol e.g. cm^3, g, °C.
- If a question asks you to give a **specific number of points**, use a **bullet point** for each to make the separate points clear.
- If you are asked to give **similarities** and **differences**, you must make it clear which points you are proposing as similarities and which points as differences. The same applies if you are asked to give **advantages** and **disadvantages**.
- **Watch the time** as you work. Know the time available for each question and stick to it.
- **Check over your answers** when you have completed all the questions.
- **Remain in the examination room** until the **end** of the examination and recheck your answers again if you have time to ensure you have done your very best. Never leave the examination room early.

Key instruction words used on examination papers

It is essential that you **fully understand** what each question is **asking you to do** before you begin to answer. The following **key instruction words** tell you the **type of detail** that you should give in your answers.

Account for: provide reasons for the information given.

Annotate: add brief notes to labels.

Apply: use knowledge and principles to solve problems.

Appraise: judge the quality or worth of the topic or issue in question.

Assess: give reasons for the importance of particular structures, relationships or processes.

Calculate: arrive at a solution to a numerical problem.

Classify: divide into groups based on observable characteristics.

Comment: state an opinion or view, giving supporting reasons.

Compare: give similarities and differences.

Construct: draw a line graph, bar graph, histogram, pie chart or table using data provided or obtained from practical investigations.

Contrast: give differences.

Deduce: make a logical connection between two or more pieces of information, or use data provided or obtained to arrive at a conclusion.

Define: state concisely the meaning of a word or term.

Demonstrate: show clearly by giving proof or evidence.

Derive: use data to determine logically a relationship, formula or result.

Describe: provide a detailed account, which includes all relevant information.

Determine: find the value of a physical quantity.

Design: plan and present ideas in a structured manner with relevant practical detail.

Develop: expand or elaborate on an idea or argument with supporting reasons.

Differentiate or **distinguish between** or **among**: give differences between or among items that place them into separate categories.

Discuss: provide a balanced argument, which considers points both for and against, or explain the relative merits of a case.

Draw: produce a line representation of a specimen or apparatus that accurately shows the relationship between the parts.

Estimate: arrive at an approximate quantitative result.

Evaluate: determine the significance or worth of the point in question.

Explain: give a clear, detailed account which makes the given information easy to understand and provides reasons for the information.

Formulate: develop and present ideas in a structured manner.

Identify: name or point out specific components or features.

Illustrate: make the answer clearer by including appropriate examples or diagrams.

Justify: provide adequate grounds for your reasoning.

Label: add names to identify structures or parts indicated by label lines or pointers.

List: itemise without detail.

Name: give only the name.

Outline: write an account which includes the main points only.

Predict: use information provided to arrive at a likely conclusion or suggest a possible outcome.

Relate: show connections between different sets of information or data.

State: give brief, precise facts without detail.

Suggest: put forward an idea.

Drawing graphs

Graphs are used to display numerical data. When drawing a graph:

- Use a **sharp HB pencil**, preferably a mechanical pencil with a 0.5 mm lead.
- Plot the **manipulated variable** on the *x*-axis (horizontal axis) and the **responding variable** on the *y*-axis (vertical axis).
  - The **manipulated variable** is the factor that is **varied** or **adjusted** by the person carrying out the investigation. It **affects** the responding variable and is given in the **left column** of the table of data.
  - The **responding variable** is the factor that **changes** as a result of the changes made to the manipulated variable. It is **affected by** the manipulated variable and is given in the **right column** of the table of data.

- Choose appropriate **scales** that are easy to work with, and which use more than half of the graph grid in both the *x*- and *y*-directions. Avoid using scales having multiples of 3, 7 or 9, which make plotting difficult.
- **Label** each axis with its correct **quantity** and **unit**, if any. To do this, use the **column headings** in the table of data.
- Give a **key** if more than one set of data is being plotted on the same axes.
- Give the graph an appropriate, underlined **title** at the top, which must include reference to the responding variable and the manipulated variable.
- When drawing a **line graph**:
    - Plot each **data point** accurately using a **small dot** surrounded by a circle or triangle, e.g. ⊙, to locate it. Alternatively a small cross (X) may be used.
    - Draw a **smooth curve** or **straight line of best fit** passing between the data points such that the mean deviation from the points is minimised.
- When drawing a **bar graph**, the height of each bar indicates the value of the responding variable:
    - Draw **vertical bars** of equal width and draw an accurately positioned **horizontal line** to show the top of each bar.
    - Ensure that a **space** is left between the *y*-axis and the first bar, and **spaces** of equal width are left between each of the bars.

School-Based Assessment (SBA)

School-Based Assessment (SBA) is an integral part of your CXC® CSEC® examination and is worth **21%** of your final examination mark. It assesses you in the **experimental skills** and the **analysis and interpretation skills** that are involved in laboratory and field work.

- The assessments are carried out at your school by **your teacher** during Terms 1 to 5 of your two-year programme.
- The assessments are carried out during **normal practical classes** and not under examination conditions. You have every opportunity to gain a high score in each assessment if you make a **consistent effort** throughout your two-year programme.
- Assessments are made of the following **five skills**:
    - Manipulation and Measurement
    - Observation, Recording and Reporting
    - Drawing
    - Planning and Designing
    - Analysis and Interpretation.
- Each skill will be assessed at **least twice** over the two-year period with the exception of Drawing, which will only be assessed once. You will be awarded a mark between **0** and **10** for each assessment made.
- You will be **taught** the skills and be given enough opportunity to **develop** them before you are assessed. You will do a minimum of **eighteen** practical experiments over the two-year period.
- As an integral part of your SBA, you will also carry out an **Investigative Project** during the second year of your two-year programme. This project assesses your **Planning and Designing,** and your **Analysis and Interpretation skills.**
- All your experimental reports are recorded in a **practical notebook**, which is subject to moderation by the CXC® Examination Board to assess the standard of marking in your school.

Module 1 – Organisms and life processes

1 Units of life

All living organisms are made of **cells**. Cells are so small that they can only be seen with a microscope and not with the naked eye. Substances move into and out of cells, and from one cell to another, by three processes: **diffusion**, **osmosis** and **active transport**.

Cells

A **cell** is the smallest unit of **life**. It is the basic structural and functional unit of all living organisms. Some organisms are **unicellular**, being composed of a single cell only, whilst others are **multicellular**, being composed of many cells. The latter includes **plants** and **animals**.

Animal and plant cells

Both **animal** and **plant cells** are surrounded by an extremely thin outer layer known as the **cell membrane**, which has jelly-like **cytoplasm** inside. The cytoplasm contains structures called **organelles**, which are specialised to carry out one or more vital functions. Organelles include the nucleus, mitochondria, ribosomes, chloroplasts and vacuoles. Most organelles are surrounded by one or two **membranes**. Cells and most organelles can be seen under a **light microscope**; however, the detailed structure of organelles can only be seen under an **electron microscope**.

The following are found in **both** animal and plant cells.

- A **cell membrane** or **plasma membrane**.
- **Cytoplasm**.
- A **nucleus**.
- **Mitochondria** (singular: **mitochondrion**).
- **Ribosomes**.

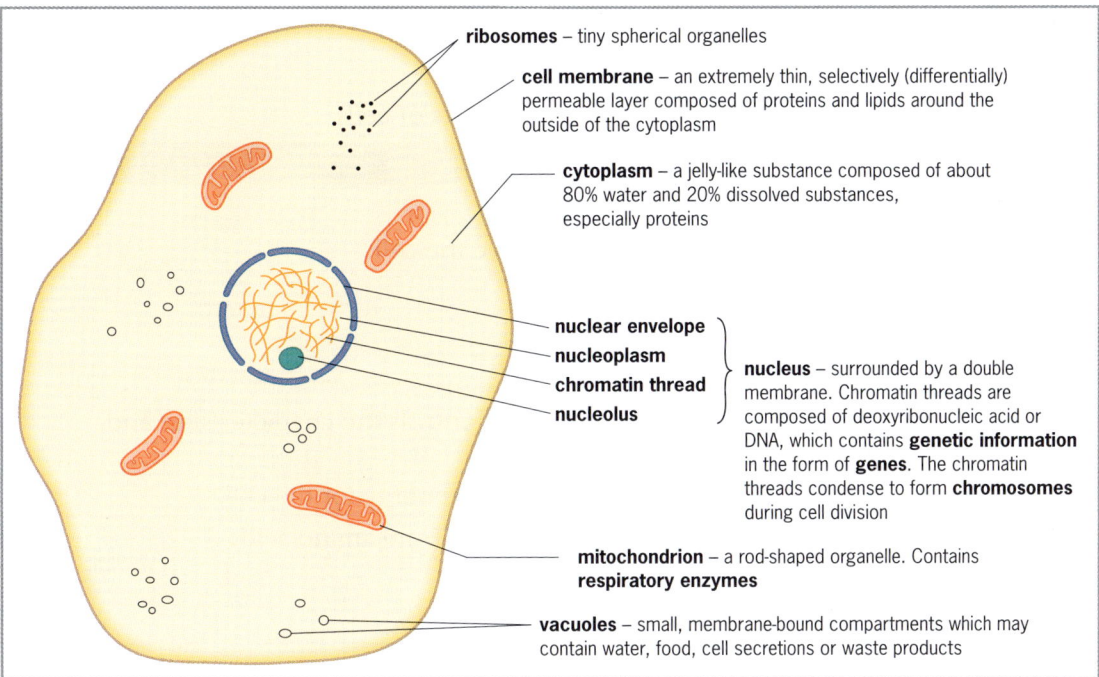

★**Figure 1.1** *Structure of a generalised animal cell*

In addition to the five components listed on page 1, **plant cells** also possess the following.
- A **cell wall**.
- **Chloroplasts**.
- A large **vacuole**.

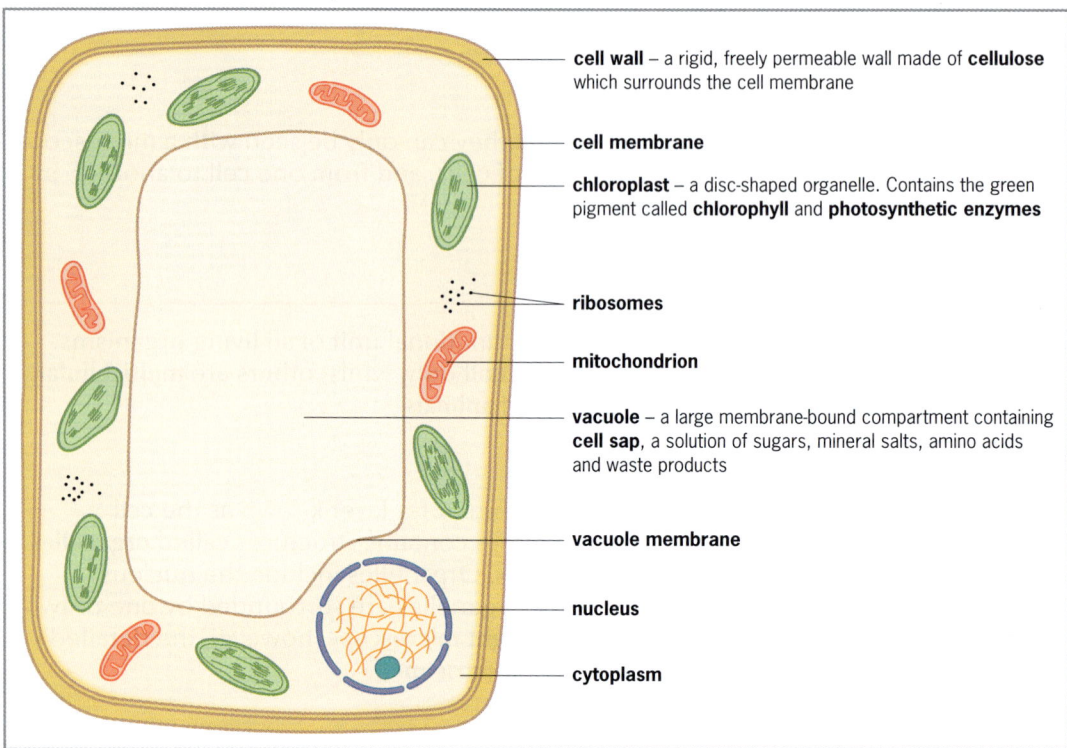

★Figure 1.2 *Structure of a generalised plant cell*

It is important to note that **plant cells** have **regular shapes**, usually round, square or rectangular, whereas **animal cells** can have a **variety of shapes**.

Table 1.1 *A summary of the functions of the different components of cells*

Cell component	Function
Cell membrane	• Keeps the cell contents inside the cell. • Controls what substances enter and leave the cell.
Cytoplasm	• Supports the organelles. • The site of many chemical reactions.
Nucleus	• Essential for cell division. • **Genetic information** in the form of **genes** controls the characteristics and functioning of the cell.
Mitochondrion	• Where **aerobic respiration** occurs to release energy for the cell.
Ribosome	• Where proteins are synthesised (produced) from amino acids.
Vacuole	• Stores food, cell secretions or cell waste. • Supports a **plant cell** when it is turgid (firm).
Cell wall	• Supports and protects a **plant cell** and gives it shape.
Chloroplast	• Where **photosynthesis** occurs in **plant cells** to produce food for the plant.

Movement of substances into and out of cells

Substances can move into and out of cells, and from cell to cell by **three** different processes: **diffusion**, **osmosis** and **active transport**. Diffusion and osmosis are **passive** processes because they do not require energy released in respiration. Active transport is an **active** process because it requires energy released in respiration.

Diffusion

Diffusion is the net movement of particles down a concentration gradient from an area of higher concentration to an area of lower concentration until the particles are evenly distributed.

Concentration gradient is the difference in the concentration of a substance between two areas.

The particles, which can be **molecules** or **ions**, are said to move **down a concentration gradient** because they are moving from a region of higher concentration to one of lower concentration. Particles in **gases**, **liquids** and **solutions** are capable of diffusing because they are in constant motion.

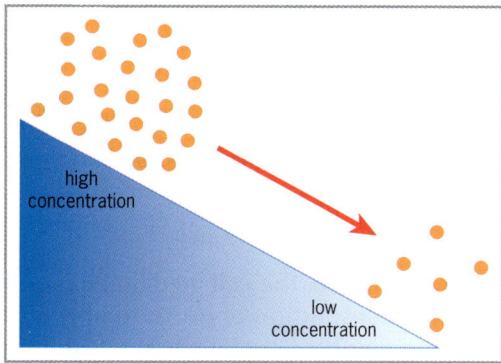

Figure 1.3 *Diffusion of molecules down a concentration gradient*

The role of diffusion in living organisms

Diffusion is **important** in living organisms because it is the way cells obtain many of their requirements and get rid of their waste products which, if not removed, would poison them.

- **Oxygen**, for use in **aerobic respiration**, moves into cells by diffusion, and **carbon dioxide**, produced in **aerobic respiration**, moves out of cells by diffusion.
- Some of the **glucose** and **amino acids** produced in **digestion** are absorbed through the cells in the walls of the small intestine and blood capillaries, and into the blood by diffusion.
- **Carbon dioxide**, for use by plants in **photosynthesis**, moves into leaves and leaf cells by diffusion, and **oxygen**, produced by plants in **photosynthesis**, moves out of leaf cells and leaves by diffusion.

Osmosis

Osmosis is the movement of water molecules through a selectively or differentially permeable membrane from a solution containing a lot of water molecules; e.g. a dilute solution (or water), to a solution containing fewer water molecules; e.g. a concentrated solution.

Osmosis is a special form of diffusion. Only **water molecules** move by osmosis, and they always move from a solution with a **higher water content** (a more dilute solution) into a solution with a **lower water content** (a more concentrated solution) through a **selectively** or **differentially permeable membrane**.

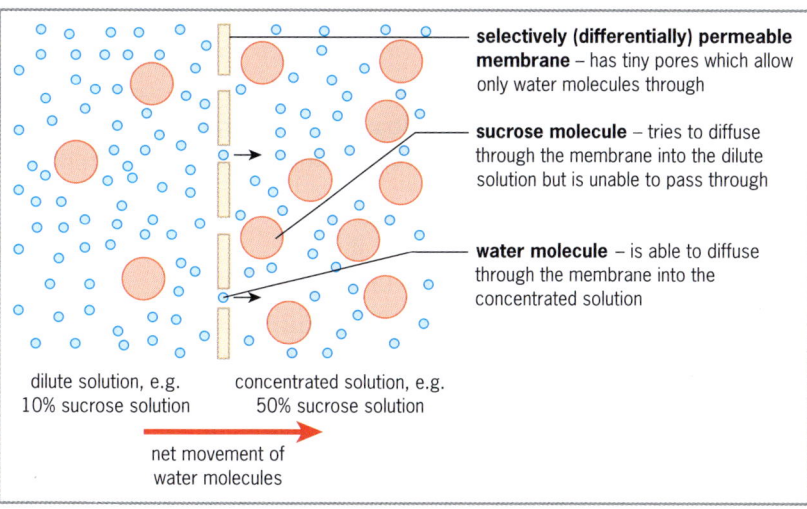

Figure 1.4 *Explanation of osmosis*

1 Units of life 3

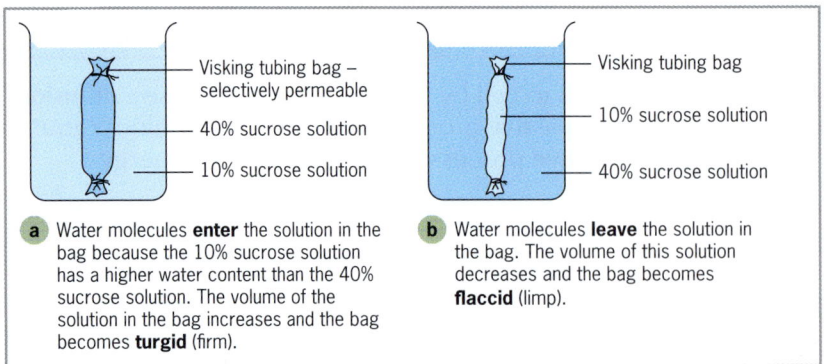

Figure 1.5 *Demonstrating osmosis using Visking tubing*

Osmosis in living cells

In any **living cell**, the **cell membrane** is selectively or differentially permeable. There is always **cytoplasm**, a solution of proteins and other substances in water, on the inside of the membrane and usually a solution, such as tissue fluid, on the outside. **Water molecules**, therefore, move into and out of living cells by **osmosis**. It is important to note that osmosis does **not** occur in **dead cells**.

The role of osmosis in living organisms

Osmosis is **important** in living organisms because it is the way **water** enters cells.
- **Water** moves into animal cells from blood plasma and body fluids by osmosis. This keeps the cells **hydrated**.
- **Water** is absorbed from the intestines into the blood by osmosis. This ensures the body obtains the water it needs from food and drink consumed.
- **Water** is reabsorbed from the filtrate in the kidney tubules into the blood by osmosis. This prevents the body from losing too much water.
- **Water** is absorbed from the soil by the root hairs of plant roots, and then moves into and through the cells of roots and leaves by osmosis. This keeps water **moving** through plants and it ensures that the leaves get water for **photosynthesis**.
- **Water** that moves into plant cells by osmosis keeps the cells **turgid**. This causes non-woody stems to stand upright and keeps leaves firm.

Active transport

Active transport is the movement of particles through cell membranes against a concentration gradient using energy released in respiration.

During active transport, **energy** released in respiration is used to move molecules and ions through cell membranes from areas of **lower** concentration, outside cells, to areas of **higher** concentration, inside cells.

The role of active transport in living organisms

Active transport is **important** in living organisms because it allows cells to accumulate high concentrations of important substances; e.g. glucose, amino acids and various ions.
- Some of the **glucose** and **amino acids** produced in **digestion** are absorbed from the small intestine into the blood by active transport.

- **Useful substances** such as glucose, amino acids, hormones and vitamins are reabsorbed from the filtrate in the kidney tubules into the blood by active transport. This prevents the loss of these substances from the body.
- **Mineral ions** move from the soil into the cells of plant roots by active transport.

Examples of diffusion in natural and artificial environments

- **Ash**, produced during a **volcanic eruption**, can **diffuse** long distances in ash clouds from the site of the eruption.
- **Smog**, a mixture of nitrogen oxides, sulfur oxides, carbon oxides, ozone, volatile organic compounds and fine particulate matter, can be formed from the **emissions** from the **exhausts** of vehicles and **industries** that burn fossil fuels. Smog **diffuses** from its source into the atmosphere, creating a hazy fog.
- **Smoke**, produced when materials **burn**, contains a variety of gases and fine particulate matter, especially carbon particles, which **diffuse** from the fire into the surrounding atmosphere.

Figure 1.6 *Smog formed from the exhausts of vehicles*

- **Perfume** contains molecules that produce a scent. When perfume is applied to the skin, these molecules evaporate and **diffuse** into the surrounding air where they can be smelt.

Revision questions

1. What is a cell?
2. Draw a simple labelled diagram to show the structure of an unspecialised plant cell.
3. Describe the structure and outline the function of EACH of the following cell structures:
   - **a** a mitochondrion
   - **b** ribosomes
   - **c** the cell membrane
   - **d** a chloroplast
   - **e** the cell wall
   - **f** vacuoles.
4. What will happen to a cell if its nucleus is removed?
5. Construct a table to give FOUR differences between the structure of a typical animal cell and the structure of a typical plant cell.
6. Provide a definition for EACH of the following terms:
   - **a** diffusion
   - **b** osmosis
   - **c** active transport.
7. **a** What is meant by the term 'concentration gradient'?
   b Distinguish between particles moving down a concentration gradient and moving against a concentration gradient.
   c Why is it important that active transport occurring in cells enables particles to move against a concentration gradient?

8. Give THREE reasons why diffusion is important to living organisms and FOUR reasons why osmosis is important to living organisms.

9. Why is the root of a plant unable to absorb mineral ions from the soil if it is given a poison that prevents respiration?

10. Provide THREE examples of diffusion occurring in natural or artificial environments.

2 Reproduction and growth in plants

All plants must **produce offspring** in order for their species to survive. They achieve this by reproduction of which there are two types: **asexual reproduction**, which requires only one parent, and **sexual reproduction**, which requires two parents. Plants also **grow**, and as they do so, they **develop**. Growth ensures that the plant is the correct size to survive in its environment.

Reproduction is the process by which living organisms generate new individuals of the same kind as themselves.

Growth is a permanent increase in the size of an organism.

Asexual and sexual reproduction compared

Asexual reproduction

Asexual reproduction involves **one** parent and offspring are produced by a type of **cell division** known as **mitosis**. All offspring produced asexually from one parent are **genetically identical**, i.e. they do not show variation, and are collectively called a **clone**. The process is **rapid** and **energy requirements** are **low** because it does not involve gamete production, finding a mate, fertilisation and embryo development. Asexual reproduction occurs in unicellular organisms such as bacteria and protozoans, e.g. an amoeba. It also occurs in fungi, some plants and a few animals.

The role of mitosis in asexual reproduction

Mitosis ensures that each daughter (new) cell receives the **same number** and **type** of chromosomes as the parent cell and that the cells are, therefore, **genetically identical** to each other and to the original parent cell. During mitosis, a cell divides once to form **two** daughter cells that contain **identical** combinations of chromosomes, therefore identical combinations of **genes**. Mitosis ensures that all **offspring** produced asexually from a single parent are **identical**.

Sexual reproduction

Sexual reproduction involves **two** parents. **Gametes**, or **sex cells**, are produced in **reproductive organs** by a type of **cell division** known as **meiosis**. A male and a female gamete fuse during **fertilisation** to form a single cell called a **zygote**. The zygote then divides repeatedly by **mitosis** to form an **embryo** and ultimately an **adult**. Offspring produced sexually receive genes from both parents, therefore they possess characteristics of both parents and they show **variation**. The process is **slow** and **energy requirements** are **high** because it involves gamete production, finding a mate, fertilisation and embryo development. Sexual reproduction occurs in most plants and animals.

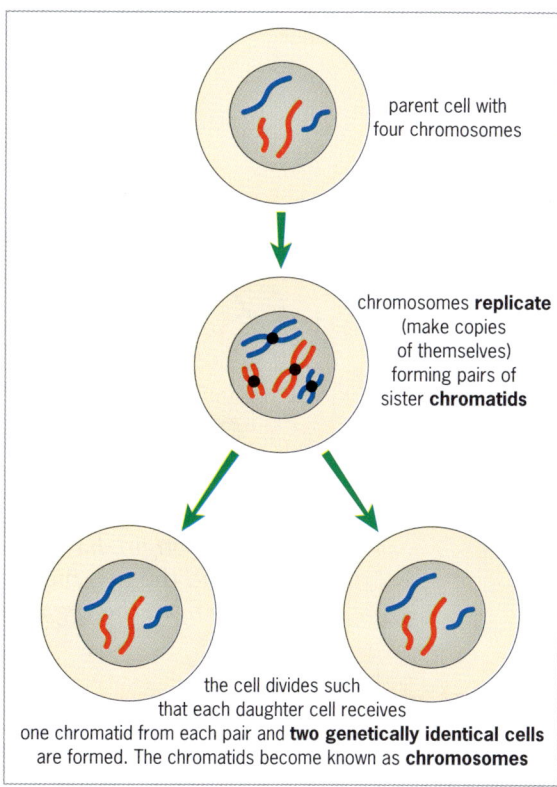

Figure 2.1 *A summary of mitosis*

The role of meiosis in sexual reproduction

Meiosis ensures that each gamete receives **half** the **number** of chromosomes as the parent cell and that all gametes are **genetically non-identical**. During meiosis, a cell divides twice to form **four** daughter cells that all have **different** combinations of chromosomes, therefore different combinations of **genes**. By halving the number of chromosomes, meiosis ensures that when gametes fuse, the zygote has the **correct number** of chromosomes for the species, and because no two gametes have the same combination of genes, all **offspring** produced sexually show **variation**.

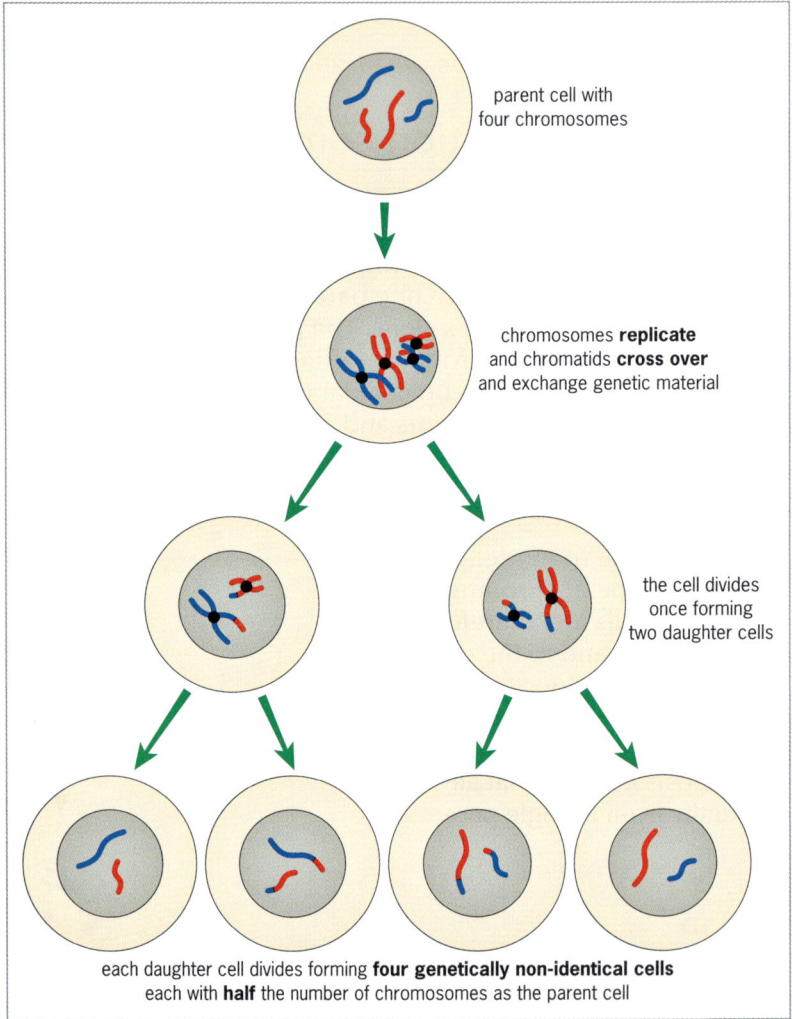

Figure 2.2 *A summary of meiosis. The two red chromosomes are of maternal origin, having been passed on from the organism's mother, and the two blue are of paternal origin, having been passed on from the organism's father*

Advantages and disadvantages of asexual reproduction

Asexual reproduction has advantages and disadvantages, some of which depend on the **environmental conditions** under which the organisms live.

Table 2.1 *Advantages and disadvantages of asexual reproduction*

Advantages	Disadvantages
• **All** offspring from a single parent will be **well adapted** to their environment and have a high chance of survival if the parent is well adapted to its environment because they are all **identical**. • **Beneficial** or **desirable characteristics** are retained within populations because all offspring from a single parent are **identical**. This is important to commercial **crop growers** because if crops are produced asexually, they can ensure the crops possess the same **desirable traits** from generation to generation (see page 11). • Population sizes can increase **rapidly** because the process is **rapid** and it does not require a lot of energy. • It enables organisms to reproduce when suitable mates are **scarce** or **absent**.	• **All** offspring from a single parent will be **adversely affected** and have a reduced chance of survival if the environmental conditions change adversely because they are all **identical**. This is detrimental to commercial **crop growers** because crops produced asexually will all be susceptible to the same adverse environmental conditions and diseases. • Species cannot change and adapt to changing environmental conditions, i.e. they **cannot improve** or **evolve** because all offspring are **identical**. • It can lead to **overcrowding** and **competition** for resources because offspring usually remain **close** to the parent.

Asexual reproduction in plants

Many **plant** species are capable of reproducing **asexually** by **mitosis** occurring in certain structures of the parent plant. This process is known as **vegetative propagation**, and humans use it to **artificially propagate (produce)** plants. Vegetative propagation is a form of **cloning**.

Cloning is the process of making genetically identical individuals by non-sexual means.

Natural vegetative propagation

Plants can propagate **naturally** by producing **perennating organs**, e.g. **stem tubers**, **rhizomes**, **corms** and **bulbs**. These are usually **underground structures** that are swollen with **stored food** at the end of a growing season, e.g. the rainy season or summer months, and they allow the plants to survive through the unfavourable season, e.g. the dry season or winter months. Several new plants can then grow from buds on a single organ when conditions improve. Plants can also propagate by producing **outgrowths** called **runners** or **stolons**.

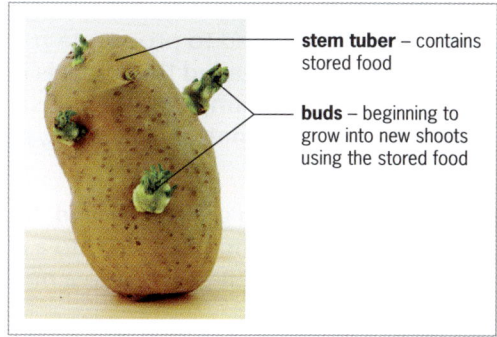

Figure 2.3 *A stem tuber with buds beginning to grow into new shoots*

- **Stem tubers** are the swollen ends of underground stems, e.g. yam and Irish potato.
- **Rhizomes** are swollen, horizontal underground stems, e.g. turmeric and ginger.
- **Corms** are short, swollen, vertical bases of stems, e.g. dasheen and eddo.
- **Bulbs** are composed of layers of fleshy, modified leaves that contain stored food and are arranged in a circular pattern, e.g. lily and onion.

Figure 2.4 *Perennating organs produced by plants for asexual reproduction*

- **Runners** or **stolons** are **horizontal stems** that grow above ground from the base of the parent plant, each with a bud at its tip, e.g. savannah grass, spider plant and strawberry.

Artificial vegetative propagation

Farmers and gardeners can **artificially** propagate plants using a variety of techniques.

- **Cuttings** are parts of stems, roots or leaves that can develop into new plants when given suitable conditions.
    - When a piece of a **sugar cane stem** with two or three **nodes** (growth rings), known as a **sett**, is planted horizontally on the soil, a new plant develops from a bud at each node.
    - When a stem of a **hibiscus** plant bearing a few leaves at the top is planted vertically into the soil, roots grow from the cut end and a new plant develops.
- **Grafting** is often used to propagate fruit trees, e.g. citrus and mango. An actively growing shoot with several buds, called the **scion**, is cut from the plant to be propagated. Its cut end is bound to the cut stem of a young plant of the same or closely related species with an actively growing root system, called the **rootstock** or **stock**, so their tissues join. The scion and stock then grow together into a new plant.
- **Budding** is similar to grafting. A **bud**, together with a small amount of bark, is cut from the plant to be propagated and inserted into a cut made into the bark of a **rootstock**. The bud is secured in place with tape and it grows into a new plant using the root system of the rootstock.

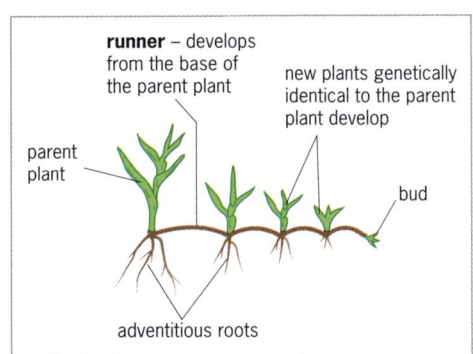

Figure 2.5 *Runners used for asexual reproduction*

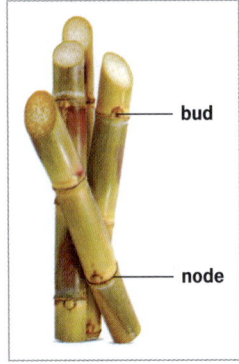

Figure 2.6 *Sugar cane setts, each with two nodes*

Figure 2.7 *Grafting two scions onto a rootstock*

- **Tissue culture** is used to artificially propagate plants such as orchids, potatoes and tomatoes. Small pieces of tissue, called **explants**, are taken from a parent plant and grown in a nutrient-rich culture medium, under sterile conditions, to form cell masses known as **calluses**. Each callus is then stimulated with the appropriate plant hormones to develop into a **plantlet**, which is transferred to the soil to grow and develop into a mature plant.

Because all the plants produced from one parent by these methods are **genetically identical**, if the **cuttings**, **scions**, **buds** or **explants** are taken from plants with **desirable characteristics**, e.g. high yield, high quality, resistance to disease or fast growth rate, then all the plants produced will have the same desirable characteristics. This is important to commercial **crop growers**.

Sexual reproduction in flowering plants

Flowering plants produce **flowers** for **sexual reproduction**. Flowers ensure that the end products of sexual reproduction, **seeds**, are produced, which can then **germinate** and produce **new plants**.

Flower structure

A **typical flower** is made up of an expanded stem tip, known as the **receptacle**, which bears **four** distinct whorls or rings of modified leaves: **sepals**, **petals**, **stamens** and one or more **carpels** in the centre. Most flowers contain both male and female reproductive parts. The male parts are the **stamens**; these produce the **pollen grains**, which contain the **male gametes**. The female parts are the **carpels**; these produce one or more **ovules**, which contain the **female gametes**.

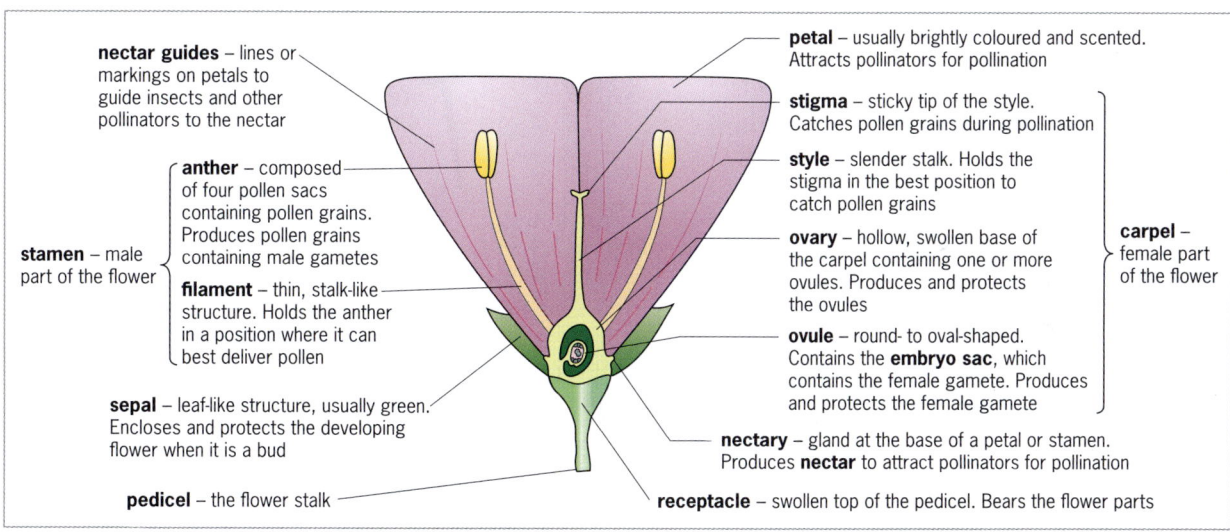

★Figure 2.8 *A longitudinal section of a typical flower showing its structure and the function of its main parts*

Pollination

***Pollination** is the transfer of pollen grains from the anther of a flower to the stigma of a flower of the same species.*

Pollination is the first step towards bringing a male gamete in a pollen grain into contact with a female gamete in an ovule for **fertilisation** to take place. Pollination can take place in **two** different ways.
- **Self-pollination** occurs when a pollen grain is transferred from an anther to a stigma of the **same flower** or to a stigma of another flower on the **same plant**.
- **Cross-pollination** occurs when a pollen grain is transferred from an anther of a flower on one plant to a stigma of a flower on a **different plant** of the **same species**.

Advantages of cross-pollination

Cross-pollination has certain **advantages** which increase the success of plant species.
- **Offspring** show **greater variation** because gametes come from two different parent plants. This increases their **chances of survival** in changing environments and helps species to adapt to changing environmental conditions, which enables them to **improve** or **evolve**.
- **Offspring** may have **superior characteristics** to both parents, which increases their chances of survival, e.g. greater resistance to disease.
- The **seeds** tend to be more **viable**, i.e. they can survive longer before they germinate and they are more likely to germinate.

Agents of pollination

Agents of pollination transport pollen grains between flowers. They include the **wind** and small **animals**, including **insects**, e.g. bees, wasps, flies, butterflies and moths, certain **bats** and some small **birds**, e.g. hummingbirds. **Flowers** are usually **adapted** to be pollinated by either **wind** or by **animals**.

- **Flowers** pollinated by **wind** are usually small and inconspicuous, and they have large anthers that hang outside the flower to produce large quantities of lightweight pollen grains, which are easily transported by the wind. They also have large, feathery stigmas to trap the pollen grains. Examples include grasses, maize and sugar cane.
- **Flowers** pollinated by **animals**, mainly **insects**, are usually large, brightly coloured, conspicuous and often scented to attract the insects to feed on their nectar. The insects pick up pollen grains on their heads and bodies when they brush against the anthers and transport them to the stigmas of other flowers as they feed. Examples include flamboyant, allamanda and pride of Barbados.

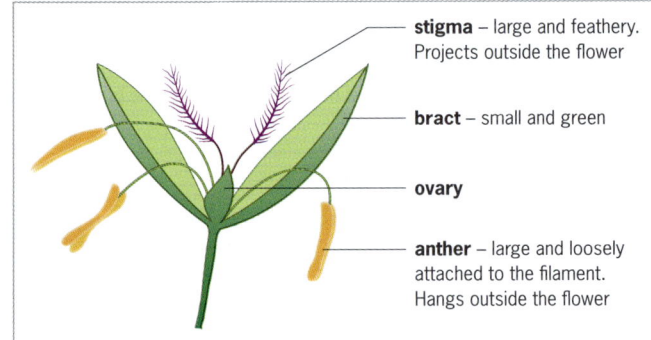

Figure 2.9 *A typical wind-pollinated flower*

Fertilisation in flowering plants

After pollination has occurred, a **male gamete** contained in a pollen grain must reach a **female gamete**, contained in the embryo sac of the ovule, for **fertilisation** to take place.

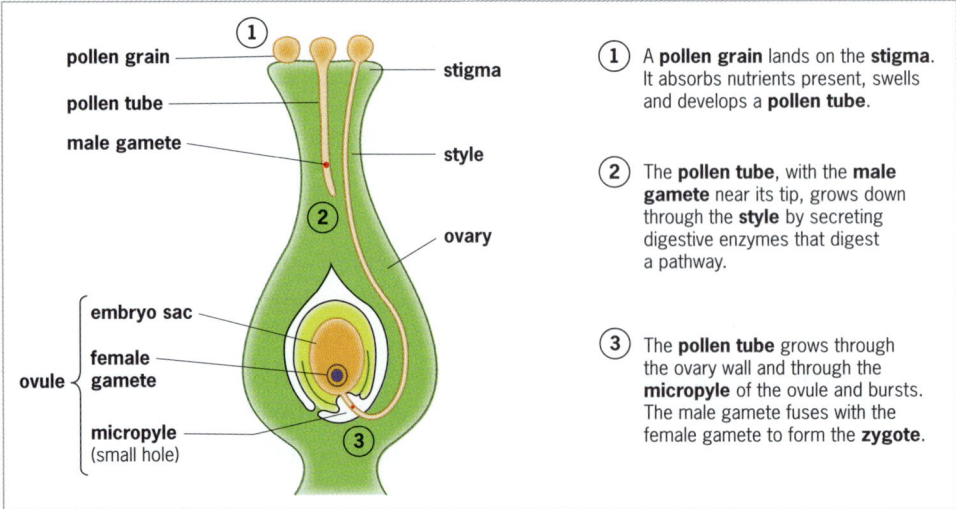

Figure 2.10 *Longitudinal section through a carpel showing the mechanism of fertilisation*

Development of seeds and fruits

After fertilisation, each **ovule** develops into a **seed** and the **ovary** develops into the **fruit**. The **stigma**, **style**, **stamens**, **petals** and **sepals** usually wither and drop off.

Seed development

The **zygote** divides by **mitosis** forming the **embryo**, which is composed of **three** parts: the **plumule** or embryonic shoot, the **radicle** or embryonic root and one or two **cotyledons** that usually store food. The embryo is surrounded by the seed coat or **testa**. Water is withdrawn from the seed and it becomes **dormant** (inactive).

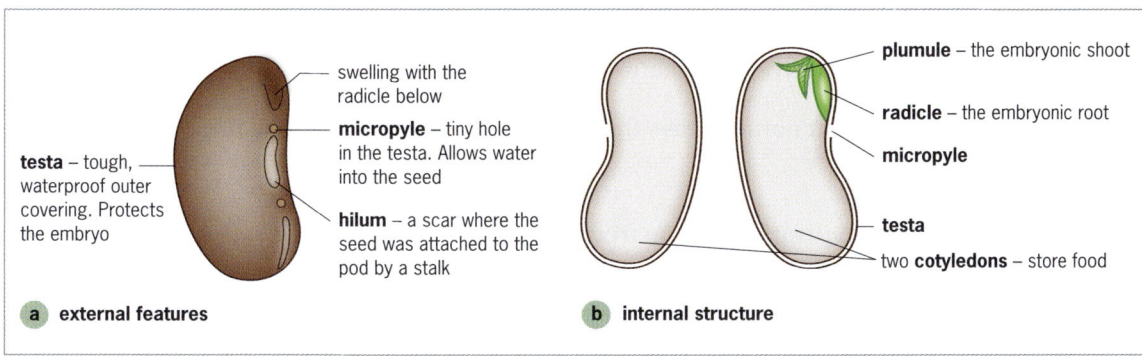

Figure 2.11 *Structure of a bean seed*

Fruit development

The **ovary wall** develops into the **fruit**. The wall may become **succulent** (**fleshy** and **juicy**), e.g. mango, guava, tomato and pumpkin, or it may become **dry** and **thin**, e.g. the pods of pride of Barbados and pigeon pea. A fruit always contains one or more **seeds**. Fruits **protect** the seeds and they help to **disperse** or **spread** the seeds they contain.

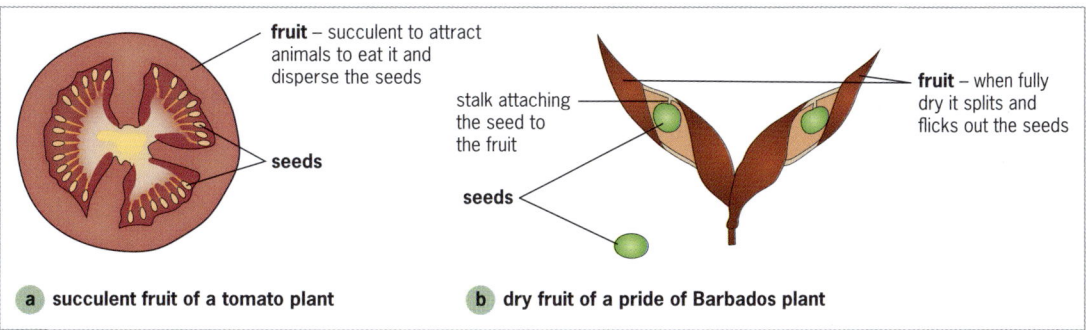

Figure 2.12 *Structure of a succulent fruit and a dry fruit*

Revision questions

1. Provide suitable definitions for:
   a asexual reproduction   b sexual reproduction.
2. Outline the roles of:
   a mitosis in asexual reproduction   b meiosis in sexual reproduction.
3. Construct a table to give THREE differences between asexual reproduction and sexual reproduction.
4. State TWO advantages and TWO disadvantages of asexual reproduction.
5. Define the term 'vegetative propagation'.
6. Outline TWO different natural ways that plants can reproduce asexually.
7. Explain how:
   a grafting is used to propagate lime trees
   b tissue culture is used to propagate tomato plants.
8. What is the importance of flowers to flowering plants?
9. Outline the structure of a typical flower.
10. Give the function of EACH of the following parts of a flower:
    a the petals   b the sepals   c the anthers
    d the stigma   e the style   f the ovule.
11. a What occurs during pollination?
    b Distinguish between self-pollination and cross-pollination, and give TWO advantages of cross-pollination.
    c Identify THREE different agents of pollination.
12. Outline the events that occur in the carpel of a flower following pollination that lead to the development of the seed and the fruit.

Growth in plants and crop production

Germination and growth in annual plants

An **annual plant** completes its life cycle, from germination of its seeds to the development of flowers, fruits and seeds, within a **single** growing season, usually the summer months or rainy season. Many **crops** are annual plants, e.g. beans, maize (corn) and other grains.

Germination

Germination is the process by which the embryonic plant in a seed grows into a seedling.

Seeds require **three conditions** to germinate.
- **Water** to activate the enzymes so that chemical reactions can occur.
- **Oxygen** for aerobic respiration to occur and release the energy required for the process.
- A **suitable temperature**, usually between about 5 °C and 40 °C, to activate enzymes.

Water is absorbed through the **micropyle** causing the seed to swell and it activates enzymes. The enzymes break down stored, insoluble food in the cotyledons into soluble food, which the radicle and plumule then use for **respiration** to release **energy** and for **growth**. As the **radicle** grows, it ruptures the testa, emerges, grows downwards into the soil and begins to form the **root system**. The **plumule** then emerges, grows upwards out of the soil and forms the **first foliage leaves**.

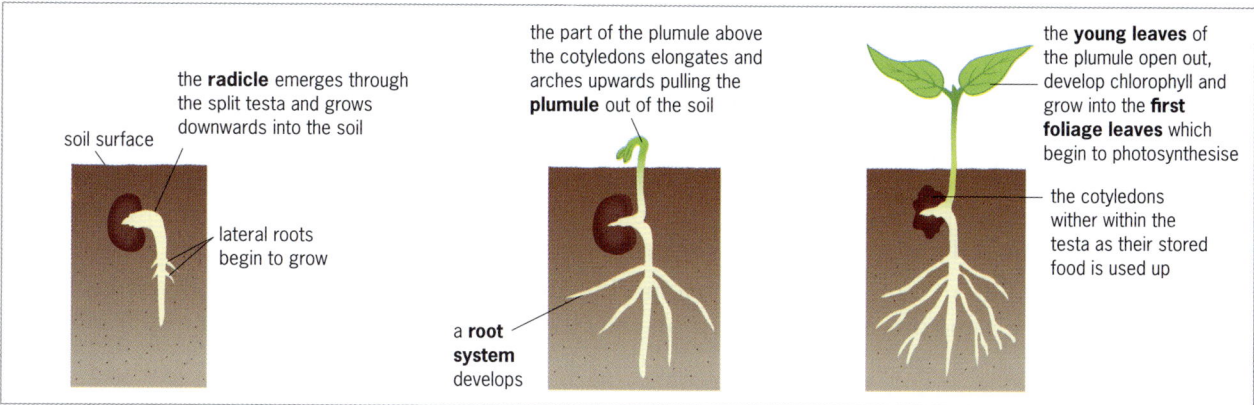

Figure 2.13 *Germination*

Growth patterns of annual plants after germination

As a seedling begins to make food by **photosynthesis**, it begins to **grow** and develop into a plant. The main stem grows upwards and begins to branch, the leaves increase in number and size, and the root system continues to develop. This period of **vegetative growth** continues until the plant approaches **maturity**, when growth slows as it begins to develop **flowers** for **reproduction**. Growth then stops as the plant enters the **reproductive phase** during which pollination and fertilisation occur and **fruits** and **seeds** develop. The seeds are then **dispersed**, the plant **dies** and the seeds remain **dormant** until the next growing season when they **germinate**.

When the **height** of an annual plant is measured at regular intervals and plotted against time, a **growth curve** is obtained that is similar in shape for most annual plants. The curve is described as being **sigmoid shaped** or **S-shaped**.

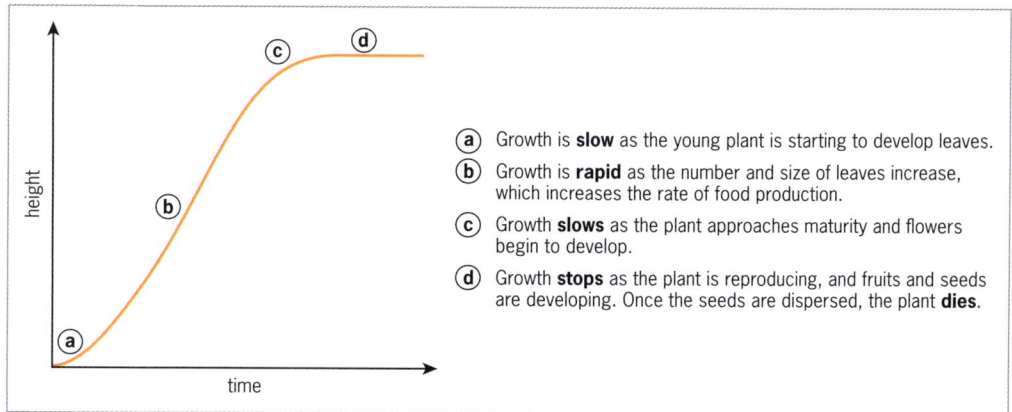

Figure 2.14 *A sigmoid growth curve for an annual plant*

2 Reproduction and growth in plants

Crop production

Farmers and gardeners use a variety of **methods** to produce **crops**. These are designed to maximise yields to meet increasing demands to provide humans with **food**, and depend on factors such as climate, availability of land and other necessary resources, soil type and type of crop being grown.

Table 2.2 *Methods used in the production of crops*

Method	Description
Strip planting	Crops are grown in **strips** or **bands** across a field or following the contours of sloping land. Strips are created by leaving unploughed strips between them or by creating raised beds. Different crops or varieties can be planted in alternating strips to help reduce pests and diseases. Strip planting reduces soil erosion, improves the use of mineral nutrients and helps manage water run-off.
Crop rotation	Different crops are grown in a **specific order** over successive growing seasons on the same piece of land. Crops usually have different mineral nutrient requirements and susceptibilities to pests and diseases. One crop is usually a **legume**, e.g. peas or beans, to replenish **nitrates** in the soil. Crop rotation maintains or enhances soil fertility, reduces soil erosion and helps control pests and diseases.
Greenhouse farming	Crops are grown in **greenhouses** with walls and roofs made of a **transparent material**, e.g. glass or plastic. The climate inside the greenhouse is carefully controlled, particularly temperature, humidity and light. Greenhouse farming allows many different kinds of crops to be grown in places where the climate would not normally be suitable, allows for year-round crop production and protects crops from adverse weather conditions, pests and diseases.
Hydroponics	Crops are grown in a **nutrient-rich solution** without soil. The solution is pumped around the plant roots, which may be suspended in it or supported in an inert growing medium, e.g. perlite, vermiculite, clay pellets, rockwool or coconut coir, ensuring the roots are well oxygenated. Hydroponics eliminates damage due to soil-borne pests and diseases, enables high yields to be produced in limited space and allows crops to be grown where soil quality is poor.
Tissue culture	Small pieces of tissue called **explants** are taken from crop plants with desirable characteristics and are grown in a sterile, nutrient-rich culture medium (see page 11). Tissue culture allows large numbers of plants with desirable characteristics to be produced in a fairly short time.
Organic farming	Crops are grown without synthetic inputs. Only **natural** pesticides and fertilisers are used, crops are rotated, organic matter (including animal manure) is recycled back into the soil, and soil conservation and preventative disease control are practised. Organic farming enhances soil fertility, minimises environmental impact and produces nutritious, healthy crops with very low levels of chemical residues.
Container gardening	Crops and herbs are grown in **containers**, e.g. half barrels, tubs and pots, with drain holes in the bottom, which contain soil or a mixture of soil and potting mix, instead of growing them in the ground. Container gardening is especially useful when outdoor space is limited and it can utilise otherwise unused spaces, e.g. flat rooftops, patios and balconies.

| a strip planting | b hydroponics | c container gardening |

Figure 2.15 *Three methods used in the production of crops*

Soil

Soil is a complex and dynamic mixture of **six** components (see Table 2.3). It forms continuously, but slowly, from the breakdown of rocks by physical weathering, chemical weathering and biological action. Soil is **important** for the survival of all terrestrial (land) organisms because it provides **support** for plant roots, and contains the **water** and **mineral nutrients (salts)** that plants need to produce **food** for themselves and for all other terrestrial organisms that rely directly or indirectly on plants for food (see pages 99 to 101).

Table 2.3 *The components of soil and their functions*

Component	Functions
Rock particles: formed from rocks by weathering.	• Provide support and anchorage for plant roots. • Provide shelter for burrowing animals.
Water: held in a thin film around rock particles.	• Provides plants with water for photosynthesis. • Dissolves mineral nutrients for plants. • Prevents soil organisms from drying out or desiccating.
Air: present in the spaces between rock particles.	• Provides plant roots and other soil organisms with oxygen for aerobic respiration.
Mineral nutrients: dissolved in the water in the soil.	• Essential for healthy plant growth.
Organic matter, including **humus**, derived from dead and decaying plant and animal material.	• Increases soil fertility. • Improves soil structure (see page 19).
Living organisms: plant roots, microorganisms, earthworms and other burrowing animals.	• Plant roots help to prevent soil erosion (see page 20). • Microorganisms, earthworms and other burrowing animals help to increase soil fertility (see page 19).

Soil profiles

Soil is composed of layers known as **soil horizons**. A **soil profile** is a vertical section through the soil showing the horizons. The soils in the horizons differ in composition, structure, texture and colour. Most soils have **three** horizons: **A**, **B** and **C**. Some also have an **O** horizon, and the parent rock is known as the **R** horizon.

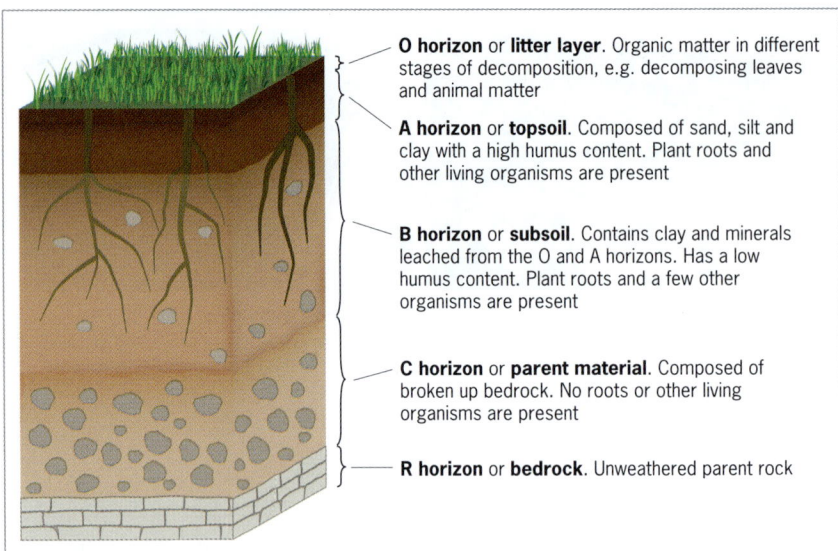

Figure 2.16 *A typical soil profile*

Soil fertility

A **fertile soil** is able to support the growth of a large number of healthy plants. A fertile soil has a high **organic matter** and **mineral nutrient** content, has a **pH** between about 6.0 and 7.5, is **well aerated**, **drains well** whilst still retaining **water** and is **loosely packed** so plant roots can penetrate and animals can burrow easily. The different **physical**, **chemical** and **biological properties** of soil all work together to determine the **fertility** of soil.

Types of soil and soil fertility

A soil can be classified as a **sandy soil**, a **clay soil** or a **loam soil** based on the **size** of the rock particles it contains. This can be determined by the **sedimentation test**, during which soil and water are shaken in a jar and the mixture is left to settle.

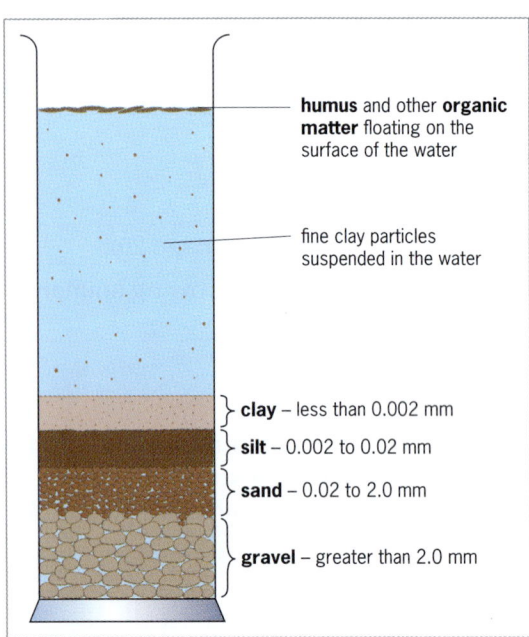

Figure 2.17 *Results of the sedimentation test*

Table 2.4 *Sandy, clay and loam soils compared*

Property	Sandy soil	Clay soil	Loam soil
Particle size	High proportion of **large** sand particles: 0.02 to 2.0 mm.	High proportion of **small** clay particles: less than 0.002 mm.	Mixture of **large** sand, **medium** silt and **small** clay particles in ideal proportions.
Air content	**High**: large particles have large air spaces between.	**Low**: small particles have small air spaces between.	**Fairly high**: different sized particles have fairly large air spaces between.
Water holding capacity	**Poor**: large particles have a small total surface area to retain water.	**Very good**: small particles have a large total surface area to retain water.	**Good**: different sized particles have a fairly large total surface area to retain water.
Drainage	**Good**: water passes through large air spaces quickly. Does not become waterlogged.	**Poor**: water passes through small air spaces slowly. Becomes waterlogged easily.	**Good**: water passes through fairly large air spaces fairly quickly. Retains water, but does not become waterlogged.
Mineral nutrient content	**Low**: rapid drainage leaches (washes) minerals through the soil easily.	**High**: slow drainage leaches minerals through the soil very slowly.	**Relatively high**: water retained around soil particles retains dissolved minerals.
pH	Typically ranges from 5.5 to 7.5.	Typically ranges from 5.0 to 8.5.	Typically ranges from 6.0 to 7.5.

A **loam soil** is the most **fertile soil**. It is **well aerated**, has good **drainage** whilst still retaining **water**, has relatively high **mineral nutrient** content and has a **pH** range that is optimum for most plants.

Soil organisms and soil fertility

Soil organisms are important in increasing **soil fertility**.

- **Microorganisms**, mainly **bacteria** and **fungi**, feed on dead and waste organic matter, e.g. dead plant and animal material, and faeces, causing it to **decompose**. This releases mineral nutrients back into the soil, which improves its **fertility**. These organisms are also known as **decomposers**.
- **Earthworms** improve **aeration** and **drainage** by burrowing through the soil, and they feed on plant debris and soil particles and egest **worm casts** that are rich in **mineral nutrients**. They also add **humus** to the soil by pulling plant debris into their burrows, which then decomposes.
- **Nematodes** (microscopic roundworms) consume bacteria and fungi in the soil and release **ammonium compounds**, which can be converted into **nitrates** for plant use (see page 20). They also **spread** useful bacteria and fungi through the soil on their bodies.

Humus and soil fertility

Humus is formed by bacteria and fungi decomposing dead and waste **organic matter** in the soil. It is a dark brown organic material that coats soil particles and improves **soil fertility**.

- Humus improves the **mineral nutrient content** of soil by adding minerals when the organic matter from which it is formed decomposes, and by absorbing and retaining minerals.
- Humus improves the **water content** of soil by absorbing and retaining water.
- Humus improves the **air content** and **drainage** of soil by binding finer soil particles together into larger **soil crumbs**. This also improves the **structure** of the soil, which makes it easier for plant roots to penetrate and animals to burrow, and makes it easier to cultivate.

Cycling of nitrogen and soil fertility

Plants require **nitrogen** to make **proteins** for healthy growth. They cannot use nitrogen (N_2) gas directly from the air; they must obtain it from the soil in the form of the **nitrate (NO_3^-) ion**. Certain **soil bacteria** cycle nitrogen within the soil, which helps maintain **soil fertility**.

- **Nitrifying bacteria** convert **ammonia (NH_3)** and **ammonium (NH_4^+) compounds**, formed from the breakdown of proteins in dead and waste organic matter, into **nitrites (NO_2^-)** and then into **nitrates**.
- **Denitrifying bacteria** convert **nitrates** into **nitrogen gas** under anaerobic conditions (without oxygen), usually in waterlogged or compact soil. The gas is returned to the air in the soil and eventually the atmosphere. This **reduces** soil fertility.
- **Nitrogen-fixing bacteria** present in root nodules of legumes, e.g. peas and beans, and free-living in the soil, convert **nitrogen gas**, present in soil air, into **ammonium compounds**. Legumes use these compounds to make proteins, and those produced by bacteria in the soil can be converted into **nitrates** by nitrifying bacteria.

Composting and soil fertility

Composting is a natural process that uses microorganisms to convert organic material, e.g. kitchen and garden waste, into a nutrient-rich **soil conditioner**, known as **compost**, which can be used to improve **soil fertility** in a similar way to humus.

Soil erosion and its impact on food production

Soil erosion is the wearing away of the upper layer of soil due to the action of wind and water.

Soil is one of the world's most **important natural resources** and the **topsoil** (upper layer) is the most fertile because it contains the most decomposing organic matter. If topsoil is removed, the remaining soil is **less fertile**.

Causes of soil erosion

Soil erosion occurs naturally and is speeded up if the soil has **no plants** growing in it to bind the particles together with their roots. Both **natural** and **human-induced** factors lead to soil erosion.

Figure 2.18 *Deforestation exposes the soil to erosion*

- **Rainfall** on the soil surface breaks down soil crumbs and separates soil particles, which can be washed away by the rainwater as it runs off the land, called **surface runoff**.
- **Wind** picks up loose soil particles and carries them away as it blows.
- **Deforestation**, where trees are cut down and not replanted, removes leaves that break the force of the rain and roots that bind the soil together. This leaves the soil barren and exposed to the rain and wind.
- **Bad agricultural practices**, e.g. leaving the soil barren after harvesting, overgrazing of animals, ploughing down hillsides and using chemical fertilisers instead of organic fertilisers such as manure, can all lead to topsoil being eroded by the rain or wind.

Effects of soil erosion

Soil erosion has many effects which go beyond those resulting directly from the loss of fertile soil.
- It reduces the number of **trees** and **other plants** that can be grown, which leads to a decline in **biodiversity**.
- It reduces **agricultural productivity**. As fertile topsoil is lost, **crop yields** decrease, leading to a decrease in food production and an increase in global food insecurity (see below).
- It can lead to **desertification** in areas where the soil has become eroded to such an extent that plants can no longer grow.
- It causes **sediment** to build up in bodies of water, e.g. lakes, rivers and streams. Sediment reduces light penetration, which reduces the growth of aquatic plants, and it clogs the gills of fish, smothers bottom-dwelling organisms and blocks waterways as it builds up.
- It leads to an increase in **flooding** as eroded lands are less able to retain water during heavy rainfall.

Prevention of soil erosion

Various measures can be implemented to **reduce soil erosion** and **conserve soil**.

- Practise **contour farming** on **sloping land**, e.g. a hillside, by ploughing the land and planting crops along **natural contours**. This helps to retain water by reducing runoff and minimises soil loss.
- Cut **terraces** out of the sides of hills or mountains to create a series of flat platforms in which crops are planted. This method, known as **terracing**, reduces rainwater runoff and minimises soil loss.
- Practise **crop rotation** (see Table 2.2, page 16) which includes cover crops (see below). The varying root structures of different crops help to improve soil structure and minimise soil loss.

Figure 2.19 *Terracing helps to minimise soil erosion*

- Plant **windbreaks**, consisting of rows of trees, shrubs or other vegetation, around the edges of fields or in well-planned positions within large areas of cropland to protect against erosion by the wind.
- Plant **cover crops**, e.g. grasses or legumes such as clover. These are planted when the main crops are not being grown or immediately after harvesting. They cover the soil to protect its surface against erosion by the wind and rain, rather than to produce crops, and they reduce water runoff and their roots bind soil particles together.

The impact of soil erosion on food production

Food security exists when all people at all times have access to sufficient, safe and nutritious food to maintain a healthy, active life. Food security depends directly on **soil productivity**, and **soil erosion** is the main cause of a decrease in soil productivity. According to the Food and Agricultural Organization of the United Nations (FAO), **soil degradation** is the single biggest threat to **global food security**.

It is estimated that soil erosion can cause up to 50% losses in **crop yields** and can lead to a decrease in the **quality** of crops and an increase in the **costs** associated with their production. These factors can lead to economic losses for farmers, food shortages, higher food prices, hunger and starvation. **Food insecurity** already affects more than 750 million people worldwide, especially in tropical and sub-tropical regions. Preventing soil erosion is crucial for ensuring global food security.

Revision questions

13 Provide a suitable definition for EACH of the following:
   a growth   **b** germination.

14 **a** Identify the THREE conditions that seeds need to germinate and outline the role of EACH condition in the germination process.
   b Describe the events that occur when a seed germinates until its first foliage leaves develop.

15 What is an annual plant and what is the name given to the shape of a typical growth curve of an annual plant?

16 Give a brief description of EACH of the following methods that can be used to produce crops and identify TWO advantages of EACH method:
   a greenhouse farming   **b** strip planting   **c** organic farming
   d hydroponics   **e** crop rotation   **f** container gardening.

17 **a** What is soil?
   b Outline the importance of soil to living organisms.

18 What are the differences among a sandy soil, a clay soil and a loam soil in terms of:
   a particle size   **b** air content   **c** water holding capacity?

19 What makes a soil fertile? Your answer must include the definition of a fertile soil.

20 Explain the importance of EACH of the following to the fertility of soil:
   a earthworms   **b** decomposers   **c** humus
   d nitrogen-fixing bacteria   **e** composting.

21 Identify THREE causes and THREE consequences of soil erosion.

22 Explain why soil erosion has a major impact on global food production.

3 Reproduction and growth in animals

Like plants, animals must also **produce offspring** in order for their species to survive, and they achieve this by reproducing. Some animals can reproduce **asexually**, however, most reproduce **sexually**. Animals also **grow** and **develop** which ensures that they are the correct size to survive in their environment.

Reproduction is the process by which living organisms generate new individuals of the same kind as themselves.

Growth is a permanent increase in the size of an organism.

Asexual reproduction in animals

A few **animal** species can reproduce **asexually** (see page 7). Several different methods are employed, including **budding**, **fragmentation** and **parthenogenesis**. Unicellular organisms such as protozoans and bacteria also reproduce asexually by **binary fission**.

- **Binary fission** involves a single parent cell dividing into two identical cells by **mitosis** (see page 7).

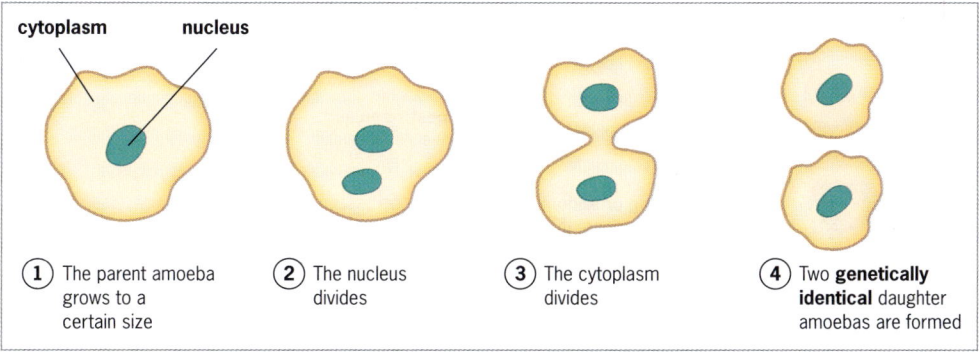

Figure 3.1 *Binary fission in an amoeba*

- **Budding** occurs in some simple animals, e.g. coral polyps, jellyfish and hydra. New individuals develop from **buds** produced by cells dividing by **mitosis** in specialised areas of the animal's body. Budding can also occur in some unicellular organisms, e.g. yeast and bacteria.

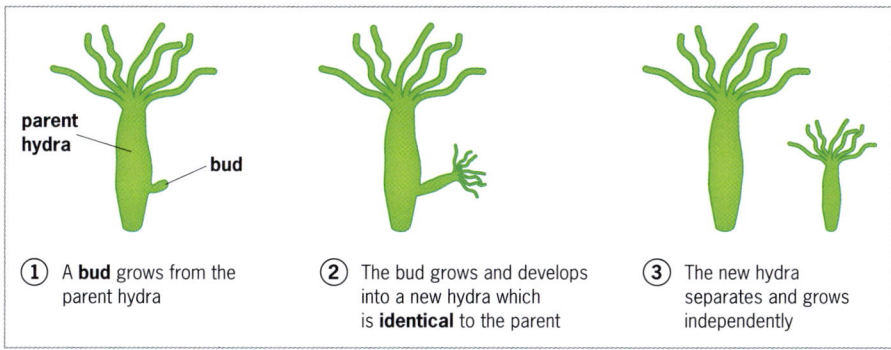

Figure 3.2 *Budding in hydra*

Note: Budding in animals and unicellular organisms is a **natural** process. It must not be confused with budding used by humans to propagate plants (see page 10), which is an **artificial** process.

- **Fragmentation** occurs in some invertebrates, e.g. certain species of flatworms and sponges. The body of the parent splits into two or more pieces or **fragments**, and each fragment then grows and develops into a new individual identical to the parent.
- **Parthenogenesis** occurs in certain insects, e.g. aphids, stick insects, and some ants, wasps and bees. It can also occur in certain reptiles, amphibians and fish. An **ovum** or **egg cell** is **activated** to divide by **mitosis** to form an embryo, without being fertilised. Activation can be caused by environmental stresses, changes in day length, hormonal changes or lack of mating opportunities.

Sexual reproduction in humans

Humans reproduce **sexually** (see page 7). This involves the fusion of male and female **gametes** or **sex cells** produced in the male and female **reproductive systems**, which consist of both internal and external **reproductive organs**.

The female reproductive system

The **female gametes** are called **ova** (singular: **ovum**) and they are produced in the **ovaries**, which form part of the female reproductive system.

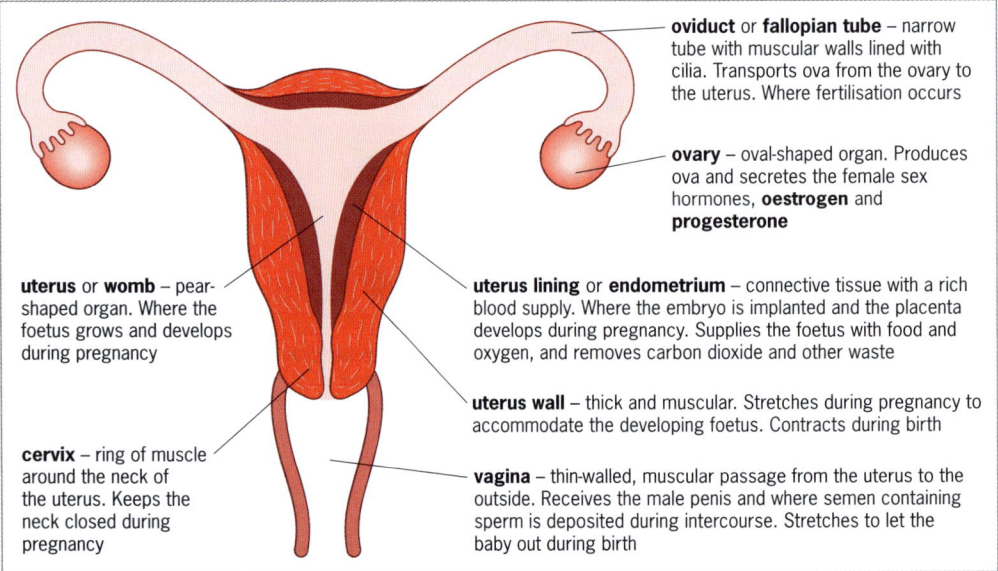

Figure 3.3 *Structure and function(s) of the parts of the female reproductive system*

The male reproductive system

The **male gametes** are called **sperm** or **spermatozoa** and they are produced in the **testes**, which form part of the male reproductive system.

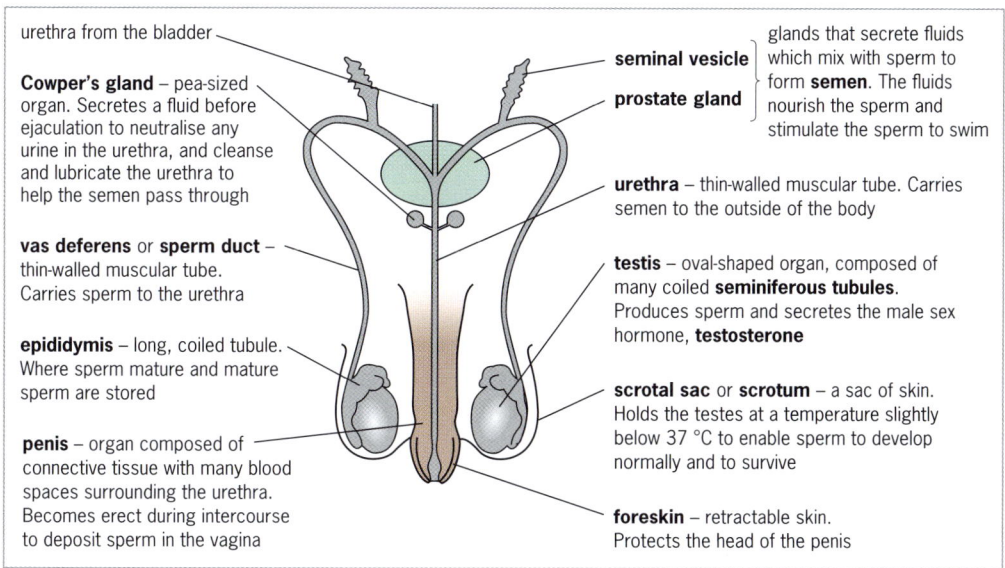

Figure 3.4 *Structure and function(s) of the parts of the male reproductive system*

Ova and sperm

At birth, each female ovary contains many thousand **immature ova**, each surrounded by a fluid-filled space that forms a **primary follicle**. Each month between **puberty** at about 11 to 13 years old, and **menopause** at about 45 to 55 years old, one immature ovum undergoes **meiosis** (see page 8) and one of the four cells develops into a **mature ovum**, which is released during **ovulation**.

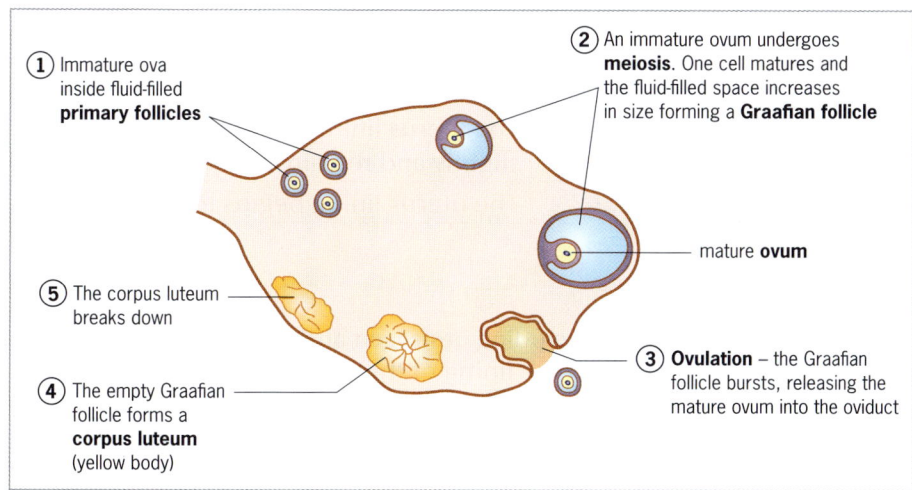

Figure 3.5 *Production of an ovum in an ovary*

Sperm cells are produced continuously from **puberty** in the **seminiferous tubules** of the **testes**. Cells in the tubule walls undergo **meiosis** and **all** the cells produced develop into sperm. These sperm mature in the **epididymis**, where they are stored until ejaculation.

3 Reproduction and growth in animals

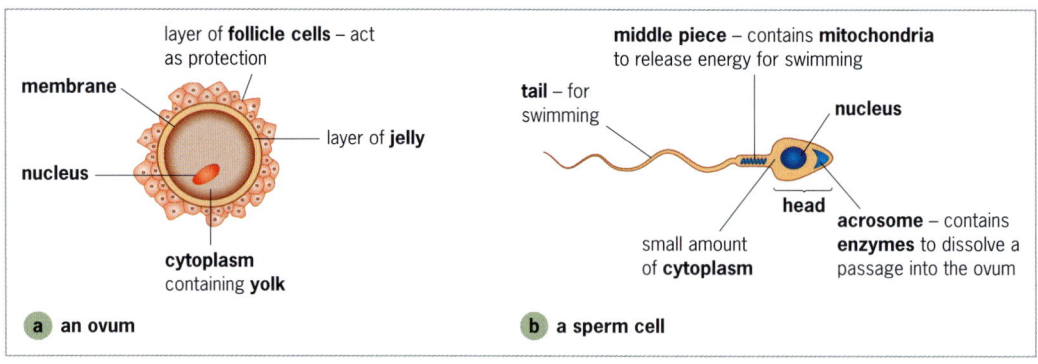

Figure 3.6 *Structure of an ovum and a sperm cell*

The menstrual cycle

The **menstrual cycle** is a cycle of about 28 days which begins when a female reaches **puberty**. It prepares the uterus lining each month to receive the embryo if fertilisation occurs. The cycle comprises **two** main events: **ovulation** and **menstruation**.

Ovulation is the release of an ovum from an ovary.

Menstruation is the loss of the uterus lining from the body.

Menstruation starts to occur about 14 days after ovulation, and the **start** of each cycle is taken from the start of **menstruation**.

Table 3.1 *A summary of the events occurring in an ovary and the uterus during the menstrual cycle*

Time	Events in an ovary	Events in the uterus
Day 1 to day 14	An immature ovum undergoes meiosis and one cell matures. The Graafian follicle develops around the ovum as it matures.	Day 1 to day 5: the uterus lining breaks down and is lost from the body. Day 6 to day 14: the uterus lining thickens and its blood supply increases.
Day 14	The mature ovum is released and the Graafian follicle forms the corpus luteum.	
Day 14 to day 25	The corpus luteum remains.	The uterus lining continues to thicken slightly and remains thick.
Day 26 to day 28	The corpus luteum breaks down.	The uterus lining begins to break down.

Hormonal control of the menstrual cycle

The **menstrual cycle** is controlled by several **hormones**, including **oestrogen** and **progesterone**. These synchronise the production of an ovum in an ovary with the lining of the uterus being ready to receive the embryo if the ovum is fertilised.

- **Oestrogen** is produced by the **Graafian follicle**, mainly during the **second week** of the cycle. It stimulates the lining of the uterus to thicken and its blood supply to increase after menstruation.
- **Progesterone** is produced by the **corpus luteum** during the **third week** of the cycle. It causes the uterus lining to increase slightly in thickness and remain thick. If fertilisation does not occur, the corpus luteum degenerates during the fourth week and reduces secretion of progesterone. The decrease in progesterone causes the uterus lining to begin to break down towards the end of the fourth week.

Menopause

Menopause is the time in a female's life that marks the end of her menstrual cycles.

Menopause is considered to have occurred when a female has not had a menstrual period for 12 consecutive months. It typically happens between the ages of 45 and 55, and marks the **end** of a **female's fertility**; the ovaries stop releasing ova and she can no longer become pregnant.

Menopause is preceded by **perimenopause**, the transitional stage leading up to menopause that can last for several years. During perimenopause, the ovaries gradually reduce secretion of **oestrogen** and **progesterone**, which leads to irregular menstrual cycles and eventually to menopause. Perimenopause is often accompanied by hot flushes, night sweats, anxiety, depression, mood swings, vaginal dryness and difficulty sleeping that can continue for several years after menopause.

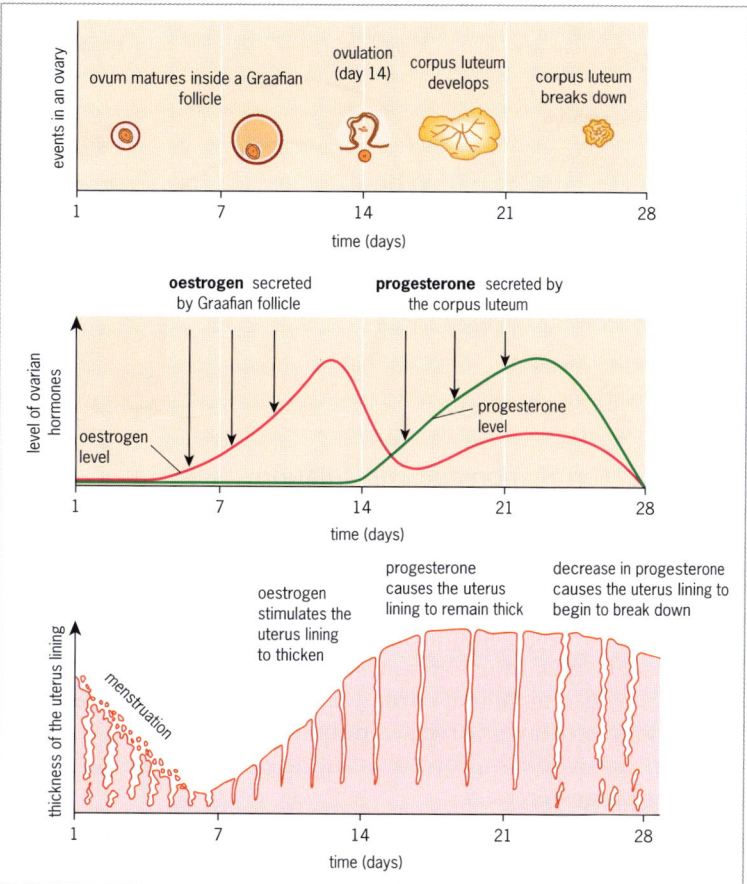

Figure 3.7 *A summary of the events occurring during the menstrual cycle*

Fertilisation, implantation and pregnancy

Bringing sperm and ova together

When a male becomes sexually excited, blood spaces in the penis fill with blood. The penis becomes **erect** and is placed into the female vagina. **Semen**, composed of sperm and secretions from the seminal vesicles and prostate gland, is **ejaculated** into the top of the vagina by muscular contractions of the tubules of the epididymis and sperm ducts. The **sperm** swim through the cervix and uterus and into the oviducts or fallopian tubes.

Fertilisation

If an **ovum** is present in one of the **oviducts**, one **sperm** enters using enzymes produced by the acrosome to digest a pathway. Its tail is left outside. A **fertilisation membrane** immediately develops around the ovum to prevent other sperm from entering, and the nuclei of the ovum and sperm fuse to form a **zygote**.

Implantation

The **zygote** divides repeatedly by **mitosis**, using **yolk** stored in the original ovum as a source of nourishment. This forms a ball of cells called the **embryo**, which moves down the oviduct and sinks into the uterus lining, a process called **implantation**. Food and oxygen diffuse from the mother's blood in the uterus lining into the embryo, and carbon dioxide and other waste diffuse back into the mother's blood.

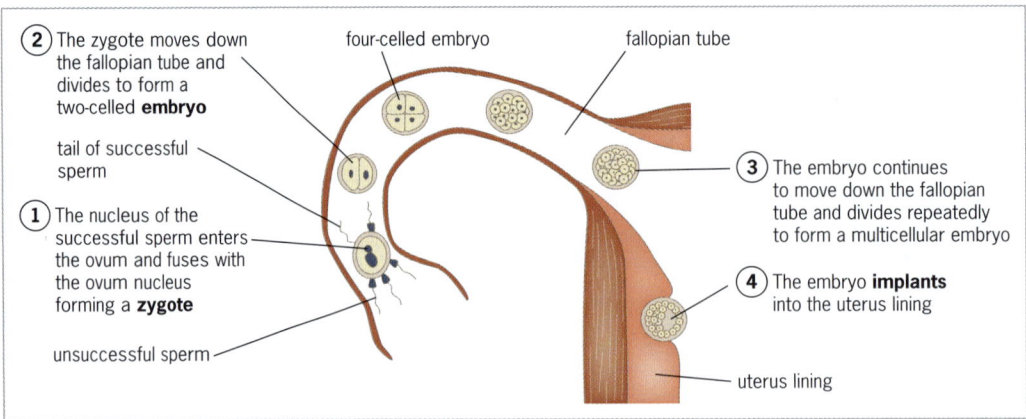

Figure 3.8 *Fertilisation and implantation*

Pregnancy and development

The cells of the **embryo** continue to divide and some develop into the **placenta**. The placenta is a disc of tissue with finger-like projections called **villi**, which project into the uterus lining and give the placenta a large surface area. Capillaries run throughout the placenta. The embryo is joined to the placenta by the **umbilical cord**, which has an **umbilical artery** and an **umbilical vein** running through. These connect the embryo's capillaries with those in the placenta. The **placenta** allows exchange of materials between the mother's and the embryo's blood, but prevents the bloods from mixing. It also secretes **progesterone**.

The developing embryo is surrounded by a thin, tough membrane called the **amnion**, which forms a sac containing **amniotic fluid** to support and protect the embryo as it develops.

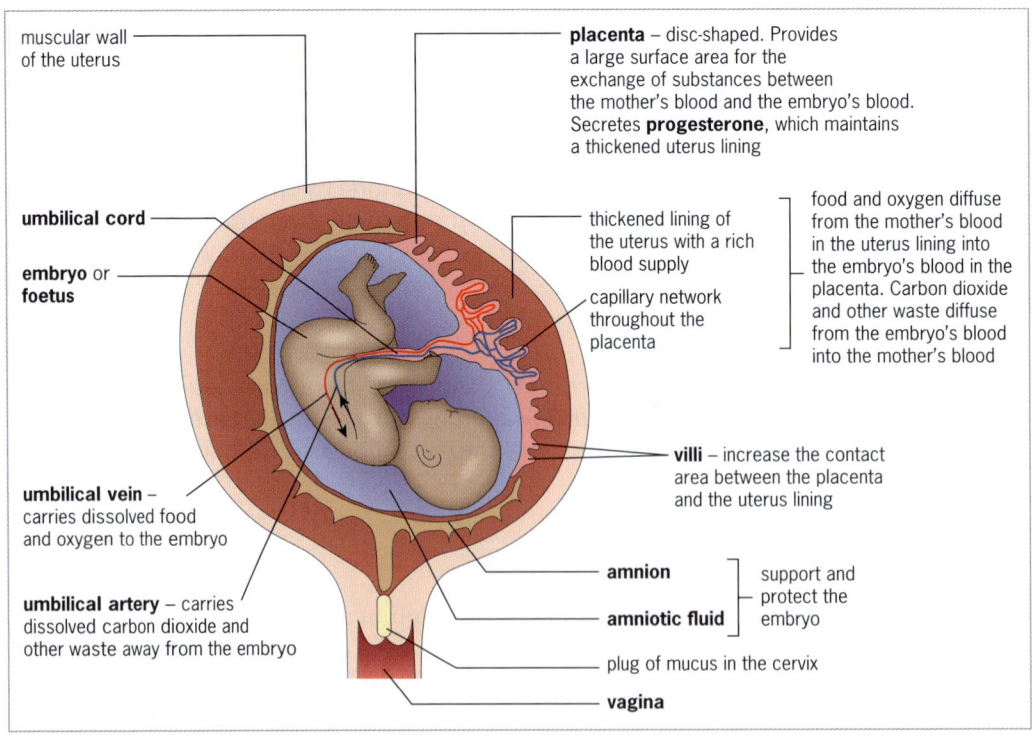

Figure 3.9 *The developing human embryo/foetus in the uterus*

Table 3.2 *A summary of the development of a human embryo/foetus*

Time after fertilisation	Characteristics
7 to 10 days	A hollow ball of cells, which is implanted in the uterus lining.
4 weeks	The brain, eyes and ears are developing along with the nervous, digestive and respiratory systems. Limb buds are forming and the heart is beginning to beat.
8 weeks	The embryo has a distinctly human appearance. All the vital organs have been formed and limbs with fingers and toes are developed.
10 weeks	The embryo is now known as a **foetus**. External genitals are beginning to appear, fingernails and toenails form, and the kidneys start to function.
11 to 38 weeks	The foetus continues to grow and the organs continue to develop and mature.
38 weeks	Birth occurs.

Note: The **gestation period (pregnancy)** is considered to last for 40 weeks or 280 days, since it is calculated from the first day of the last menstrual cycle and not from the time of fertilisation.

Birth

Towards the end of the pregnancy, the baby turns to lie **head down**. As it continues to grow, it stretches the uterus wall to a point where **stretch receptors** are stimulated. This inhibits secretion of progesterone by the placenta and stimulates the **pituitary gland** to secrete the hormone **oxytocin**. Oxytocin begins the **birthing process**, which is divided into **three** stages.

Dilation of the cervix

Oxytocin stimulates the muscles in the wall of the uterus to start **contracting**, i.e. **labour** begins. At some point, the amnion bursts and the amniotic fluid is released, referred to as **water breaking** or **waters breaking**. The contractions strengthen and cause the **cervix** to gradually **dilate (widen)**, and they push the baby towards the cervix. This stage ends when the cervix has dilated to **10 cm**.

Expulsion of the baby or birth

Pressure of the baby's head on the **cervix** gives the mother the urge to **push**. She aims to push with each contraction by contracting her abdominal muscles. This pushes the baby, head-first, through the cervix into the vagina. **Crowning** occurs as the top of the baby's head emerges through the opening of the vagina and is visible. The rest of the baby's head then emerges, followed by its body. Once outside, the baby stats to **breathe** and the **umbilical cord** is clamped and cut.

Expulsion of the placenta

After the baby is born, the **placenta** detaches from the uterus lining and is expelled from the mother's body as the **afterbirth** by further **contractions** of the uterus wall.

Methods of birth control (contraception)

Birth control is used to **prevent unintended pregnancies**. Various methods are available that aim to **prevent fertilisation** or **prevent implantation**, and they can be **natural**, **barrier**, **hormonal** or **surgical**. Two methods, **abstinence** and the **condom**, also protect against the spread of sexually transmitted infections (STIs), e.g. HIV/AIDS. When choosing a method, its reliability, availability, side effects and if both partners are comfortable using it must be considered.

Table 3.3 *Methods of birth control*

Method	How the method works	Advantages	Disadvantages
Abstinence	• Refraining from sexual intercourse.	• Completely effective.	• Relies on self-control from both partners.
Withdrawal	• Penis is withdrawn before ejaculation.	• No artificial device needs to be used or pills taken – it is **natural**, therefore acceptable to all religious groups.	• Very unreliable since some semen is released before ejaculation. • Relies on self-control.
Rhythm method	• Sexual intercourse is restricted to times when ova should be absent from the oviducts.	• No artificial device needs to be used or pills taken – it is **natural**, therefore acceptable to all religious groups.	• Unreliable since the time of ovulation can vary. • Restricts the time when intercourse can occur. • Unsuitable for women with irregular menstrual cycles.
Billings method	• Refraining from sexual intercourse when the cervical mucus is slippery and elastic, and for three days afterwards. Mucus is slippery from a few days before ovulation until ovulation.	• No artificial device needs to be used or pills taken – it is **natural**, therefore is acceptable to all religious groups.	• Restricts the time when intercourse can occur. • Requires commitment, accurate observations of the cervical mucus and a proper understanding of the method. • Reliability varies.
Condom	• A latex rubber or polyurethane sheath placed over the erect penis or into the female vagina before intercourse. • Acts as a physical **barrier** to prevent sperm entering the female body.	• Very reliable if used correctly. • Easy to use. • Readily available. • Protects against sexually transmitted infections.	• May reduce sensitivity, so interferes with enjoyment. • Condoms can tear, allowing sperm to enter the vagina. • Latex may cause an allergic reaction.
Diaphragm and **cervical cap**	• **Diaphragm** – a dome-shaped latex or silicone cup inserted over the cervix and surrounding area before intercourse. • **Cervical cap** – a silicone cap that fits snugly over the cervix only, inserted before intercourse. • Both are used with a spermicide to kill sperm. • Both act as physical **barriers** to prevent sperm entering the uterus.	• Both are fairly reliable if used correctly. • They are not felt, therefore do not interfere with enjoyment. • Both are easy to use once the female is taught.	• Must be left in place for 6 hours after intercourse, but no longer than 24 hours in the case of the diaphragm and 48 hours in the case of the cervical cap. • Latex in latex diaphragms may cause an allergic reaction. • Both may slip out of place if not fitted properly.

Method	How the method works	Advantages	Disadvantages
Intra-uterine device (IUD or coil)	• A T-shaped plastic device, usually containing copper or **progestin** (synthetic form of **progesterone**), inserted into the uterus by a doctor. • Prevents sperm reaching the ova or prevents implantation.	• Very reliable. • Once fitted, no further action is required except an annual check-up. • Long-acting; can work for 3 to 10 years. • Few, if any, side effects.	• Must be inserted by a medical practitioner. • May cause menstruation to be heavier, longer or more painful.
Contraceptive pill	• A **hormone** pill, taken daily, which contains **oestrogen** and **progestin** or **progestin**. • Prevents ovulation. • Thickens the cervical mucus making it harder for sperm to swim through. • Thins the uterus lining reducing the likelihood of implantation.	• Almost totally reliable if taken daily. • Menstruation is lighter, shorter and less painful.	• Ceases to be effective if one pill is missed. • May cause side effects in some women, especially those who smoke.
Contraceptive patch	• A small, sticky patch attached to the skin and replaced weekly. • Releases **oestrogen** and **progestin** into the bloodstream. • Works in the same way as the contraceptive pill.	• Almost totally reliable if replaced every 7 days. • Menstruation is lighter, shorter and less painful.	• Ceases to be effective if not changed for 9 days or longer. • May cause side effects in some women, especially those who smoke.
Contraceptive injection	• An injection containing **progestin** given by a doctor every 3 months. • Works in the same way as the contraceptive pill.	• Almost totally reliable. • Once injected, no further action is required for 3 months. • Can reduce heavy, painful menstruation.	• May cause side effects in some women. • Can take up to 1 year for fertility to return to normal after injections are discontinued.
Contraceptive implant	• A small, flexible plastic rod placed under the skin of the upper arm by a doctor every 3 years. • Releases **progestin** into the bloodstream. • Works in the same way as the contraceptive pill.	• Almost totally reliable. • Once fitted, no further action is required for 3 years. • Can be removed at any time and fertility returns immediately. • Can reduce heavy, painful menstruation.	• May cause side effects in some women.

Method	How the method works	Advantages	Disadvantages
Surgical sterilisation (**vasectomy** in males, **tubal ligation** in females)	• The sperm ducts or oviducts are **surgically** cut and tied off. • Prevents sperm leaving the male body or ova passing down the oviducts in females.	• Totally reliable. • No artificial device needs to be used or pills taken. • No need to think further about contraception.	• Usually irreversible.
Hysterectomy	• The uterus, and usually the cervix, are **surgically** removed. • Usually only recommended when the individual is experiencing a serious medical condition, e.g. fibroids or cancer.	• Totally reliable. • No artificial device needs to be used or pills taken. • No need to think further about contraception. • No more menstrual periods.	• A major surgical procedure that has risks associated with it. • Irreversible.

Note: One **disadvantage** of all methods, except abstinence and condoms, is that they do not protect against sexually transmitted infections.

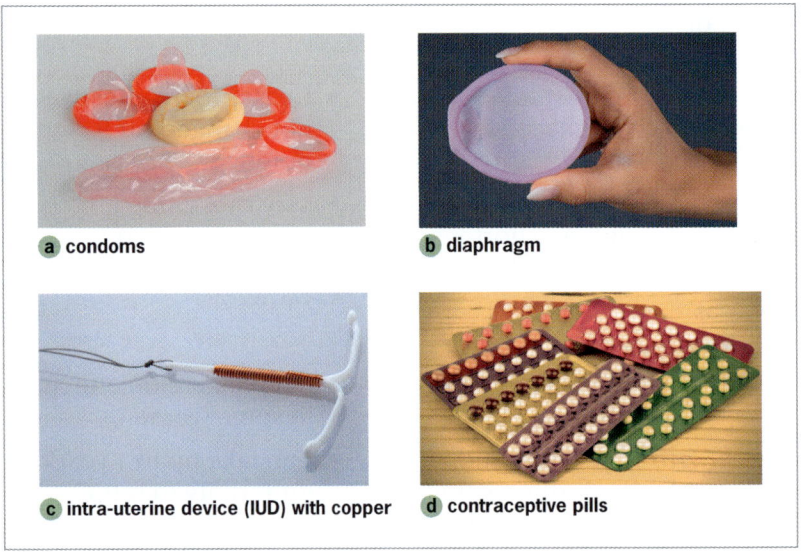

a condoms b diaphragm c intra-uterine device (IUD) with copper d contraceptive pills

Figure 3.10 *Methods of birth control*

The importance of prenatal (antenatal) care

Prenatal care, or care before birth, is essential to ensure the foetus **grows** and **develops normally** and **healthily**, and that the mother **remains healthy** throughout her pregnancy. During pregnancy, the mother should:

- Attend regular **prenatal check-ups** with her doctor or clinic to monitor her health and the development of her baby.
- Have two **ultrasound scans** if possible at about 6 to 8 weeks and at about 18 to 20 weeks. These use high-frequency sound waves to create an **image** of the baby to monitor its growth and development, and to detect any abnormalities.
- Eat a **balanced diet** that contains adequate quantities of proteins, carbohydrates, vitamins and minerals, especially folic acid or vitamin B_9, vitamin D, calcium and iron, to ensure the foetus obtains all the nutrients it needs to grow and develop.

- **Exercise** regularly to maintain fitness.
- Not use **unprescribed drugs** of any kind, especially alcohol, cigarettes and illegal drugs, which can interfere with normal foetal development and damage foetal organs.
- Protect herself against harmful **X-rays** due to risks radiation poses to her developing foetus.
- Protect herself against **infectious diseases** because certain diseases pose serious risks to both the mother and her foetus, e.g. rubella (German measles) and Zika fever.
- Be tested for **sexually transmitted infections (STIs)** and treated, if necessary, because they can pose significant health risks to her baby.

The importance of postnatal care

Postnatal care, or care after birth, is essential to ensure the baby **grows** and **develops healthily**, and that the mother remains both physically and emotionally **healthy**.

- The **newborn baby** should be **breastfed**, if possible, for a minimum of 6 months, because:
  - Breast milk contains all the **nutrients** the baby needs in the correct proportions.
  - Breast milk contains **antibodies** which protect the baby against bacterial and viral diseases.
  - Breast milk is **sterile**, so it reduces the risk of infection, it is at the correct **temperature**, and it is **available** whenever needed.
  - Breastfeeding lowers the baby's risk of developing **asthma**, **allergies** and other **non-communicable diseases** as it grows older.
  - Breastfeeding creates a strong **emotional bond** between mother and baby.
- The **newborn baby** must be kept **warm** and **clean**, have plenty of **interaction** with both parents and its surroundings, and be taken for regular **check-ups** with the doctor. As the baby grows, it must be **weaned** onto semi-solid and solid food, **cared for** physically and emotionally, and given continual **teaching**.
- The **baby** must be **vaccinated** to **immunise** it against a variety of different infectious diseases, following a vaccination or immunisation programme.

Growth in animals

Growth patterns

When the height or weight of a typical girl or boy is plotted against age, a **height** or **weight curve** is obtained, which shows the girl's or boy's **pattern** of **growth**. Both girls and boys have a similar growth pattern, however there are some differences.

- **Girls** and **boys** both grow **rapidly** during **infancy** and **very early childhood**, from birth to the age of about 3. Their growth rates then slow slightly during **childhood**, with **boys** in general being slightly **taller** and **heavier** than girls at all ages.
- As a child approaches **puberty**, typically between ages 8 and 13 in girls, and ages 9 and 14 in boys, **growth rate** begins to

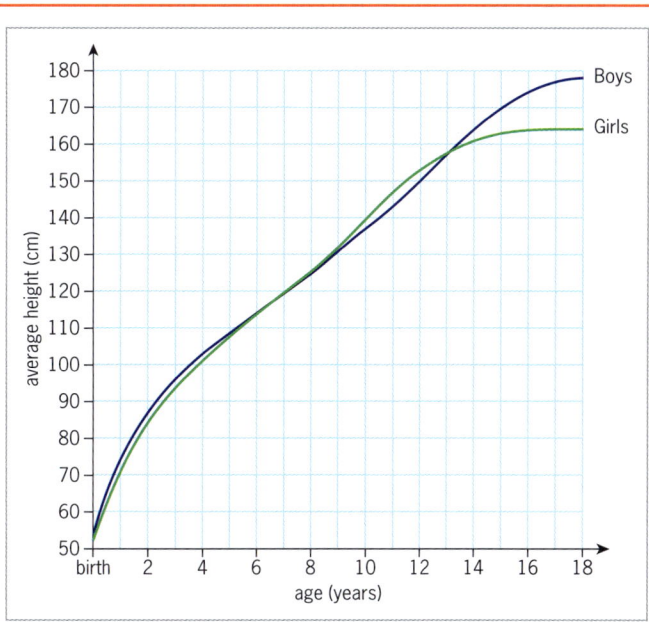

Figure 3.11 *Average height against age of boys and girls up to age 18*

increase. The average age for **girls** to start puberty is **earlier** than boys, therefore their growth rate begins to increase at a **younger age**, such that they become **taller** and **heavier** than boys of their age during this period.

- Towards the **end** of **puberty**, the growth rate of both girls and boys **decreases**. This happens at a **younger age** in girls than boys and, once again, **boys** become **taller** and **heavier** than girls of their age.
- **Girls** typically **stop growing** earlier than boys. Most girls reach their maximum size and stop growing by the age of about 14 to 16, while boys usually reach their maximum size at about 16 to 18. At this age, **boys** are, in general, significantly **taller** and **heavier** than girls.

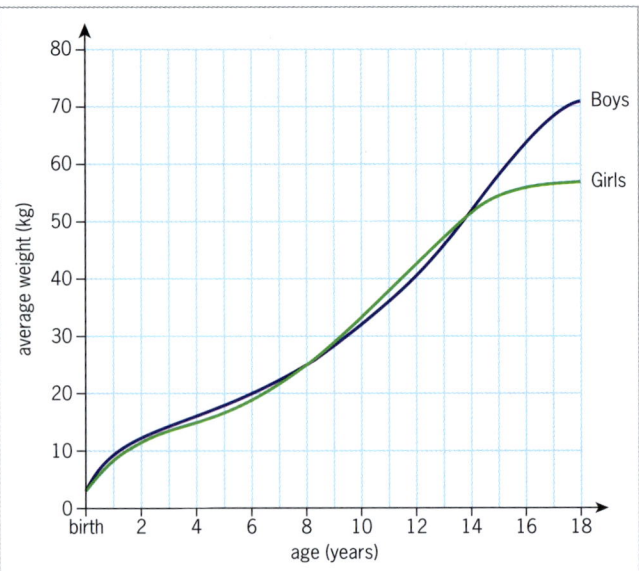

Figure 3.12 *Average weight against age of boys and girls up to age 18*

Human population growth and control

Populations grow when the **birth rate** is **higher** than the **death rate**. The **human population** is currently growing **rapidly** or **exponentially**. Estimates show that the world population was approximately 5 million in the year 8000 BC. It then took 9804 years to reach 1 billion in 1804 AD, and only 218 years to reach 8 billion in 2022.

Current estimates are that the human population will continue to increase, but at a **slower rate** compared to the recent past, reaching 9 billion by 2037 and 10 billion by 2058.

Problems arising from human population growth

Human population growth can lead to countries becoming **overpopulated**, which can cause various interconnected issues that impact the environment, society and economy of countries.

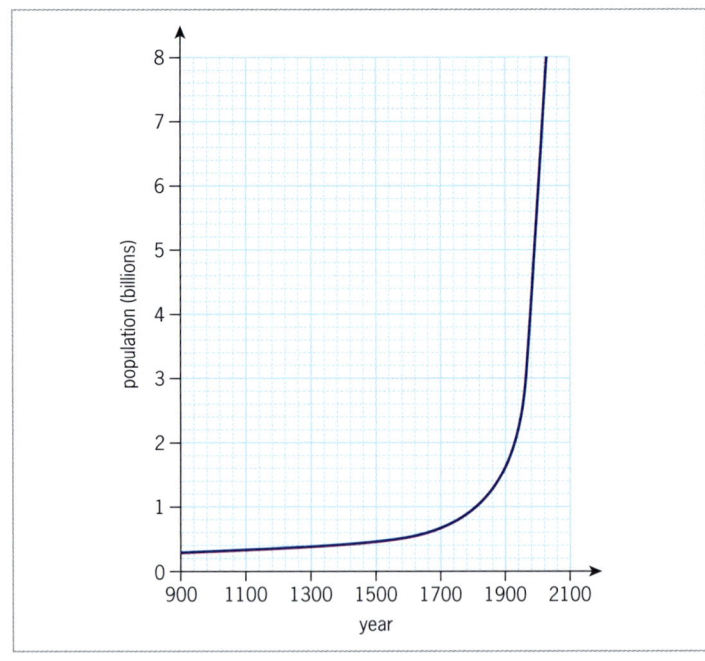

Figure 3.13 *The growth of the human population since 900 AD*

- It is causing a **depletion** of **natural** and **material resources**. Natural resources can be either non-renewable or renewable.
  - **Non-renewable resources** include fossil fuels, e.g. coal, crude oil and natural gas, and mineral resources, e.g. iron ore, bauxite (aluminium ore), zinc ore, copper, tin and gold. These cannot be replenished and some are **running out** rapidly.
  - **Renewable resources** include land, wildlife, forests, soil and water. These can be replenished, but many are being **depleted** at a faster rate than they can be replenished.

- **Material resources** are produced from natural resources, e.g. glass, plastics, textiles, processed foods, fertilisers and construction materials such as cement, concrete blocks, steel and lumber. Their production is **depleting** the natural resources from which they are made.

As the human population increases, the **demand** for these resources **increases**, which increases their rate of depletion and will, ultimately, lead to worldwide **resource shortages**.

- It can lead to **food insecurity**, as a rapidly growing human population has the potential to outpace agricultural production leading to food shortages, increased food prices, hunger and starvation, especially in developing countries with limited resources.
- It can lead to various other issues including increased **pollution** and the build-up of **waste**, **degradation** of the **environment**, **overcrowding**, increased **unemployment**, **social inequality**, **poverty** and **resource-based conflict**, and an increase in the **spread of disease**, all leading to decreased **living standards**.

Controlling and **reducing** human population growth should help to **reduce** the depletion of resources, pollution, environmental degradation, overcrowding, unemployment, social inequality, poverty, resource-based conflict and the spread of disease, and **increase** worldwide food security, standards of living and the overall quality of people's lives.

Effects of teenage pregnancy on human population growth

Teenage pregnancy contributes significantly to the **growth** of human populations in various ways, especially in developing countries.

- Girls who become **pregnant** as **teenagers** have many **more years** to continue having children than women who become pregnant for the first time later in life. Therefore they can have **more children** during their reproductive life, which **increases birth rates** within populations.
- The **cycle** of girls born to teenage mothers becoming pregnant themselves as teenagers can continue through generations if proper interventions are not put in place to break the cycle, and this contributes to **higher birth rates** within populations.
- High teenage pregnancy rates affect the **age distribution** within populations and create the 'youth bulge' phenomenon in which populations have a **higher proportion** of **young individuals** compared to older individuals. This affects education systems, labour markets, healthcare services and social welfare programmes, as they need to accommodate a younger population.

Effects of birth control on human population growth

Using **birth control** can play a significant role in **reducing** human population growth.

- Most birth control methods are highly effective in preventing **unintended pregnancies**. This allows individuals to control their own **fertility**, which helps to **reduce birth rates** in populations.
- Birth control allows parents to control **when** they have children, enabling them to **delay** having children until later in life, and this can **lower birth rates** in populations.
- Using birth control allows parents to control the **number** of children they have, which enables them to **restrict the size** of their family, thus **reducing birth rates** within populations.

Revision questions

1. Outline THREE different methods by which animals can reproduce asexually.

2. Construct a table to give the function(s) of EACH of the following parts of the female reproductive system: the uterus, the oviducts, the cervix, the vagina and the ovaries.

3. Describe the structure and state the function(s) of EACH of the following parts of the male reproductive system:
   a the penis   b the prostate gland   c the testes.

4. a Outline the events that occur in the uterus and ovaries during one complete menstrual cycle.
   b Explain the roles of oestrogen and progesterone in the menstrual cycle.
   c What happens during menopause?

5. Explain how ova and sperm are brought together and fertilisation occurs.

6. Outline the structure of Adrianna's placenta and explain its role in the development of her baby.

7. Describe the stages involved in the birthing process.

8. Construct a table that explains how EACH of the following methods of birth control prevents pregnancy, and gives ONE advantage and ONE disadvantage of EACH method: the contraceptive pill, surgical sterilisation, the Billings method and the condom.

9. Jacia finds out that she is pregnant. Outline some of the steps she should take to ensure the health of her developing baby during her pregnancy.

10. Describe growth patterns of girls and boys from birth until 18 years of age.

11. Outline the growth pattern of the human population from 900 AD to the present.

12. Discuss the need for human population growth to be controlled.

13. What effects does EACH of the following have on the growth of the human population?
    a Teenage pregnancy.   b Birth control.

4 Transport systems

Living organisms need to exchange substances constantly with their environment. They need to take in useful substances and get rid of waste. **Transport systems** provide a means by which these substances are moved between the surfaces where they are exchanged and body cells.

Transport systems in multicellular organisms

The need for a transport system in multicellular organisms

The absorption and transport of substances in living organisms are affected by **two** factors.

- The **limitations of diffusion**. Diffusion is a relatively slow process, therefore it is only effective over short distances.
- The **surface area to volume ratio** of an organism. This determines the effectiveness of diffusion occurring through the surface of the organism.

As organisms increase in size, their **surface area** increases at a slower rate than their **volume**. In other words, the **ratio** of the surface area to the volume of their body **decreases**.

Large, **multicellular organisms** have a **small** surface area to volume ratio. Because of this, diffusion through their limited body surface is not adequate to supply all their body cells with their requirements and remove their waste. Also, most of their body cells are too far from the body surface for substances to move through them quickly and efficiently enough by diffusion. These organisms have, therefore, developed **transport systems** to carry **useful substances** from specialised organs that absorb them, e.g. the lungs and small intestine, to body cells, and to carry **waste substances** from body cells to specialised organs that excrete them, e.g. the lungs and kidneys.

Materials transported in animals and plants

Nutrients, **gases**, **excretory products** and **metabolic products** are transported around the bodies of **animals**.

- **Products of digestion**, mainly **glucose**, **amino acids**, **vitamins** and **minerals**, are carried from the intestines, where they are absorbed, to the body cells.
- **Oxygen** is carried from the lungs, where it is absorbed, to the body cells which use it in respiration.
- **Carbon dioxide** is carried from the body cells, where it is produced in respiration, to the lungs which excrete it.
- **Nitrogenous waste**, mainly **urea**, is carried from the liver, where it is produced, to the kidneys, which excrete it.
- **Hormones** are carried from the glands that produce them, to their target organs.
- **Heat energy** is carried mainly from the liver and muscles, where it is produced predominantly by respiration, to the rest of the body.
- Other substances carried include **antibodies** and other **blood** or **plasma proteins**, and **water**.

The following materials are transported in **plants**.

- **Water** and **mineral nutrients (salts)** are carried from the roots, where they are absorbed, to the leaves.
- **Sucrose** and **amino acids** are carried from the leaves, where they are made, to all other plant organs.

The circulatory system in humans

The **circulatory system**, also known as the **cardiovascular system**, transports substances around the human body. The system consists of **three** basic components.
- **Blood**, which serves as the **medium** to **transport** substances around the body.
- **Blood vessels**, which are **tubes** through which the blood flows to and from all parts of the body.
- The **heart**, which **pumps** the blood through the blood vessels.

The composition of blood

Blood is a **tissue** composed of **three** types of cells: **red blood cells**, **white blood cells** and **platelets**. These cells are suspended in a fluid called **plasma**. The cells make up about 45% by volume of the blood and the plasma makes up about 55%.

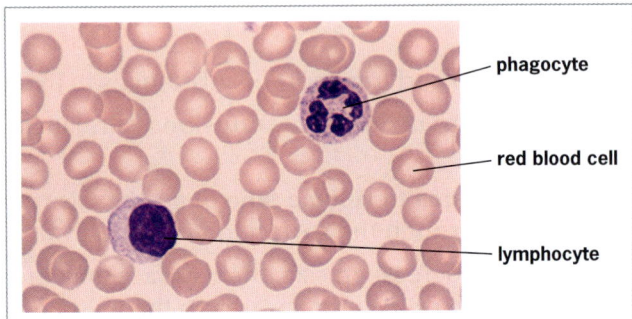

Figure 4.1 *Red and white blood cells (a phagocyte and a lymphocyte) under the microscope*

Plasma

Plasma is a yellowish fluid composed of about 90% **water** and 10% **dissolved substances**. The dissolved substances include the following.

- **Products of digestion**, mainly glucose, amino acids, vitamins and minerals.
- **Waste products**, mainly dissolved carbon dioxide and urea.
- **Hormones**, e.g. insulin and thyroxine.
- **Blood** or **plasma proteins**, e.g. fibrinogen, prothrombin, albumin and antibodies.

The main function of plasma is to **transport** all the dissolved substances around the body.

Blood cells

Table 4.1 *Structure and functions of blood cells*

Cell type and structure	Formation of cells	Functions
Red blood cells (erythrocytes) cell membrane / cytoplasm rich in **haemoglobin**, an iron-containing protein • **Biconcave discs** with a thin centre and relatively large surface area to volume ratio so gases can diffuse in and out easily. • Have **no nucleus**, therefore they cannot divide and can only live for about 3 to 4 months. • Contain the red pigment **haemoglobin**. • Slightly **elastic**, which allows them to squeeze through the narrowest capillaries.	• Formed in the red bone marrow found in flat bones, e.g. the pelvis, scapula, ribs, sternum, cranium and vertebrae, and in the ends of long bones, e.g. the humerus and femur. • Broken down mainly in the liver and the spleen.	• Transport **oxygen** as **oxyhaemoglobin** from the lungs to the body cells. • Transport small amounts of **carbon dioxide** from the body cells to the lungs.

Cell type and structure	Formation of cells	Functions
White blood cells (leucocytes) These are slightly larger than red blood cells and less numerous. Blood contains approximately 1 white blood cell to 600 red blood cells. There are two main types: 25% are **lymphocytes** and 75% are **phagocytes**.		
Lymphocytes *cell membrane* *large, round nucleus* *non-granular cytoplasm* • Have a **rounded** shape. • Have a large, **round nucleus** which controls the production of antibodies. • Have only a small amount of cytoplasm.	• Develop from cells in the red bone marrow and mature in other organs, e.g. lymph nodes, spleen and thymus gland.	• Produce **antibodies** to destroy disease-causing bacteria and viruses, also known as pathogens. • Produce **antitoxins** to neutralise toxins produced by pathogens.
Phagocytes *cell membrane* *lobed nucleus* *granular cytoplasm* • Have a **variable** shape. • Move by **pseudopodia** or **false feet**. They can move out of capillaries by passing between the cells in their walls and engulf pathogens using their pseudopodia by a process known as **phagocytosis**. • Have a **lobed nucleus**.	• Formed in the red bone marrow.	• Engulf and destroy pathogens. • Engulf pathogens destroyed by antibodies.
Platelets (thrombocytes) *cell membrane* *cytoplasm* • Cell **fragments**. • Have **no nucleus** and only live for about 10 days.	• Formed from cells in the red bone marrow.	• Help the blood to **clot** at a cut or wound.

Blood doping

Blood doping is an unethical and illegal method to **artificially** boost the blood's ability to carry oxygen to muscle cells, usually by increasing the number of **red blood cells** in the blood. It is carried out by athletes to improve their endurance and stamina, thereby improving their **athletic performance**. It can be done by **transfusing blood** into the athlete before a competition or by injecting the hormone **erythropoietin (EPO)** into the athlete to stimulate his or her body to produce higher than normal numbers of red blood cells. **Synthetic oxygen-carrying chemicals** can also be injected into the athlete's blood.

Blood vessels

There are **three** main types of blood vessels in the circulatory system: **arteries**, **capillaries** and **veins**.

Arteries carry blood **away** from the heart. On entering an organ, an artery branches into smaller arteries called **arterioles** that branch into a network of **capillaries**, which run throughout the organ. Capillaries join into small veins called **venules** that join to form a **vein**, which leads back from the organ **towards** the heart. The **exchange** of substances between the blood and body cells occurs in the **capillaries**.

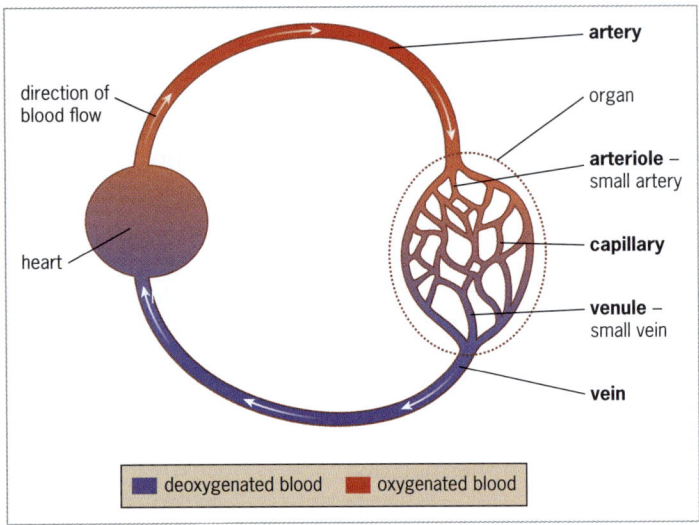

Figure 4.2 *The relationship between the different blood vessels*

Arteries carry blood which is under **high pressure**, therefore they have thick muscular walls to withstand the pressure. **Veins** carry blood which is under **low pressure**, therefore they have **thinner** muscular walls and **valves** to prevent the blood from flowing backwards. **Capillaries** have walls which are **one cell thick** so that nutrients and oxygen can easily pass from the blood into body cells, and carbon dioxide and other waste can pass back into the blood.

The **major arteries** in the human body are the **aorta** and the **pulmonary artery**, which carry blood directly out of the heart. The **major veins** are the **anterior vena cava**, **posterior vena cava** and the **pulmonary veins**, which carry blood directly into the heart.

The heart

The **heart** acts as a **pump** to maintain a constant circulation of blood around the body. The walls of the heart are composed of **cardiac muscle**, which contracts without nerve impulses and does not get tired.

The heart is divided into **four** chambers. The two on the right contain **deoxygenated blood** and are completely separated from the two on the left, which contain **oxygenated blood**, by the **septum**.

- The top two chambers, called **atria**, have thin walls and collect blood entering the heart via the **anterior** and **posterior vena cavae** and **pulmonary veins**. Their walls are **thin** because they only have to pump blood a short distance into the ventricles below.
- The bottom two chambers, called **ventricles**, have thick walls and pump blood out of the heart via the **pulmonary artery** and **aorta**. Their walls are **thick** because they have to pump blood longer distances to the lungs and around the body, therefore they have to pump with more force.

Valves are present between each atrium and ventricle, and in the pulmonary artery and aorta as they leave the ventricles, to ensure that blood flows through the heart in **one direction**. The **coronary arteries** branch from the aorta as it leaves the heart and supply the muscle of the heart with oxygen.

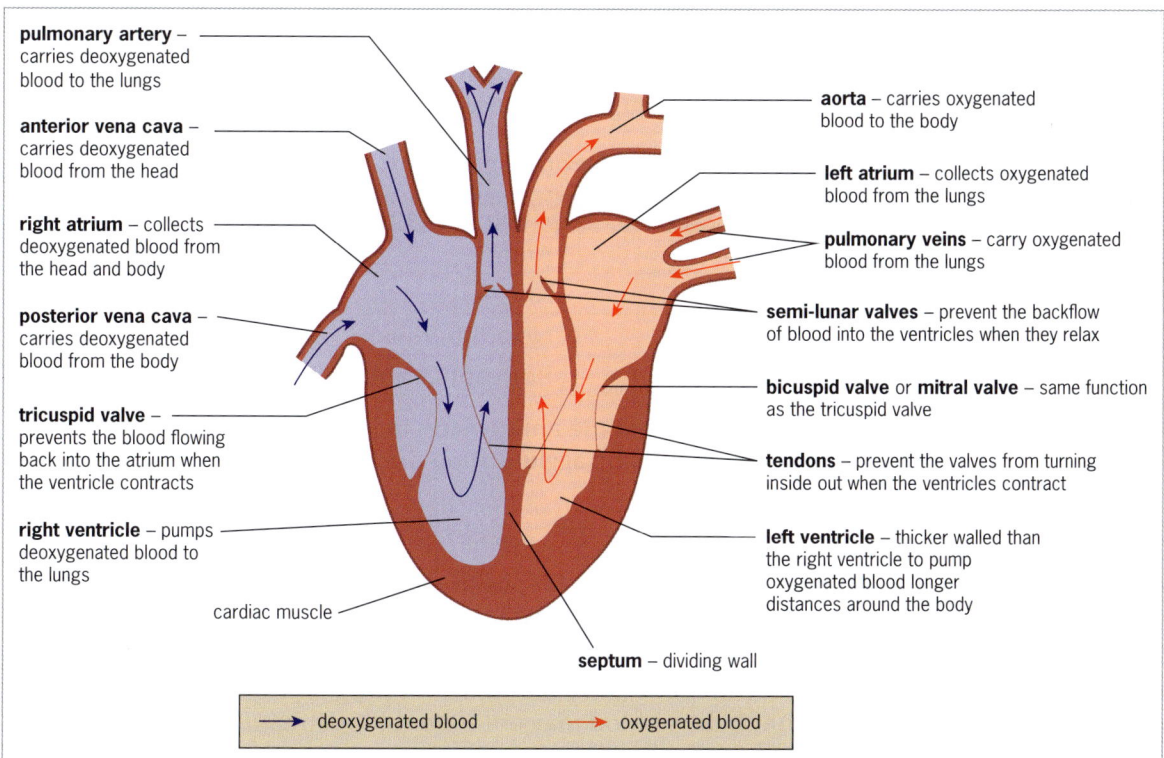

Figure 4.3 *The structure of the human heart and the functions of the different parts*

Heartbeat and circulation

As the **heart beats**, the atria and ventricles at both sides contract and relax together. A group of **specialised cells** in the wall of the **right atrium**, called the **pacemaker**, controls the rhythm of a normal heartbeat. The **contraction** of a chamber is called **systole** and its **relaxation** is called **diastole**. One **cardiac cycle** or **heartbeat** involves the following.

- **Diastole** – the **atria** and **ventricles relax** together, the semi-lunar valves close, the atria fill up with blood from the anterior and posterior vena cavae and pulmonary veins, and the blood flows into the ventricles. This takes 0.4 seconds.
- **Atrial systole** – the **atria contract** together forcing any remaining blood through the tricuspid and bicuspid valves into the ventricles. This takes 0.1 second.
- **Ventricular systole** – the **ventricles contract** together, the tricuspid and bicuspid valves close and blood is forced through the semi-lunar valves into the pulmonary artery and aorta. This takes 0.3 seconds.

During one complete **circulation** around the body, the blood flows through the heart **twice**. It travels from the **right ventricle** through the **pulmonary arteries** to the **lungs** to pick up oxygen and lose carbon dioxide, and travels back via the **pulmonary veins** to the **left atrium**. It then travels from the **left ventricle** through the **aorta** to the **body** where it gives up oxygen to the body cells and picks up carbon dioxide, and travels back via the **anterior** or **posterior vena cavae** to the **right atrium**.

Blood groups

Blood can be classified into different **blood groups** or **types** based on chemicals present on the surface of red blood cells known as **antigens**. A person's blood type is **inherited** from his or her parents and there are **two** main grouping systems: the **ABO blood group system** and the **Rhesus blood group system**.

The ABO blood group system

The **ABO system** divides blood into **four** groups: **group A**, **group B**, **group AB** and **group O**. These are determined by the presence or absence of two **antigens** and also two **antibodies** in the plasma.

Table 4.2 *Antigens and antibodies of the ABO blood grouping system*

Blood group	Antigen on the surface of red blood cells	Antibody in the plasma
Group A	A	Anti-B
Group B	B	Anti-A
Group AB	Both A and B	No antibodies
Group O	No antigens	Both anti-A and anti-B

The antibodies in the plasma must be **different** from the antigens on the red blood cells. If they are the same, the antibodies bind to the antigens causing **agglutination** or **clumping** of the red blood cells.

The Rhesus (Rh) blood group system and risks in pregnancy

The **Rhesus** or **Rh system** divides blood into **two** groups: **Rh-positive** and **Rh-negative**. These are determined by the presence or absence of an **antigen** known as the **Rh antigen** or **Rh factor**. If the antigen is **present**, the person has Rh-positive blood; if it is **absent**, the person has Rh-negative blood.

The **Rh antigen** poses a **risk** to a woman with **Rh-negative** blood who is pregnant with a baby that has **Rh-positive** blood. A small amount of her baby's blood may enter her bloodstream, especially during labour, and cause her to produce **anti-Rh antibodies**. During any **subsequent pregnancies** with Rh-positive babies, these antibodies can pass across the placenta and attack the baby's red blood cells causing **Rhesus disease**, characterised by anaemia and jaundice. To **prevent** this, the mother is given an **injection** of **anti-D** at about 28 weeks into her pregnancy and again immediately after delivery if her baby is Rh-positive. This stops her from making anti-Rh antibodies.

Precautions for blood transfusions

A **blood transfusion** involves giving blood from a healthy person to a person who needs blood. Certain **precautions** have to be followed when **handling** and **transfusing** blood.

- Persons handling the blood must **avoid direct contact** with it, e.g. by wearing medical gloves.
- Blood should **not** be taken from a person who is pregnant or has anaemia.
- Donated blood must be **screened** for pathogens, e.g. HIV, hepatitis B and C.
- Blood from the donor must be **cross-matched** with the recipient's blood to ensure that their blood groups are **compatible**. This prevents **agglutination** of red blood cells in the **donated** blood. If agglutination occurs, blood vessels may become blocked and the agglutinated cells **disintegrate**, which can be fatal. To ensure compatibility, the recipient's blood must not contain antibodies against the antigens in the donor's blood.

Type **O Rh-negative** blood or **O–** is known as the **universal donor** type because it has no **A**, **B** or **Rh** antigens, so it can be given to anybody. Type **AB Rh-positive** blood or **AB+** is known as the **universal recipient** type because it has no **A**, **B** or **Rh** antibodies, so a person with it can receive blood of any type.

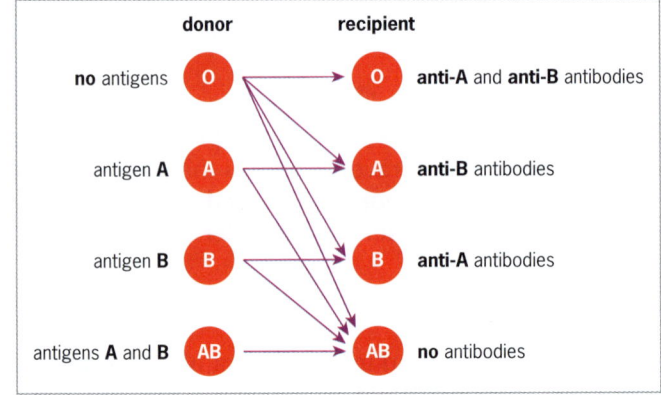

Figure 4.4 *ABO blood group system donors and recipients*

Revision questions

1. Explain why the human body needs a transport system.
2. Name FIVE materials that are transported around the human body and FOUR materials that are transported in plants.
3. a Name the liquid part of blood and state its MAIN function.
   b Identify the THREE types of cells found in blood, describe the structure of EACH type and state the function(s) of EACH type.
4. Explain the relationship between arteries, capillaries and veins.
5. Name the major blood vessels connected to the heart and state which chamber EACH leads into or out of.
6. Explain:
   a Why the walls of the ventricles of the heart are thicker than the walls of the atria.
   b Why the wall of the left ventricle is thicker than the wall of the right ventricle.
   c How blood flow through the heart is maintained in one direction.
7. Distinguish among diastole, atrial systole and ventricular systole.
8. Why do different blood groups exist? Your answer must include the names of the TWO main blood grouping systems in humans.
9. Explain the potential risk to a woman with Rh-negative blood who wishes to have children, and what precautions she must take if she is found to be carrying a baby with Rh-positive blood.
10. Identify THREE precautions that should be taken when handling and transfusing blood.

Transport systems in plants

Structure and functions of xylem and phloem

Substances are transported around plants by **vascular tissue** which is composed of **two** different tube-like tissues: **xylem tissue** and **phloem tissue**. Vascular tissue runs throughout roots and stems of plants, and it makes up the midrib and veins of leaves.

Xylem tissue

Xylem tissue is made up of **xylem vessels** grouped closely together. **Xylem vessels** are long, extremely narrow, hollow, non-living tubes and they transport **water** and dissolved **mineral salts** from the roots, where they are absorbed, up stems and into the leaves for use in photosynthesis. The main process responsible for moving water upwards through the xylem vessels is **transpiration** (see pages 44 to 45).

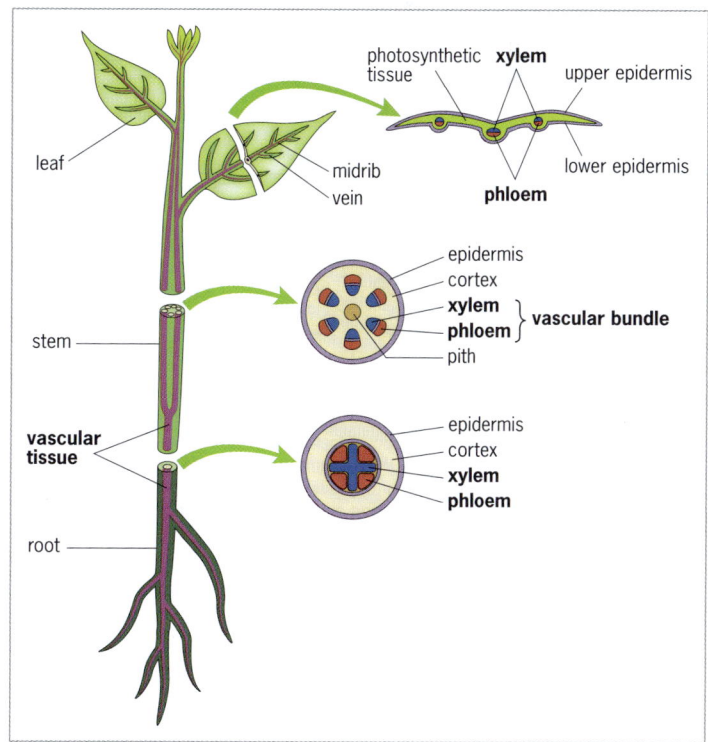

Figure 4.5 *The arrangement of vascular tissue in a plant*

4 Transport systems 43

Phloem tissue

Phloem tissue is made up of **phloem sieve tubes** and **companion cells**, grouped closely together. **Phloem sieve tubes** are long, narrow tubes containing living cytoplasm, but no nuclei, and they have perforated cross walls along their length. They transport **soluble organic food**, mainly **sucrose** and some **amino acids**, from leaves where it is made to all other plant parts. The food moves into and out of the sieve tubes by **active transport**.

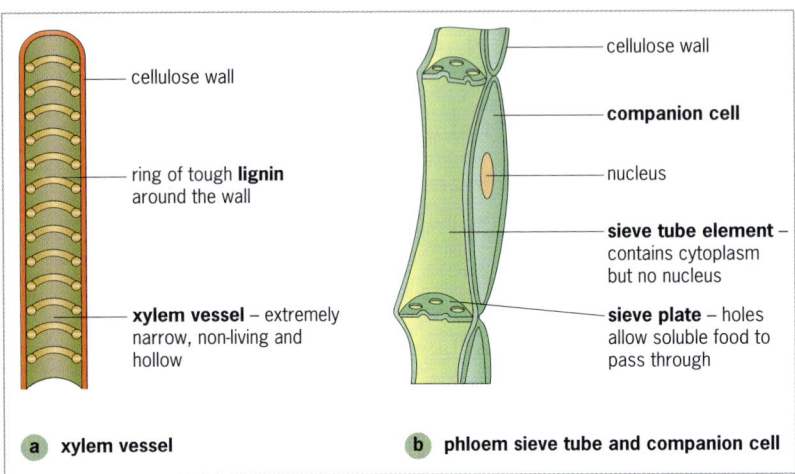

Figure 4.6 *Longitudinal section through a xylem vessel and a phloem sieve tube and companion cell*

Transpiration and its role in plants

Transpiration is the loss of water vapour from the surface of leaves.

Water moves in a **continuous stream** through the xylem vessels of a plant from the roots, where it is absorbed, up the stem and into the leaves. **Transpiration** provides the suction, known as the **transpiration pull**, which is necessary to draw the water up into the leaves.

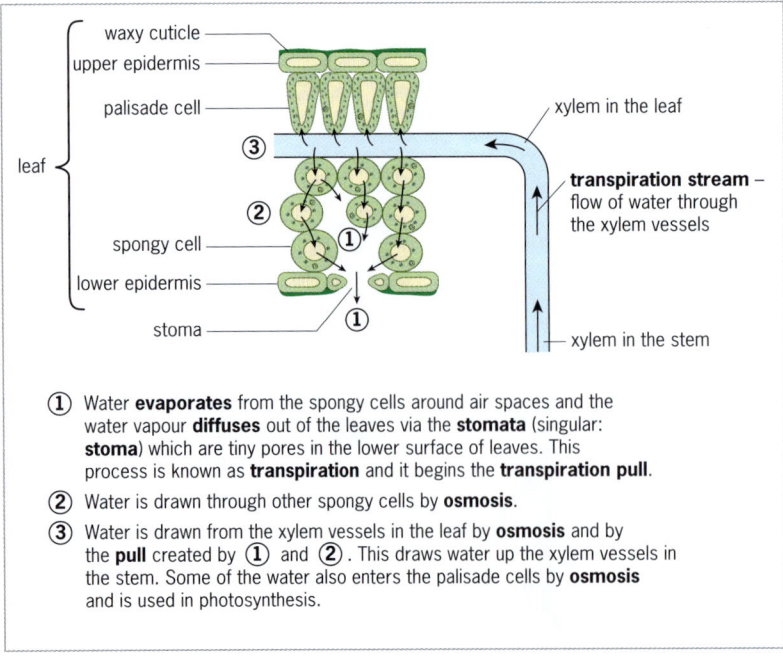

Figure 4.7 *The role of transpiration in drawing water up xylem vessels*

Transpiration is important to plants for the following reasons.
- It draws water from the soil up to leaves for use in **photosynthesis**.
- Moving water carries dissolved **mineral salts** up to the leaves for healthy growth and development.
- It supplies plant cells with water to keep them **turgid**. This **supports** non-woody stems and leaves.
- Evaporation of water from the surface of leaves **cools** the leaves as it removes heat energy.

Environmental factors affecting the rate of transpiration

If the water supply in the soil is **plentiful**, the **rate of transpiration** is controlled by the interaction of various **environmental factors** that affect the rate at which water **evaporates** and water vapour **diffuses**. These factors include **temperature**, **humidity**, **wind speed** and **light intensity**.

- As **temperature increases** the rate of transpiration **increases**. The rate at which water evaporates and water vapour diffuses, and the amount of water vapour the air can hold, increase as temperature increases.
- As **humidity decreases** the rate of transpiration **increases**. The amount of water vapour in the air decreases as humidity decreases, which increases the concentration gradient between the water vapour in the air spaces of leaves and in the surrounding air, and increases the rate of diffusion.
- As **wind speed increases** the rate of transpiration **increases**. The increase in wind speed increases the speed at which water vapour is carried away from the leaves.
- As **light intensity increases** the rate of transpiration **increases**. The stomata open more fully as light intensity increases, allowing more water vapour to diffuse out.

Revision questions

11 Construct a table that names the TWO types of tissue that transport substances around plants, outlines the structure of EACH tissue, identifies the substance(s) carried by EACH tissue, and states where EACH substance is carried from and its destination.

12 a What do you understand by the term 'transpiration'?
b Explain the role played by transpiration in drawing water up the stem of a hibiscus plant.

13 Outline the role played by transpiration in the life of a flowering plant.

14 Account for the fact that transpiration is slower on a cool, humid, overcast day than on a hot, dry, sunny day.

5 Excretion

Chemical reactions occurring in living organisms constantly produce **waste** and **harmful** substances which the organisms must get rid of from their bodies. **Excretion** is responsible for getting rid of these substances. It is also essential that the body's internal environment is kept constant, and excretion plays a part in this by helping to keep certain conditions surrounding cells constant.

Excretion and egestion in living organisms

Excretion is the process by which waste and harmful substances, produced by the body's metabolism, are removed from the body.

The **body's metabolism** refers to all the **chemical reactions** occurring within cells, many of which produce waste and harmful substances. Excretion is **important** for the following reasons.

- It prevents **toxic** metabolic waste from building up in the body and damaging or killing cells.
- It is important in **homeostasis**, i.e. keeping conditions surrounding body cells **constant** so that they are **optimum** for enzyme action (pages 109 to 110) and for cells to function properly.

Egestion is the process by which undigested dietary fibre and other undigested materials are removed from the body as faeces.

Excretion and **egestion** must **not** be confused. The undigested dietary fibre and other indigestible materials that are removed during egestion are **not** produced by the body's metabolism, so their removal cannot be classed as excretion.

Excretory products and organs in humans

Humans produce the following waste products during metabolism.

- **Carbon dioxide** is produced by **respiration**, which occurs in all body cells.
- **Water** is produced by **respiration**, which occurs in all body cells.
- **Nitrogenous waste**, mainly **urea**, is produced by the liver from excess **amino acids** in the diet (see page 112).
- **Bile pigments** are produced by the **breakdown of haemoglobin** from red blood cells in the liver.
- **Heat** is produced in general metabolism, especially **respiration**.

Table 5.1 *Excretory organs in humans and their excretory products*

Excretory organ	Products excreted
Kidneys	• Water, urea and salts are excreted in **urine** (see Figure 5.3, page 48).
Lungs	• Carbon dioxide and water vapour are excreted in **exhaled air** during **exhalation** (see page 117).
Skin	• Water, urea and salts are excreted in **sweat** (see Figure 5.5, page 50). • Heat is also excreted.
Liver	• Bile pigments are excreted in **bile**, which passes into the intestines and is excreted in faeces. • Urea is made in the liver, carried by the blood to the kidneys and excreted.

The kidneys

Humans have two **kidneys** which form part of the **urinary system**. Kidneys have **two** main **functions**.

- They **excrete** metabolic waste, mainly urea, from the body as **urine**.
- They **regulate** the volume and concentration of blood plasma and body fluids by regulating the amount of water they contain, a process known as **osmoregulation** (see page 48).

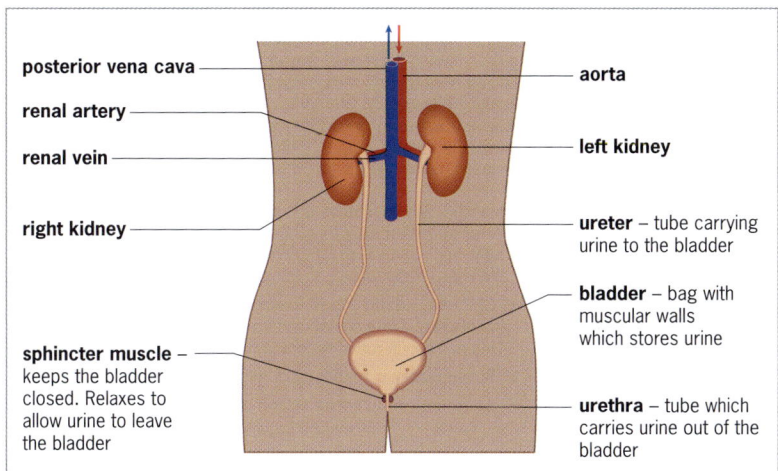

Figure 5.1 *Structure and functions of parts of the urinary system in a human*

The kidneys and excretion

Each **kidney** is divided into **three** regions: an outer **cortex**, an inner **medulla** and a central hollow region, the **pelvis**, which leads into the top of the **ureter**. Each kidney is composed of about 1 million thin-walled **kidney tubules** or **nephrons** that produce urine. A **renal artery** carries oxygenated blood containing urea from the aorta to each kidney and a **renal vein** carries deoxygenated blood lacking urea back to the posterior vena cava.

Each **nephron** begins with a cup-shaped **Bowman's capsule** in the cortex, which surrounds an intertwined cluster of capillaries, called a **glomerulus** (plural: **glomeruli**). After the Bowman's capsule, each nephron is divided into **three** sections.

- The **first convoluted (coiled) tubule** or **proximal convoluted tubule** in the cortex.
- The **loop of Henle** in the medulla.
- The **second convoluted (coiled) tubule** or **distal convoluted tubule** in the cortex.

An **afferent arteriole**, which branches from the renal artery, leads into each glomerulus. An **efferent arteriole** leads out of each glomerulus and branches to form a **network of capillaries** that wrap around each nephron and then join into a venule, which leads into the renal vein. Nephrons join into **collecting ducts** in the cortex and these ducts lead through the medulla and out into the pelvis.

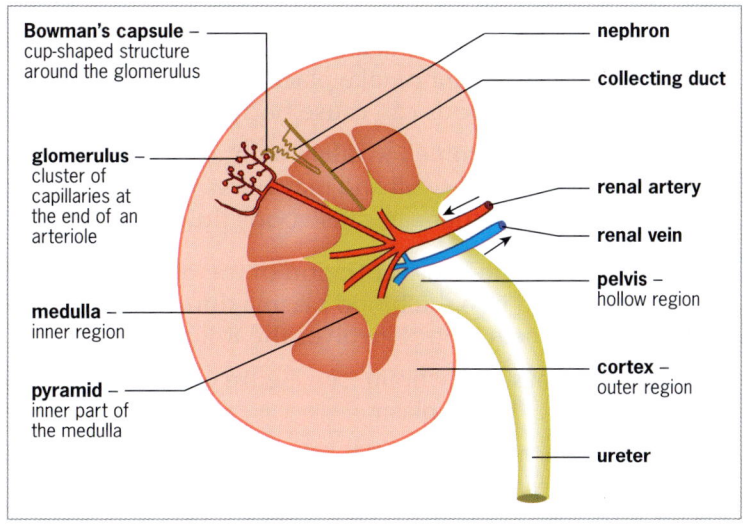

Figure 5.2 *A longitudinal section through a kidney showing the position of a nephron*

5 Excretion

Urine is produced in the nephrons by **two** processes.
- **Ultra-filtration** or **pressure filtration**.
- **Selective reabsorption**.

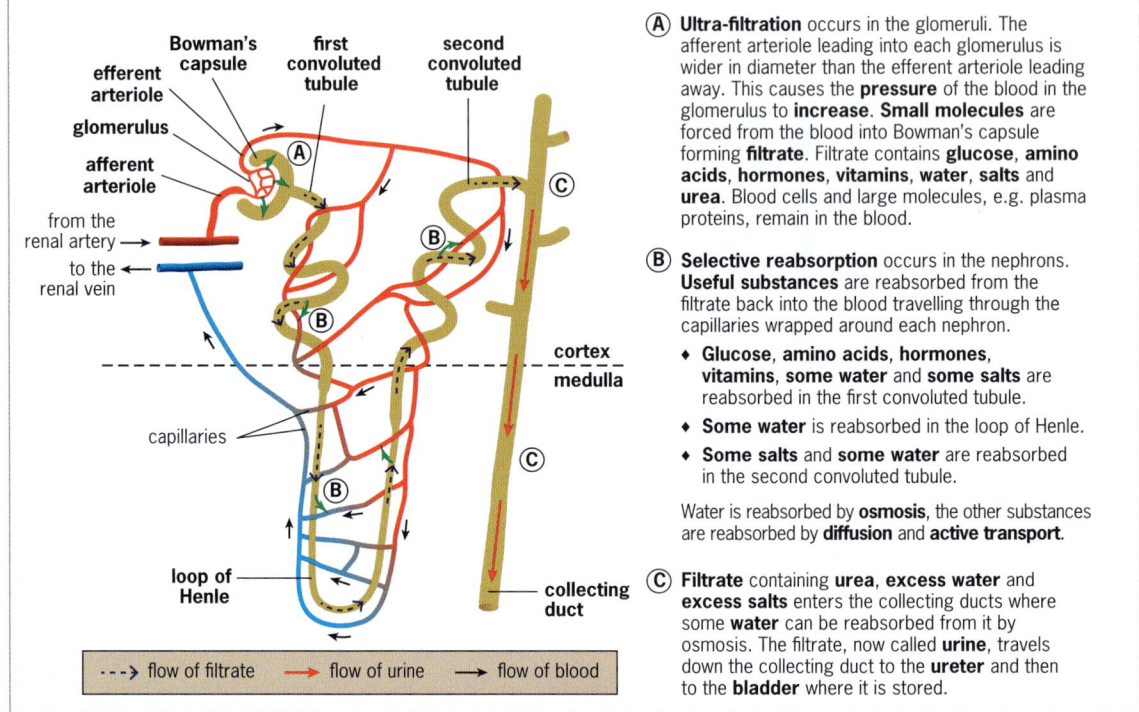

Figure 5.3 *Detailed structure of a nephron explaining how urine is formed*

The kidneys and osmoregulation

Osmoregulation is the regulation of the concentration of blood plasma and body fluids.

The concentration of blood plasma and body fluids must be kept **constant** to prevent water moving into and out of body cells unnecessarily.

- If the body fluids become **too concentrated** (contain too little water), water will **leave** body cells by osmosis. The cells will shrink and the body will become **dehydrated**. If too much water leaves cells, metabolic reactions cannot take place and cells die. Not drinking enough, excessive sweating or eating a lot of salty foods can cause body fluids to become too concentrated.
- If the body fluids become **too dilute** (contain too much water), water will **enter** body cells by osmosis. The cells will swell and may burst. Drinking a lot of liquid or sweating very little because of being in cold weather can cause body fluids to become too dilute.

The **kidneys** regulate the concentration of the blood plasma and body fluids by controlling how much **water** is reabsorbed into the blood plasma during **selective reabsorption**. This determines how much water is lost in urine. Control involves the following.

- The **hypothalamus** of the **brain** (see Figure 6.15, page 61), which detects changes in the concentration of blood plasma.
- The **antidiuretic hormone (ADH)**, which is released by the **pituitary gland** at the base of brain in response to messages from the hypothalamus. ADH is carried by the blood to the kidneys where it controls the **permeability** of the walls of the second convoluted tubules and collecting ducts to water.

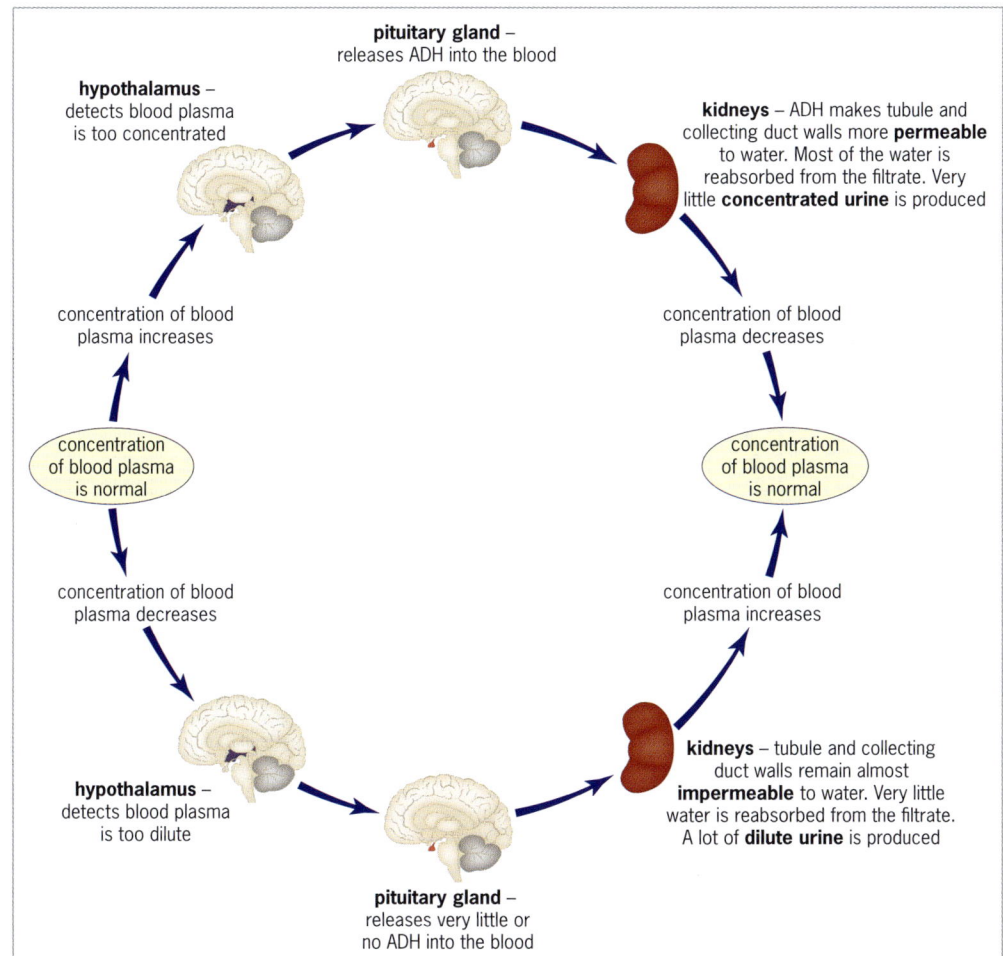

Figure 5.4 *Osmoregulation explained*

Dialysis for malfunctioning kidneys

When nephrons stop functioning properly, the kidneys are unable to **remove waste** from the blood and regulate the **volume** and **composition** of blood plasma and body fluids. Harmful waste, especially urea, builds up in the blood and can reach toxic levels, resulting in death. Malfunctioning kidneys can be treated by a **kidney transplant** or **renal dialysis**.

During **renal dialysis**, blood from a blood vessel, usually in the arm, flows through a **dialysis machine** or **dialyser** and is returned to the body. In the machine, **waste products**, mainly **urea**, together with **excess water** and **excess salts**, are removed from the blood, and the volume and composition of the blood plasma and body fluids are regulated. Dialysis must occur at regular intervals; most people require three sessions per week, each lasting 4 hours.

The skin

The **skin** is the largest organ in the human body and it has a variety of important functions. These include **excreting waste** in the form of **sweat** and **regulating** the **body temperature**.

The skin and excretion

The **skin** is made up of **three** layers:

- The **epidermis** is the outermost layer. It is waterproof and made of **three** layers (see Figure 5.5).
- The **dermis** is below the epidermis. It is made of **connective tissue** and has **nerve endings**, **hair follicles**, **sweat glands** and **blood vessels** throughout.
- The **subcutaneous layer** is the innermost layer. It is composed of **adipose tissue**, which is composed mainly of **fat cells**.

Sweat is produced by the **sweat glands** of the skin when the body temperature rises above 37 °C.

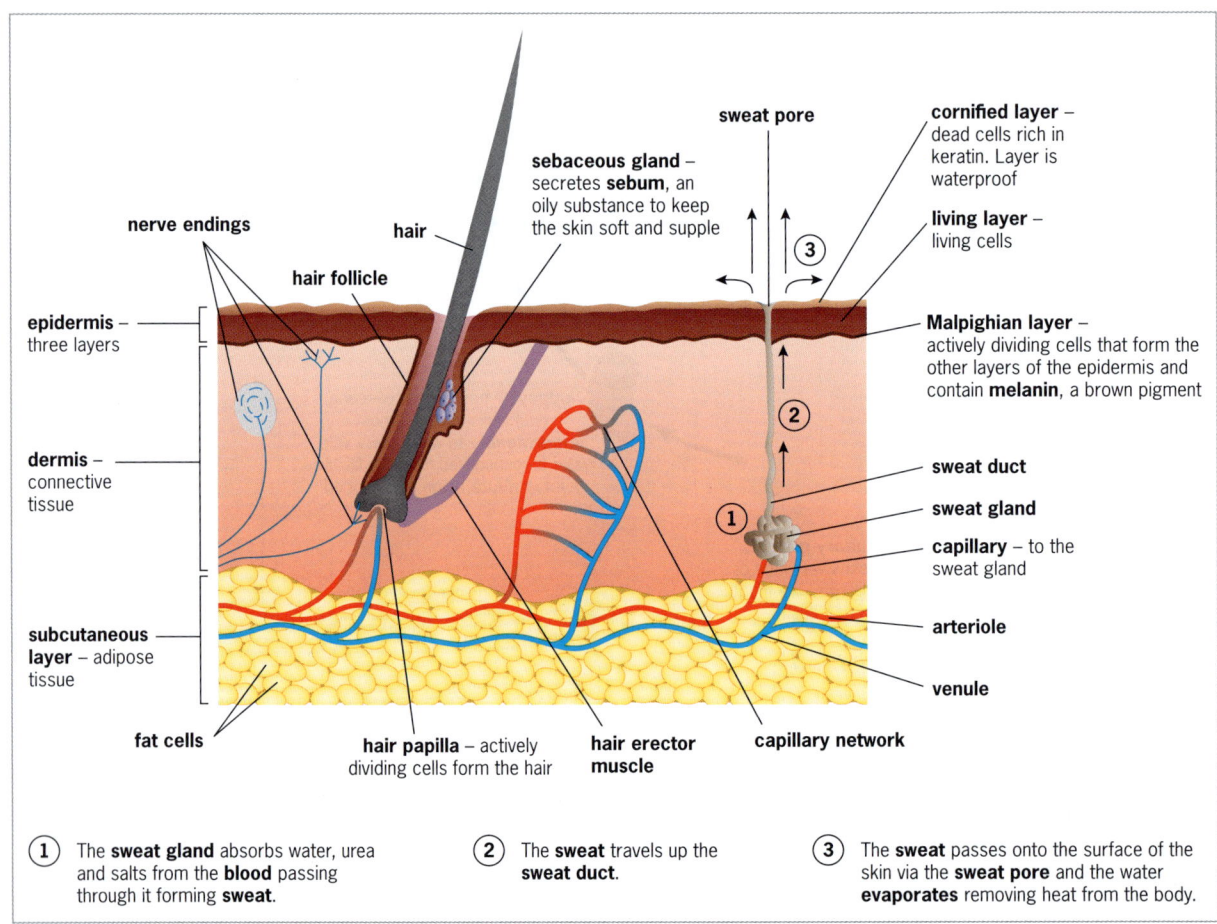

★**Figure 5.5** *A section through the human skin showing its structure and the mechanism of sweat formation*

The skin and temperature control

Humans must maintain a **constant** internal body temperature of about **37 °C** for **enzymes** to function properly (see page 110). Most heat is **gained** from internal **metabolic processes**, mainly respiration, and the blood carries this heat around the body. Heat is **lost** mainly by conduction, convection and radiation through the skin, and evaporation of water during **sweating**.

The **human skin** plays a major role in regulating body temperature. The **hypothalamus** of the brain detects changes in the temperature of the blood flowing through the brain and sends messages along nerves to appropriate structures in the skin, causing them to respond.

Table 5.2 *How the skin helps to maintain a constant body temperature*

Body temperature rises above 37 °C	Body temperature drops below 37 °C
Sweating occurs: water in the sweat **evaporates** and removes heat from the body.	**Sweating stops**: there is no water to evaporate and remove heat from the body.
Vasodilation occurs: blood vessels in the dermis of the skin **dilate** so more blood flows through them and more heat is lost to the environment from the blood.	**Vasoconstriction occurs**: blood vessels in the dermis of the skin **constrict** so very little blood flows through them and very little heat is lost. The heat is retained by the blood flowing through vessels deeper inside the body.
Hair erector muscles relax: this causes the hairs to **lie flat** so no insulating layer of air is created.	**Hair erector muscles contract**: this causes the hairs to **stand up** and trap a layer of air next to the skin, which acts as **insulation**. This prevents heat loss in hairy mammals and creates 'goose bumps' in humans.

Excretion in flowering plants

Plants produce the following **waste substances** during metabolism.

- **Oxygen** is produced in **photosynthesis**, which occurs during daylight hours. It is excreted during the **day** when the rate of photosynthesis is higher than the rate of respiration, and more oxygen is being produced than is being used in respiration.
- **Carbon dioxide** is produced in **respiration**, which occurs at all times. It is excreted during the **night** when no photosynthesis is occurring to use the carbon dioxide produced.
- **Water** is produced in **respiration**. It is excreted as water vapour during the **night** when no photosynthesis is occurring to use the water produced.

Plants do not have any specialised excretory organs. **Oxygen**, **carbon dioxide** and **water vapour** are all excreted by **diffusing** out through tiny pores in the undersurface of the leaves, called **stomata**, and through small areas of loosely packed cells in the bark of bark-covered stems and roots, called **lenticels**.

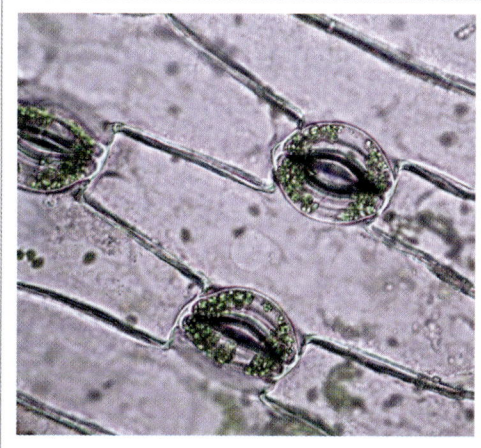

Figure 5.6 *Stomata in the lower surface of a leaf seen under the microscope*

Figure 5.7 *Lenticels in the bark of a tree trunk*

Revision questions

1. **a** Distinguish between excretion and egestion.
   b Explain why egestion is not considered to be a form of excretion.

2. Explain what will happen to an organism if excretion does not occur.

3. List the different excretory organs in humans and state what EACH organ excretes.

4. Identify the MAIN functions of the human kidneys.

5. Explain how urine is produced in Mario's kidney.

6. **a** Name the process by which the water content of body fluids is regulated.
   b Why is it important to regulate the water content of body fluids?

7. Tamesha plays tennis all day in the hot sun and drinks very little. What effect does her behaviour have on the quantity and concentration of her urine? Provide an explanation for your answer.

8. A person suffering from kidney failure can be treated using dialysis. Outline what happens during dialysis and explain why the treatment must occur at regular intervals.

9. **a** Why should humans maintain a constant internal body temperature?
   b Explain the changes that occur in Kia's skin if her body temperature rises above 37 °C.

10. **a** Explain how plants excrete their metabolic waste.
    b Over a 24-hour period, plants excrete both oxygen and carbon dioxide, but animals only excrete carbon dioxide. Provide an explanation for these facts.

6 Sense organs and coordination

Humans must constantly monitor their environment and **respond** appropriately to any changes that they detect in this environment to help them survive. To do this, two systems are involved: the **nervous system** and the **endocrine** or **hormonal system**, and the **sense organs** play a critical role in detecting environmental changes.

Some important definitions

- A **stimulus** is a change in the internal or external environment of an organism that initiates a response.
- A **response** is a change in an organism or part of an organism that is brought about by a stimulus.
- A **receptor** is the part of the organism that **detects** the stimulus.
- An **effector** is the part of an organism that **responds** to the stimulus.

Sense organs in humans

Humans have **five sense organs** (see Table 6.1) which contain specialised **receptor cells** that detect changes in the environment or **stimuli**. The cells turn these stimuli into **electrical impulses** or **nerve impulses**. The impulses travel along nerves to the **brain**, which interprets them as **sensations** of **seeing**, **hearing**, **smelling**, **tasting** and **touching**.

Table 6.1 *Sense organs in the human body*

Sense organ	Specialised receptor cells	Stimuli detected
Eyes	Rods and cones (photoreceptors) in the retina.	• Light. Light intensity and colour are detected.
Ears	Hair cells (mechanoreceptors) in the inner ear.	• Sound waves. • The position and movement of the head.
Nose	Olfactory cells (chemoreceptors) in the top of the nasal cavities.	• Chemicals in the air.
Tongue	Taste receptor cells (chemoreceptors) in the taste buds on its upper surface.	• Chemicals in food. Five tastes are detected: sweet, sour, salty, bitter and umami (savoury).
Skin	Touch receptor cells.	• Touch and texture.
	Pressure receptor cells.	• Pressure.
	Pain receptor cells.	• Pain and itching.
	Temperature receptor cells.	• Hot and cold.

The eye

The **eye** detects **light** that has been reflected from an object and converts it into **electrical impulses**. The impulses are transmitted along the **optic nerve** to the brain and the brain translates them into a precise picture of the object.

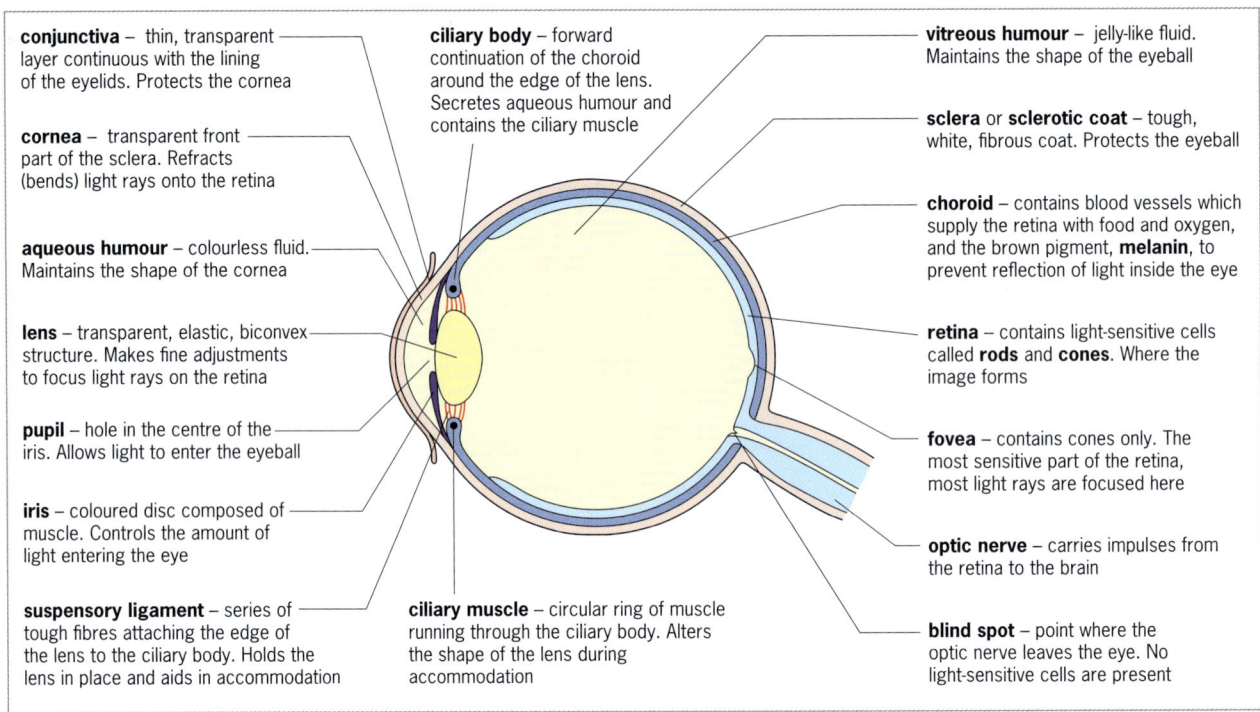

★Figure 6.1 *Structure and functions of the parts of the human eye, as seen in longitudinal section*

Image formation

In order to see, light rays reflected from an object must be **refracted** (bent) as they enter the eye so that they form a clear **image** of the object on the receptor cells of the retina. Being **convex** in shape, both the **cornea** and the **lens** refract the light rays; the cornea refracts them to the greatest extent and the lens then makes fine adjustments to focus them onto the retina.

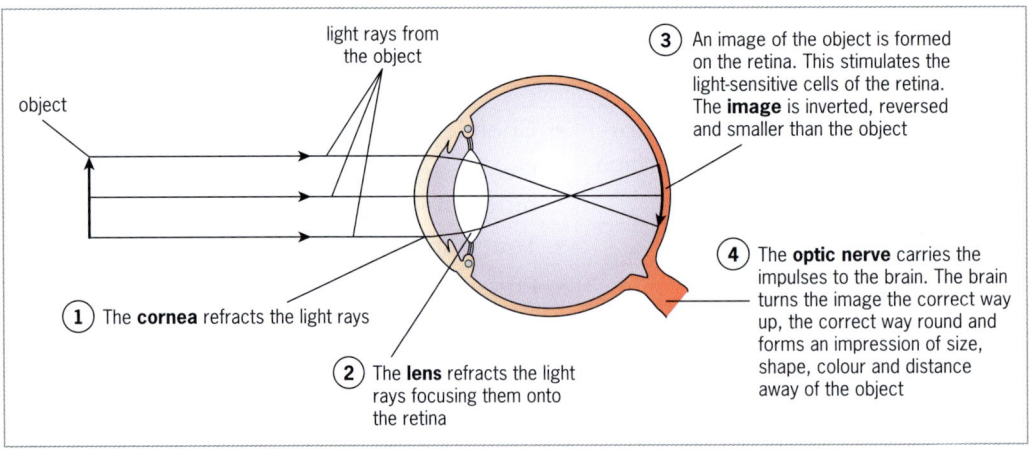

★Figure 6.2 *Formation of an image in the eye*

Detection of light intensity and colour by the eye

The **retina** is composed of **two** types of specialised **light-sensitive cells** or **photoreceptors**.
- **Rods** function in **low light intensities**. They detect the **brightness** of light and are mainly located around the sides of the retina. Images falling on the rods are seen in **shades** of black and white.

- **Cones** function in **high light intensities**. They detect **colour** and **fine detail**, and are mainly located around the back of the retina. The **fovea** is composed entirely of cones packed closely together. Light rays focusing onto the fovea produce the **sharpest** image. There are **three** types of cones which detect either the **red**, **green** or **blue** wavelengths of light.

Control of the amount of light entering the eye

The size of the **pupil** controls the amount of light entering the eye, and the **circular** and **radial muscles** of the **iris** control the pupil size.

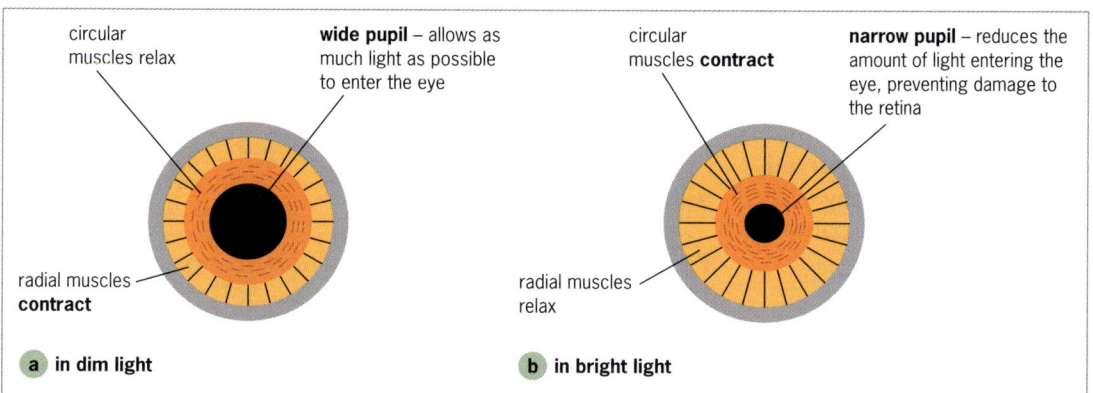

★ Figure 6.3 *Controlling the amount of light entering the eye*

Focusing light onto the retina – accommodation

Accommodation is the process by which the shape of the lens is changed to focus light coming from different distances onto the retina.

By its **shape** changing, the **lens** makes fine adjustments to focus light rays onto the retina. This change in shape is brought about by the **ciliary muscle** in the ciliary body and the **elasticity** of the **lens**.

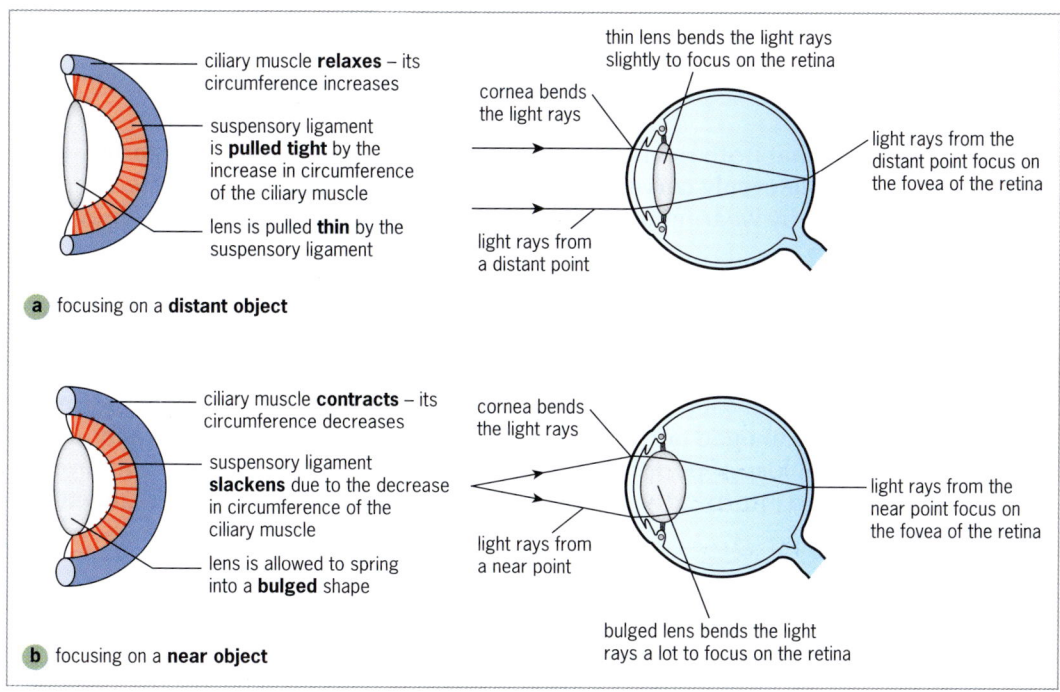

★ Figure 6.4 *Accommodation*

6 Sense organs and coordination

Sight defects and their corrections

Any condition that prevents light rays from focusing properly on the retina causes a person's sight to be **defective**.

Short-sightedness (myopia)

A person with **short sight** can see **near** objects clearly, but distant objects are out of focus. Light rays from near objects focus on the retina; light rays from **distant** objects focus **in front** of the retina. It is caused by the eyeball being too **long** from front to back, or by either the cornea or the lens being too **curved**. It is corrected by wearing **diverging (concave) lenses** as spectacles or contact lenses.

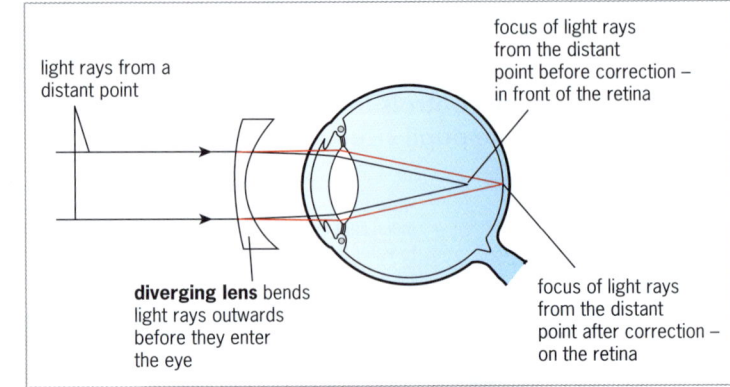

Figure 6.5 *The cause and correction of short sight*

Long-sightedness (hyperopia or hypermetropia)

A person with **long sight** can see **distant** objects clearly, but near objects are out of focus. Light rays from distant objects focus on the retina; light rays from **near** objects focus **behind** the retina. It is caused by the eyeball being too **short** from front to back, or by either the cornea or the lens being too **flat**. It is corrected by wearing **converging (convex) lenses** as spectacles or contact lenses.

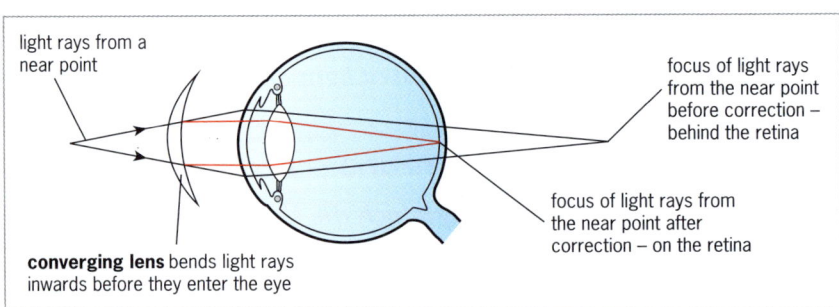

Figure 6.6 *The cause and correction of long sight*

Astigmatism

A person who has **astigmatism** finds that both near and distant objects appear **blurry** or **distorted**. It occurs if either the cornea or lens is **unevenly curved**, so not all light rays are equally refracted and not all focus on the retina. It is corrected by wearing **unevenly curved lenses** that counteract the uneven curvature of the cornea or lens.

Glaucoma

Glaucoma is a condition in which the **pressure** of the fluid within the eye **increases** because the drainage channels that allow aqueous humour to flow from the eye become blocked. If left untreated, the optic nerve becomes damaged and it can lead to permanent **blindness**. It usually develops slowly and causes a gradual loss of peripheral (side) vision. Glaucoma is treated with **eye drops** to reduce fluid production or to improve the flow of fluid from the eye, or by **laser treatment** or **surgery** to open the drainage channels.

Cataract

A **cataract** is a **cloudy** area that forms in the lens. It develops slowly and, as it increases in size, it leads to cloudy or blurred vision, halos forming around lights, colours appearing faded and difficulty seeing in bright light and at night. It is usually caused by **ageing** and is usually corrected by **surgery** to remove the clouded lens and to replace it with an **artificial lens**.

Colour blindness

Colour blindness occurs if one or more of the **cone types** in the retina do not function properly and the sufferer is unable to distinguish differences between certain colours. The commonest form is red-green colour blindness, where red and green cannot be distinguished. Most forms are **inherited**.

Damage to the eyes

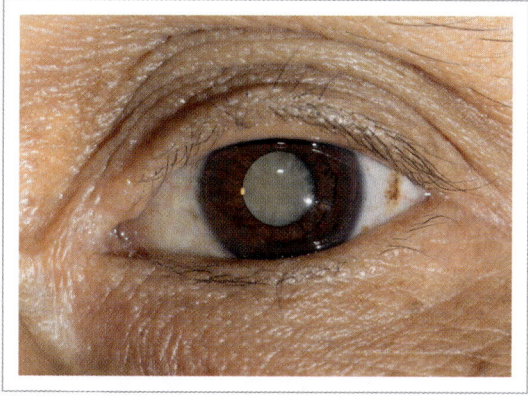

Figure 6.7 *An eye with a cataract*

Staring directly at the **Sun** or very **bright lights** can damage the **retina** at the back of the eye and cause blind spots to develop. Extended exposure to **ultraviolet light** from the Sun can lead to **cataracts** developing and can also damage the **cornea** or the **retina**. Damage to the retina may lead to **blindness**.

Eyes can also be **physically injured** by being **poked**, e.g. with a stick or a finger, by **flying objects** getting into the eye, e.g. small pieces of metal or sand, by being **hit**, e.g. by a sports ball or a fist during a fistfight, or by **harsh chemicals** that can burn the surface of the cornea.

The ear

The **ear** detects **sound waves**, which are produced when objects vibrate, and converts them into **electrical impulses**. These impulses are transmitted along the **auditory nerve** to the brain and the brain interprets them as **sound**. The ear also detects the **position** and **movement** of the head, and this helps to control **balance** and **posture**.

The **ear** is divided into **three** regions: the **outer ear** consisting of the **pinna** and **ear canal**, the **middle ear** which is a small, air-filled cavity in the skull containing the **ear bones** or **ear ossicles**, and the **inner ear** which is filled with fluid and composed of three **semicircular canals** and the **cochlea**.

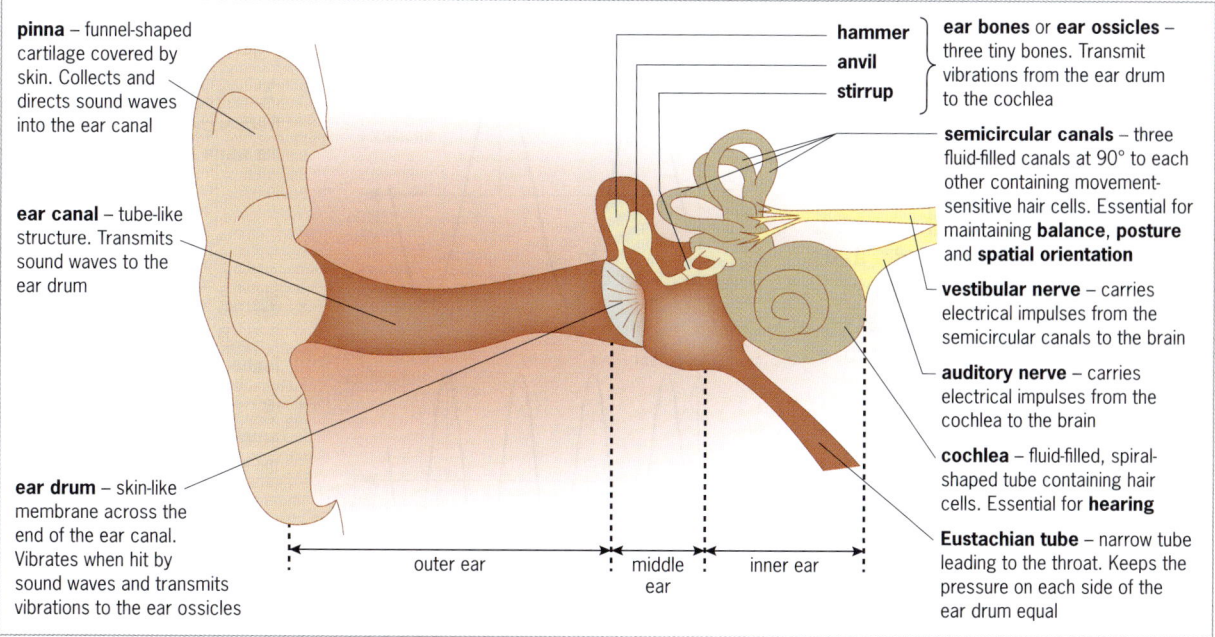

Figure 6.8 *Structure and functions of the parts of the human ear*

6 Sense organs and coordination

The mechanism of hearing

The **mechanism of hearing** is explained in Figure 6.9.

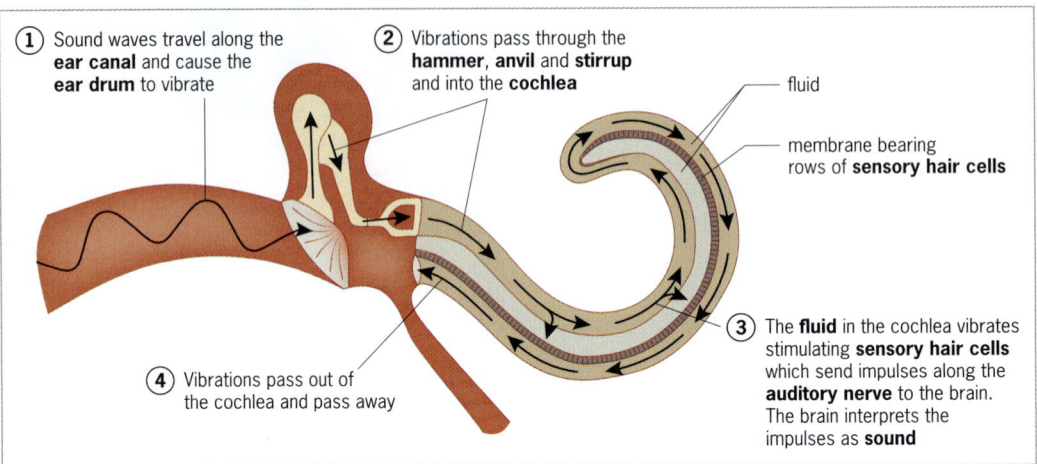

Figure 6.9 *The mechanism of hearing*

Controlling balance and posture

The **semicircular canals** help control **balance** and **posture**. When the head moves, the fluid inside the canals moves and stimulates the **hair cells**. Impulses travel along the **vestibular nerve** to the **brain**. The brain processes the impulses and integrates the information with information received from the eyes and other parts of the body, mainly muscles and joints. It then sends impulses back to muscles to help maintain balance and posture, coordinate movement and provide an awareness of spatial orientation.

Sound

Sound waves travelling through the air are made of areas of high pressure alternated with areas of low pressure. A sound can be described in terms of its **loudness** and its **pitch**.

- The **loudness** of a sound depends on the **amplitude** or **height** of the sound wave. The greater the amplitude, the louder the sound. Loudness is measured in **decibels** or **dB**. Humans can **safely** listen to sounds up to **70 dB**. However, prolonged or repeated exposure to sounds of **85 dB** and higher can permanently damage the hair cells of the cochlea, leading to permanent **hearing loss**. Sounds exceeding **120 dB** can cause **immediate** damage and hearing loss, and sounds above **150 dB** can cause the ear drums to rupture and damage the ear ossicles.

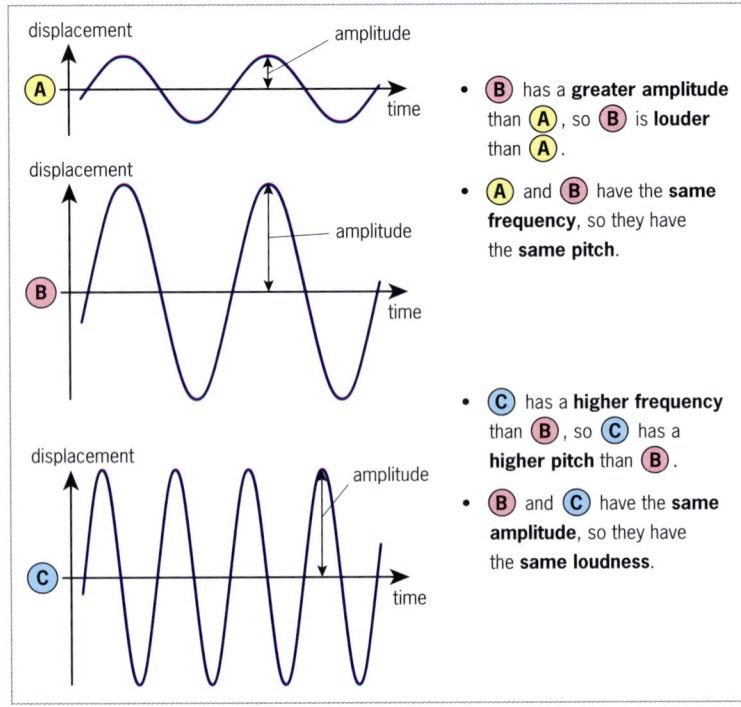

Figure 6.10 *Sound waves with different amplitudes and frequencies*

- The **pitch** of a sound is how high or low the sound is, and this depends on the **frequency** of the sound waves or **number** of waves per second. The higher the frequency, the higher the pitch. Frequency is measured in **hertz** or **Hz**. The approximate **audio frequency range** or **spectrum** of human hearing is **20 Hz** to **20 000 Hz**, and the ear is most sensitive to sounds between **2000 Hz** and **5000 Hz**. As a person **ages**, the range **decreases** and a person often finds it harder to hear high-pitched sounds.

Revision questions

1. Make a list of the sense organs of the human body and identify the stimulus or stimuli detected by EACH.

2. Construct a table to give the function of EACH of the following parts of the eye: the cornea, the iris, the pupil, the lens, the choroid, the ciliary muscle, the suspensory ligament, the optic nerve and the vitreous humour.

3. Suggest why an object is not seen if light rays reflected from that object fall on the blind spot.

4. Explain in detail how Omari is able to see the tree at the end of his garden.

5. Distinguish between the rods and cones of the retina.

6. Explain how Jason's eyes adjust when:
   a   he walks from his dimly lit kitchen into his sunny garden
   b   he watches his dog, Gizmo, run from his feet to chase a monkey at the far end of his garden.

7. Nadira has her eyes tested and is told that she is long-sighted. Explain the possible cause of her sight defect and how it can be corrected.

8. Discuss the cause and treatment of EACH of the following sight defects:
   a   glaucoma   b   a cataract.

9. Explain how the ear functions to:
   a   detect sound   b   control balance and posture.

10. What does EACH of the following depend on?
    a   The pitch of a sound.
    b   The loudness of a sound.

11. What is the approximate audio frequency spectrum of the human ear?

The nervous system

The **nervous system** helps coordinate and control activities of the body. It is made up of **neurones** or **nerve cells** which transmit messages as **electrical impulses** throughout the system. Neurones link **receptors**, which are present in the sense organs, to **effectors**, which are muscles and glands. The system is divided into **two** parts.

- The **central nervous system (CNS)** consists of the **brain** and the **spinal cord**.
- The **peripheral nervous system (PNS)** consists of **cranial** and **spinal nerves** that connect the central nervous system to all parts of the body. Cranial nerves emerge from the brain and spinal nerves emerge from the spinal cord.

Neurones

Neurones are specialised cells that conduct nerve impulses throughout the nervous system.

All neurones have a **cell body** with thin fibres of cytoplasm extending from it called **nerve fibres**. **Dendrites** are the nerve fibres that carry impulses **towards** the cell body and **axons** carry impulses **away**; each neurone has only **one** axon. There are **three** types of neurones.

- **Sensory neurones** transmit impulses from **receptors** to the **CNS**.
- **Motor neurones** transmit impulses from the **CNS** to **effectors**.
- **Relay** or **intermediate neurones** transmit impulses throughout the **CNS**. They link sensory and motor neurones and their nerve fibres lack myelin sheaths.

Nerves are cord-like bundles of nerve fibres of sensory and/or motor neurones surrounded by connective tissue. The **brain** and **spinal cord** are made up mainly of relay neurones and cell bodies of motor neurones.

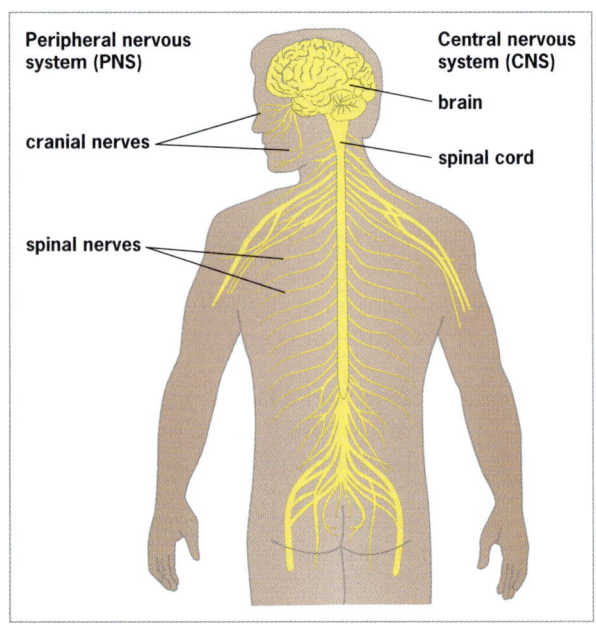

Figure 6.11 *Organisation of the human nervous system*

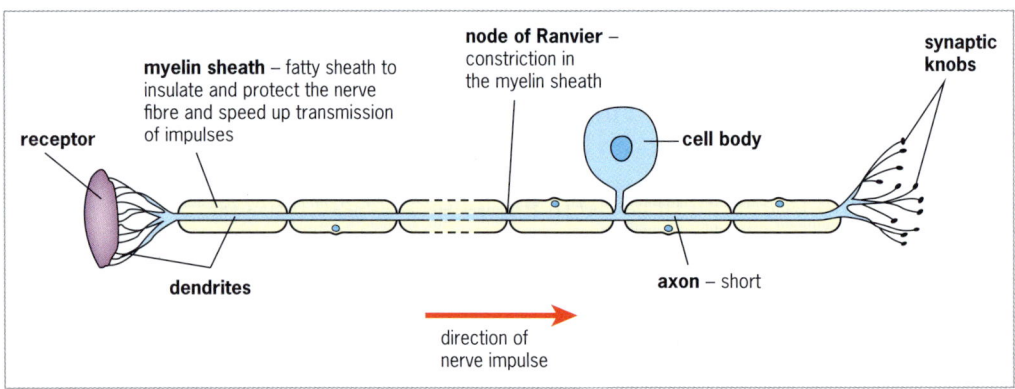

Figure 6.12 *Structure of a sensory neurone*

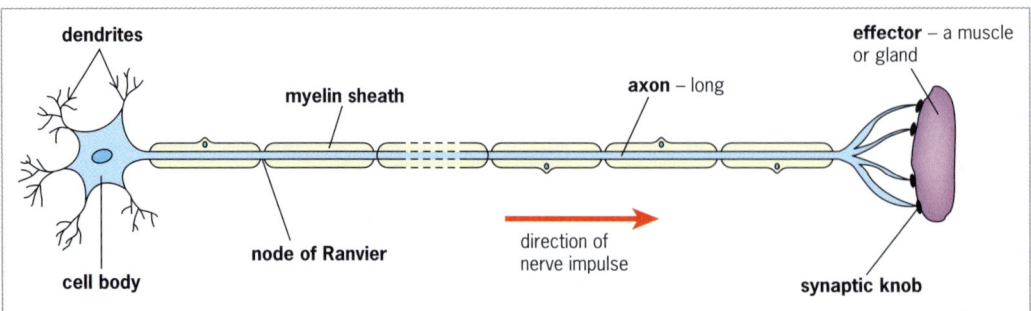

Figure 6.13 *Structure of a motor neurone*

When a **receptor** is stimulated, impulses pass from the receptor along **sensory neurones** into the CNS, where they pass into **relay neurones**. The impulses then pass into **motor neurones** which carry them out of the CNS to **effectors**.

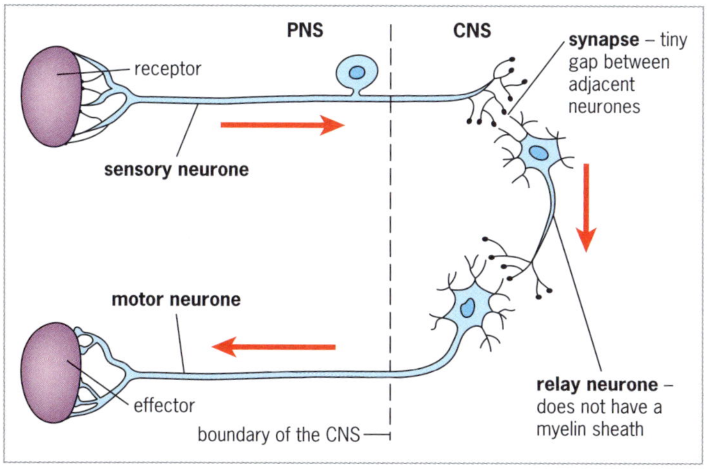

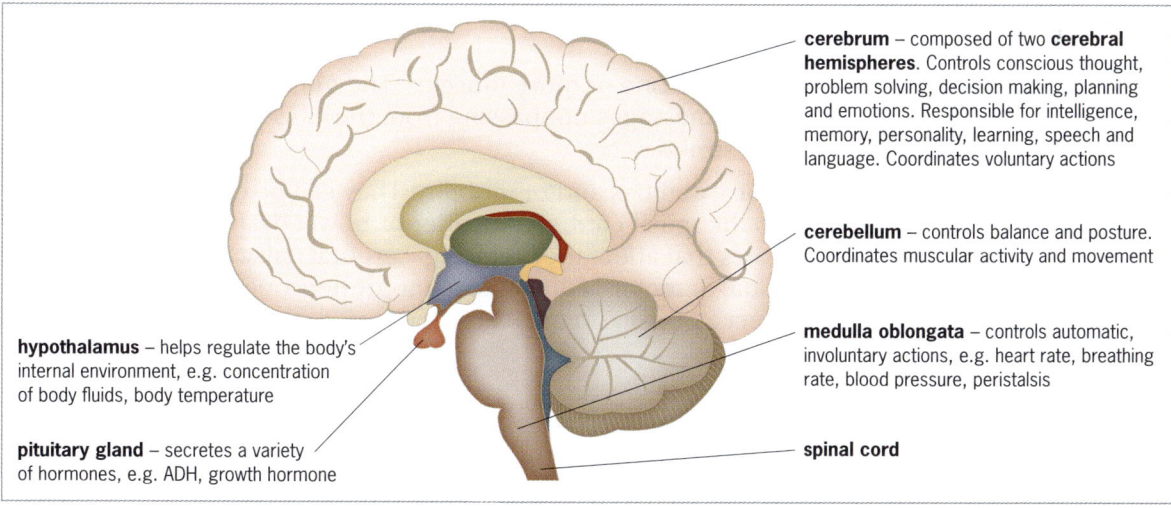

Figure 6.14 *The connection between a receptor and an effector*

The brain

The **human brain** is an extremely complex organ composed of billions of interconnected **neurones**. It has **five** main regions, each concerned with different functions.

Figure 6.15 *Functions of the main parts of the human brain as seen in longitudinal section*

The spinal cord

The **spinal cord** is a long, thin structure running through the vertebral column from the medulla oblongata of the brain almost to the bottom of the column. It is protected by the **vertebrae**, is made up of interconnected **neurones**, and transmits impulses between the **brain** and the **rest of the body**.

Sensory neurones in **spinal nerves** carry impulses from receptors into the spinal cord. Impulses then travel to and from the brain through **relay neurones**. **Motor neurones** in **spinal nerves** then carry impulses back out to effectors. In this way, the spinal cord helps coordinate and control body movements and other body processes. It also coordinates certain reflex actions without involving the brain.

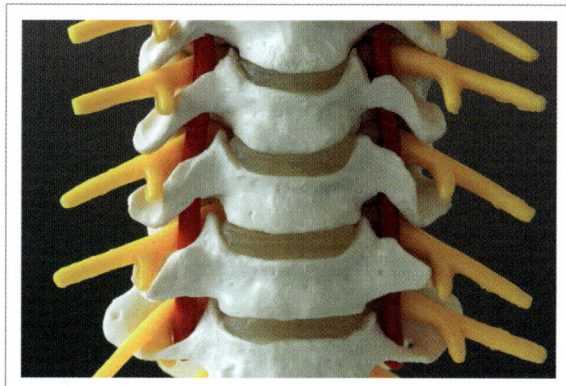

Figure 6.16 *Model showing pairs of spinal nerves exiting the spinal cord between the vertebrae*

6 Sense organs and coordination

Voluntary and involuntary actions

*A **voluntary action** is an action that is consciously controlled by the **brain**.*

The **cerebrum** of the brain **initiates** voluntary actions in one of two ways, both of which involve **conscious thought**:
- The cerebrum can receive **incoming information** from sensory neurones, **process** this information and then initiate an action, e.g. deciding whether or not to accept an apple offered by a friend.
- The cerebrum can **spontaneously** initiate an action without receiving any incoming information, e.g. deciding to go for a walk.

Voluntary actions are **learned**, they are relatively **slow** and they are **complex** because a variety of **different responses** can result from **one stimulus**.

Other **examples** include talking, writing, running, eating, reading, watching TV and playing football.

*An **involuntary action** is an action that occurs without conscious thought.*

Involuntary actions are **not learned**, they are **rapid** and they are **simple** because the **same response** always results from the **same stimulus**.

Examples include breathing, digestion, peristalsis, control of blood pressure and **reflex actions**, e.g. the withdrawal reflex in response to pain, the knee jerk reflex, the pupil reflex, blinking, sneezing, coughing and saliva production.

Malfunctioning of the nervous system

Paralysis occurs when a person loses control of one or more muscles. The muscles stop being able to contract and bring about movement due to messages not passing from the CNS to them. It is mainly caused by **injury** to the spinal cord or by a **stroke**.

Physical disabilities affect a person's physical abilities and/or mobility. They can be caused by **injuries** to the **brain**, e.g. visual impairment, or to the **spinal cord**, e.g. bladder or bowel dysfunction. They can also be caused by **conditions** or **diseases** that affect the **brain**, e.g. cerebral palsy and Parkinson's disease, or the **spinal cord**, e.g. spina bifida, or **motor neurones**, e.g. motor neurone disease. Damage to the **cerebrum** of the brain can also affect a person's mental ability, memory and personality.

The endocrine (hormonal) system

The **endocrine system** is composed of **endocrine glands** or **ductless glands** which secrete chemicals called **hormones** directly into the blood. Hormones are known as **chemical messengers** and they help coordinate and control the body's activities.

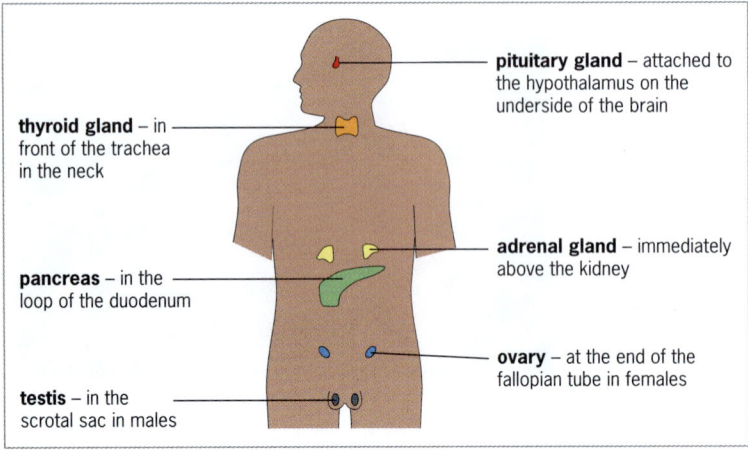

★**Figure 6.17** *The position of the main endocrine glands*

The role of the main hormones in the human body

Hormones are carried around the body by the **blood**. Some affect specific organs, known as **target organs**, e.g. ADH. Others affect cells and tissues throughout the body, e.g. thyroxine.

Table 6.2 *Hormones secreted by the main endocrine glands and their functions*

Endocrine gland	Hormone(s)	Function(s)
Pituitary gland	Antidiuretic hormone (ADH)	Controls the **water content** of blood plasma and body fluids by controlling water reabsorption in the **kidneys** (see page 48).
Thyroid gland	Thyroxine	Controls the **rate of metabolism** and energy production in cells, and **physical growth** and **mental development**, especially in children.
Adrenal glands	Adrenaline (flight, fright or fight hormone)	Released in large quantities when frightened, excited or anxious. Speeds up **metabolism**, mainly respiration, and increases blood sugar levels, heart rate, breathing rate and blood supply to muscles, i.e. it triggers the **fight-or-flight response** and gives the feeling of **fear**.
Pancreas	Insulin	Regulates **blood glucose** levels. Secreted when blood glucose levels rise. Stimulates body cells to absorb glucose for respiration and liver cells to convert excess glucose to glycogen (animal starch), causing blood glucose levels to drop.
Ovaries	Oestrogen (produced by the Graafian follicle)	Controls the development of female **secondary sexual characteristics** at puberty, i.e. development of breasts, pubic and underarm hair, and a broad pelvis. Helps regulate the **menstrual cycle** by stimulating the **uterus lining** or **endometrium** to thicken each month after menstruation (see page 26).
	Progesterone (produced by the corpus luteum)	Helps regulate the **menstrual cycle** by maintaining a thickened **uterus lining** after ovulation each month (see page 26).
Testes	Testosterone	Controls the development of male **reproductive organs** and **secondary sexual characteristics** at puberty, i.e. development of a deep voice, facial and body hair, muscles and broad shoulders. Controls **sperm** production in the testes.

Revision questions

12. **a** What is the importance of the nervous system to humans?
    b Describe the main divisions of the human nervous system.

13. Identify the THREE types of neurones that make up the nervous system and explain the relationship between them.

14. Construct a table to give the functions of the following regions of the brain: the hypothalamus, the cerebrum, the cerebellum and the medulla oblongata.

15. Outline the role of the spinal cord.

16. Give THREE differences between a voluntary action and an involuntary action and give THREE examples of EACH.

17. Jan pricked her finger and immediately withdrew her hand from the source of the pain. Outline the events that occurred in Jan's nervous system to bring about her response.

18. Fabian is involved in a car crash and his right leg is paralysed. Explain the possible cause of his paralysis.

19. What is the endocrine system and what role does it play in the human body?

20. For EACH of the following endocrine glands, identify where the gland is located in the body, name the hormone it produces and outline the functions of the hormone:
    a the thyroid gland **b** the adrenal glands **c** the testes
    d the pancreas.

7 Health

The World Health Organization (WHO) is a specialised agency of the United Nations (UN) responsible for **international public health**. The Constitution of the WHO states that its main objective is 'the attainment by all people of the highest possible level of health'. According to the WHO, **healthy people** are able to function well physically, mentally and socially.

***Health** is a state of complete physical, mental and social wellbeing and not merely an absence of disease and infirmity.*

Microbes

Microbes or **microorganisms** are extremely small organisms, including **viruses**, **bacteria**, **protozoans** and some **fungi**. They play an **important role** in maintaining life on Earth; however, they can also be extremely **harmful**.

- **Viruses** are considered to be **particles** and not cells because they lack a cellular structure. Each **virus particle** is composed of **nucleic acid (DNA or RNA)** surrounded by a coat of protein. All viruses live **parasitically** inside living cells and they can only reproduce inside these cells. Viruses often cause **disease** in plants and animals.
- **Bacteria** are unicellular organisms; however, their cells lack true nuclei and other membrane-bound organelles. Their **DNA** is found in the **nucleoid**, which lacks a nuclear membrane. Bacteria are present in almost every environment found on Earth. Many are **beneficial** to their environment and other living organisms, others cause **disease**.
- **Fungi** considered to be microbes include **yeasts**, which are unicellular, and **moulds**, which are multicellular. In multicellular fungi, the cells form thread-like, branching filaments called **hyphae** (singular: **hypha**), and a network of hyphae, known as a **mycelium**, makes up the body of the fungus. Many fungi are **beneficial** to their environment and other living organisms, a few cause **disease**.

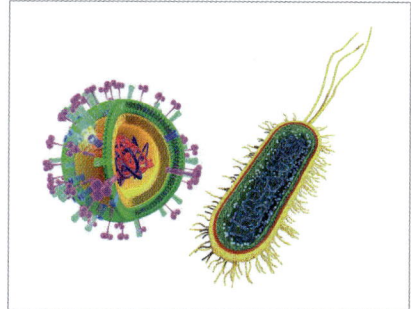

Figure 7.1 *The influenza virus (left) and the E. coli bacterium (right)*

Positive effects of microbes

- Some **bacteria** and **fungi** contribute to **soil fertility** by feeding on dead and waste plant and animal material (organic matter), causing it to **decompose**. This releases **mineral nutrients** into the soil, which increase its fertility (see page 19).
- Certain **bacteria** and **fungi** are used in **food production** and **processing**. Yeast is used to make bread and alcoholic beverages (see page 114). **Lactic acid bacteria** are used to make **fermented products**, e.g. yoghurt, sour cream, buttermilk, kimchi and sauerkraut, by producing **lactic acid** from sugars in the raw ingredients. **Acetic acid bacteria** are used to produce **vinegar** by converting ethanol in fermented beverages into **acetic acid** or vinegar.
- Certain **bacteria** are used to break down domestic and industrial organic waste during the **treatment** of **sewage** in sewage treatment plants. This forms **fertiliser** from harmful waste.
- Certain **bacteria** and **fungi** found in the digestive system of herbivores, e.g. cows and rabbits, enable the herbivores to **digest** their food by breaking down the cellulose in the food.

Negative effects of microbes

- **Viruses**, certain **bacteria** and some **fungi** cause **disease** in both plants and animals. Microbes that cause disease are known as **pathogens** (see page 66).
- Certain **viruses**, **bacteria** and **fungi** cause **damage** to **food crops** by growing parasitically on various parts of the plants and causing **disease**. This reduces **yields**, affects the **quality** of crops and can lead to food insecurity.

- Some **bacteria** and **fungi** cause **food** to **spoil**. They break down food using enzymes, and some produce acids or gases. This causes the appearance, texture, flavour and odour of the food to change. Others produce **toxic chemicals** that can cause food poisoning.

Figure 7.2 *Mould causes bread to spoil*

Communicable or infectious diseases

*A **disease** is a condition that impairs the normal functioning of part or all of an organism and leads to a loss of good health.*

*A **communicable disease** is a disease that can be passed from one person to another and is caused by a **pathogen**.*

Pathogens are **microscopic parasites** that cause disease in their hosts and include **viruses**, **bacteria**, **fungi** and **protozoans**. Communicable diseases are also known as **pathogenic diseases** or **infectious diseases**, and pathogens can be referred to as **causative agents** of the disease.

Some communicable diseases are **contagious** because they are spread by **direct** or **indirect contact** with an infected person. Not all communicable diseases are contagious, e.g. dengue is spread by mosquitoes.

Note: When **treating** a disease, the aim is to **relieve the symptoms** and to **cure** it, if possible. When **controlling** a disease, the aim is to **prevent further development** and **spread** of the disease so its incidence in the population is reduced. Treating a disease is always one method to control it.

Sexually transmitted infections (STIs)

Sexually transmitted infections or **STIs** are passed on from one person to another during **unprotected sexual contact**. Some can also be transmitted from mother to child during pregnancy, childbirth and breastfeeding, or through infected blood and blood products during transfusions.

Table 7.1 *Causes, signs/symptoms and treatment of some common STIs*

Infection	Cause	Signs/symptoms	Treatment
AIDS – acquired immune deficiency syndrome	Virus known as the human immuno-deficiency virus or HIV	• **Primary infection:** Flu-like symptoms lasting 1 to 2 weeks may develop 2 to 6 weeks after infection. Some people have no symptoms. • **Asymptomatic stage:** Usually no symptoms are experienced for 10 years or more, but the virus is damaging the immune system and the person can pass it on without knowing. • **Symptomatic stage:** Symptoms of **AIDS** begin to develop, including weight loss, prolonged fever, severe tiredness, night sweats, chronic diarrhoea, swollen glands and skin rashes. Severe damage to the body's immune system also leaves the person vulnerable to **opportunistic infections**, e.g. pneumonia, tuberculosis and certain cancers.	• **Antiretroviral drugs** to prevent the virus from replicating and reduce the level of the virus in the body. • **Drugs** to treat opportunistic infections. • **No cure** exists.

Infection	Cause	Signs/symptoms	Treatment
Genital herpes	Virus known as the herpes simplex virus or HSV	• Recurrent painful blisters on the genitals and surrounding areas. • Flu-like symptoms may accompany the initial appearance of the blisters and become reduced in severity during subsequent outbreaks.	• Antiviral drugs to reduce symptoms. • No cure exists.
Hepatitis B	Virus known as the hepatitis B virus or HBV	• Fatigue, jaundice (yellowing of the skin and whites of the eyes), dark coloured urine, joint pain, abdominal pain, loss of appetite, nausea, vomiting and fever. Some people have no symptoms. • Can lead to serious liver damage if left untreated, e.g. cirrhosis and liver cancer.	• Antiviral drugs to reduce symptoms. • No cure exists.
Gonorrhoea	Bacterium known as *Neisseria gonorrhoeae*	• **In females**: abnormal vaginal discharge, pain or burning sensation when urinating, pain or tenderness in the lower abdomen and bleeding between periods. • **In males**: abnormal discharge from the tip of the penis, pain or burning sensation when urinating and pain or swelling in the testes. If left untreated, it can lead to infertility.	• Antibiotics to kill the bacterium.
Chlamydia	Bacterium known as *Chlamydia trachomatis*	• **In females**: abnormal vaginal discharge, burning sensation when urinating, pain during sexual intercourse, itching or soreness in or around the vagina and bleeding between periods. • **In males**: abnormal discharge from the tip of the penis, burning sensation when urinating, itching or burning around the opening of the penis and pain or swelling in the testes. Some people have no symptoms in the early stages. If left untreated, it can damage the reproductive organs and cause infertility, especially in females.	• Antibiotics to kill the bacterium.
Syphilis	Bacterium known as *Treponema pallidum*	• **Primary syphilis**: painless, round sores called **chancres** develop on the genitals at the point of infection and last for 3 to 6 weeks. • **Secondary syphilis**: a red, non-itchy rash spreads over the body and may be accompanied by patchy hair loss, fever, sore throat and swollen lymph glands. These symptoms eventually go away. • **Latent syphilis**: no symptoms occur for years. • **Tertiary syphilis**: if left untreated, damage can occur to the brain, nerves, heart, blood vessels and other organs years after the original infection.	• Antibiotics to kill the bacterium.

Infection	Cause	Signs/symptoms	Treatment
Candidiasis, yeast infection or **thrush**	A **yeast-like fungus** – *Candida albicans*	• **In females**: white vaginal discharge, itching or soreness of the vagina and surrounding area, and vaginal burning during intercourse or urination. • **In males**: irritation or burning at the head of the penis, soreness during intercourse, white discharge from the penis and redness of the penis.	• **Antifungal medications**, e.g. tablets and creams, to kill the fungus.

Note: Hepatitis A and **hepatitis C** have similar symptoms to hepatitis B and they can also be transmitted by sexual contact, however, this means of transmission of both is rare.

Prevention and control of STIs

To **prevent** or **control** the spread of **all STIs**, the following methods can be employed.

- **Abstain** from sexual intercourse or keep to **one**, uninfected sexual partner.
- Use **condoms** during sexual intercourse.
- Visit a doctor or healthcare facility to be **tested** if an infection is suspected and, if an STI is diagnosed, undergo the correct **treatment**.
- **Trace** and **treat** all sexual contacts of infected persons.
- Set up **education programmes** to educate populations about STIs and how to prevent their spread.

Currently, the only STI referred to in Table 7.1 that can be controlled by **vaccination** is **hepatitis B**.

Prevention and control of HIV/AIDS and hepatitis B

In **addition** to the above, other **specific measures** can be taken to control the spread of HIV/AIDS and hepatitis B.

- **Do not** use intravenous drugs or share cutting instruments.
- Use **sterile needles** for all injections and dispose of all used needles following strict guidelines.
- **Test** all human products to be given intravenously for HIV and hepatitis B.
- **Prevent** mother to baby transmission by ensuring pregnant women who are HIV or hepatitis B positive receive antiretroviral drugs and their babies receive the appropriate medication after birth.
- Give **pre-exposure prophylaxis** or **PrEP**, a daily **HIV** drug, to persons at a high risk of HIV infection, e.g. those with an HIV positive partner, to reduce their risk of being infected with HIV.
- **Vaccinate** all babies against **hepatitis B**, and any children and adults who have not been vaccinated.

The immune system and immunity

The **immune system** provides the body with **immunity** by protecting it against **pathogens**. It also helps protect against **abnormal cells**, e.g. cancer cells, and **harmful substances**, e.g. toxins and allergens.

Immunity is the body's temporary or permanent resistance to a disease.

Immunity can be **innate** or **acquired**.

- **Innate immunity** refers to the **inborn ability** of the body to resist disease. It is provided by physical barriers, e.g. the **skin** and **mucous membranes**, and **phagocytes** in the blood. It is present at **birth**.
- **Acquired immunity** or **adaptive immunity**, is **highly specific**. It develops **over time** and involves **lymphocytes** in the blood (see pages 69 to 70).

Impact of diseases on the immune system

A variety of **diseases** can affect the way the immune system functions.

- **Immunodeficiency diseases** occur when the immune system is **suppressed** or **weakened**, making those affected more vulnerable to infections. They can be **inherited**, e.g. severe combined immunodeficiency (SCID), or caused by another **disease**, e.g. HIV/AIDS, or by taking certain **medications**, e.g. chemotherapy and other drugs to treat cancer or prevent organ rejection.
- **Allergic diseases**, e.g. **asthma**, **hay fever** and **eczema**, occur when certain substances in the environment, known as **allergens**, cause the immune system to respond in an **overactive** way (see page 71).
- **Autoimmune diseases** occur when the immune system attacks the body's own normal, healthy cells and tissues, e.g. **type 1 diabetes**, **rheumatoid arthritis** and **lupus** (see page 71).

Acquired immunity

Much of the body's **acquired immunity** develops in response to specific **antigens** entering the body. It can also be acquired by specific **antibodies** or **antitoxins** entering.

*An **antigen** is a substance that is recognised as being foreign to the body and it stimulates the lymphocytes to produce antibodies.*

*An **antibody** is a specific protein that is produced by lymphocytes in response to the presence of a specific antigen.*

*An **antitoxin** is an **antibody** that is produced in response to a specific toxin, and it can neutralise this toxin.*

Antigens are usually **proteins** on the surface of **pathogens** or **toxins** produced by pathogens. Antigens are **specific** to a particular pathogen, and when they are detected in the body, **lymphocytes** produce specific **antibodies**, including **antitoxins**, against them. The **antibodies** bind to the **antigens** and cause the pathogens to clump together so that phagocytes can engulf them, or they cause the pathogens to disintegrate. The **antitoxins** neutralise any **toxins** produced.

There are **two** types of acquired immunity: **natural immunity** and **artificial immunity**.

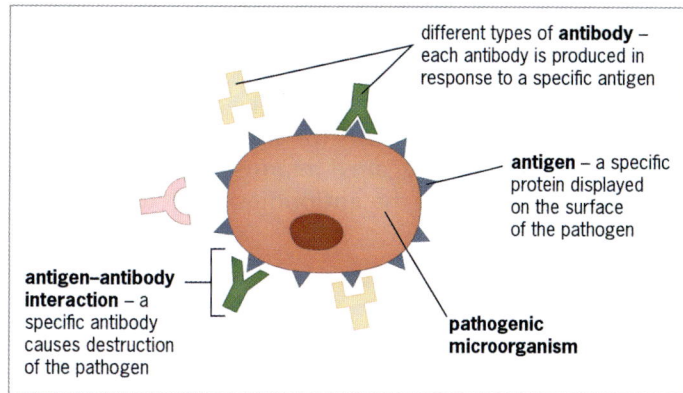

Figure 7.3 *The antigen–antibody interaction*

Natural immunity

Natural immunity is **acquired** by a person being exposed to a pathogenic disease. When the pathogen enters the body, lymphocytes produce specific antibodies in response to the specific antigens of the pathogen. These **antibodies** destroy the pathogens or neutralise their toxins. This takes time and the person usually experiences **symptoms** of the disease. When enough antibodies have been produced to destroy the pathogens or neutralise their toxins, the person recovers. The antibodies gradually disappear from the blood and some lymphocytes develop into **lymphocyte memory cells** which remember the specific antigens.

When the same pathogen re-enters the body, the memory lymphocytes recognise the antigens, multiply and produce **large quantities** of the specific antibodies **rapidly**. The antibodies destroy the pathogen, or neutralise its toxins so quickly that the person does not develop symptoms. The person has become **immune** to the disease. Depending on the type of pathogen, immunity may last for a **short** time, e.g. against the common cold, or a **long time**, e.g. chicken pox is rarely caught twice.

A **baby** acquires natural immunity by receiving **antibodies** that pass across the **placenta** before birth or from **breast milk** during breastfeeding. Immunity lasts a **short time** because the baby's body does not produce lymphocyte memory cells and the antibodies gradually disappear from the blood.

Artificial immunity

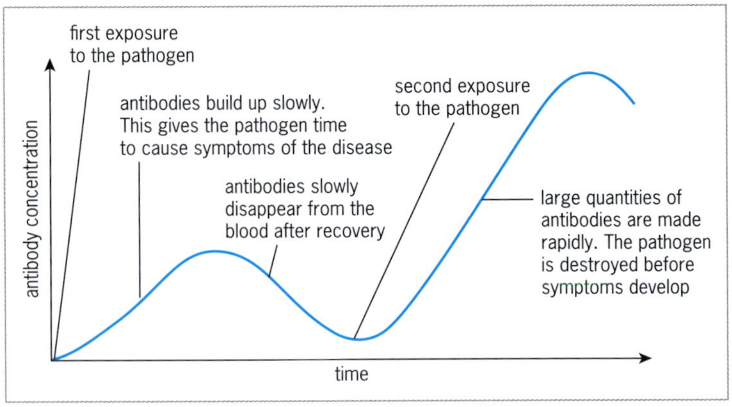

Figure 7.4 *Antibody production during the acquisition of natural immunity*

Artificial immunity is acquired by the deliberate introduction of the specific **antigens** into a person's body. This process is known as **vaccination** and the biological preparation containing the antigens is known as a **vaccine**. A **vaccine** may contain **weakened** or **dead** pathogens, e.g. polio vaccines; **antigens** from the coats of pathogens, e.g. hepatitis B vaccine; **toxins** that have been made harmless, e.g. tetanus vaccine; harmless **viral vectors** that contain genetic material from the pathogen, e.g. certain COVID-19 vaccines; or **messenger RNA (mRNA)** from the pathogen enclosed in microscopic fat particles, which provides instructions to certain body cells to produce specific antigens, e.g. certain COVID-19 vaccines.

A **vaccine** does not cause the disease, but it stimulates lymphocytes to make the specific **antibodies** or **antitoxins** needed. **Lymphocyte memory cells** also develop so that an **immune response** is set up whenever the specific pathogen enters the body. Vaccines may provide **short-term** protection, e.g. against cholera, or **long-term** protection, e.g. against tuberculosis.

Artificial immunity can also be acquired when **antiserum** containing **specific antibodies** or **antitoxins** is injected into a person's body for **immediate relief** of symptoms of a specific disease, e.g. antiserum containing antibodies against tetanus. Immunity lasts a **short time** because lymphocyte memory cells are not produced, and the antibodies gradually disappear from the blood.

Non-communicable or non-infectious diseases

*A **non-communicable disease** is a disease that cannot be passed from one person to another and is not caused by pathogens.*

Many **non-communicable diseases** or **NCDs** are long-term medical conditions that can worsen over time. They are often influenced by genetic, environmental and lifestyle factors, require ongoing medical attention and may limit a person's daily activities. There are several different **types** of NCDs.

- **Nutritional deficiency diseases** are caused by the shortage or lack of a particular **nutrient** in the diet, e.g. vitamin and mineral deficiency diseases, and protein-energy malnutrition (see page 107).
- **Degenerative diseases** are caused by a gradual **deterioration** of body tissues or organs over time that prevents their normal functioning, e.g. Alzheimer's and Parkinson's diseases and osteoporosis.
- **Inherited disorders** are passed on from one generation to the next via genes and are caused by an **abnormal gene**, e.g. sickle cell anaemia, cystic fibrosis and Huntington's disease.
- **Mental health problems** affect how a person feels, thinks, behaves and interacts with other people. These include depression, anxiety disorders, neurosis, stress, schizophrenia and eating disorders.
- **Immune system diseases** disrupt the normal functioning of the immune system, e.g. immunodeficiency diseases, allergic diseases and autoimmune diseases (see page 71).
- **Lifestyle diseases** are linked to the way people **live** their lives, e.g. type 2 diabetes, hypertension and obesity (see pages 71 to 72 and 107).

Allergies

Allergies are caused by an **overactive response** of the immune system to certain **allergens** in the environment, e.g. pollen, dust, animal dander (skin flakes), various air pollutants, certain components of food, certain drugs and the venom in insect stings. Allergens stimulate the immune system to release **inflammatory substances** which cause allergic reactions in the affected parts of the body. The following are some common allergic reactions.

- **Asthma** is a chronic inflammation of the walls of the bronchi and bronchioles of the lungs. This makes the airways narrower, which causes breathing to become extremely difficult.
- **Allergic rhinitis** or **hay fever** mainly affects the nasal passages and eyes, causing sneezing, a runny or blocked nose, itchy or watery eyes and pain around the temples.
- **Eczema** affects the skin causing it to become dry, itchy, cracked, scaly, red and inflamed.
- **Food allergies** occur when the immune system reacts to certain components in food, mainly proteins, and their effects vary from mild itching or swelling to severe **anaphylaxis**, a life-threatening allergic reaction.

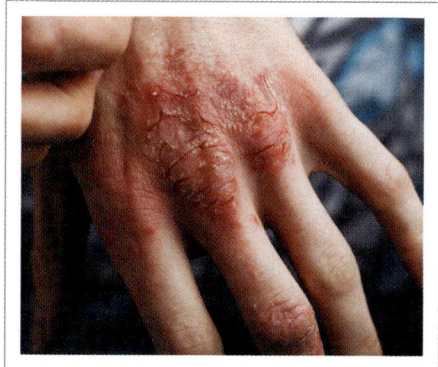

Figure 7.5 *A person scratching an eczema rash on his hand*

Autoimmune diseases

Autoimmune diseases are caused by the immune system developing an immune response to the antigens of its own, healthy cells, tissues and organs as if they were foreign. These include **psoriasis, coeliac disease, rheumatoid arthritis, lupus** and **type 1 diabetes.**

- **Rheumatoid arthritis** is an inflammatory disorder that mainly affects the joints, causing pain, swelling, stiffness and loss of function, leading to decreased mobility.
- **Lupus** can affect various tissues and organs including the skin, joints, kidneys, heart, lungs and brain. Symptoms vary and include fatigue, joint pain, skin rashes, fever, hair thinning, chest pain and organ dysfunction. These often appear and disappear in cycles.

Diabetes

Diabetes is a condition in which the **blood sugar level** is consistently **high** over a prolonged period.

Causes of diabetes

There are **two** types of diabetes which have different **causes**.

- **Type 1** or **insulin-dependent diabetes** is the most common type in people under 30 years old. It is **caused** by the pancreas not producing any insulin because the insulin-producing cells have been damaged, usually by the body's own immune system.
- **Type 2** or **non-insulin dependent diabetes** occurs most often in people that are over 40 years old. It is **caused** by the pancreas not producing enough insulin and/or by the body cells developing resistance to the insulin so they do not respond to it. Being overweight or obese, or having a family history of type 2 diabetes, increases a person's chances of developing the condition.

Effects of diabetes

Symptoms of diabetes include glucose in the urine, frequent urination, excessive thirst, fatigue, blurred vision, excessive hunger, unexplained weight loss, tingling or numbness of the hands and feet, and slow healing of wounds. There is **no cure** and if diabetes is not managed properly, it can lead to various **complications** over time, e.g. **heart disease**, **nerve damage**, **vision loss**, **kidney disease**, **foot complications** and an increased risk of **skin infections**.

Hypertension

Hypertension or **high blood pressure** is a condition in which the pressure of blood in the **arteries** is consistently higher than normal; in other words, **140/90 mm Hg** or **higher**.

Causes of hypertension

A number of **factors** can put a person at risk of developing hypertension. These include the following.
- Being **overweight** or **obese**.
- Being **physically inactive**.
- Consuming too much **saturated fat** and **salt** in the diet.
- **Smoking**.
- Drinking too much **alcohol**.
- Being under **stress**.

Effects of hypertension

Hypertension usually has **no symptoms**; it is known as the 'silent killer'. However, if it is not treated with appropriate medications and properly managed by making lifestyle modifications, it can lead to various **complications** including **heart disease**, **stroke**, **kidney failure**, **impaired vision** and **dementia**.

Physiological effects of exercise

During exercise, **respiration** speeds up in muscles cells, and heart rate and breathing rate increase to supply the muscles with the extra oxygen and glucose they need, and to remove the extra carbon dioxide produced. **Exercise** has many **benefits**.

- It improves the **circulatory system** because it strengthens the heart muscles and increases cardiac output, i.e. the amount of blood pumped by the heart per minute. It also lowers the resting heart rate and blood pressure, improves circulation, and reduces the risk of hypertension, heart disease and stroke.
- It improves the **respiratory system** because it strengthens the diaphragm and intercostal muscles (see page 115) and increases lung capacity. It also increases the efficiency of gaseous exchange and improves the body's ability to use oxygen.
- It helps to control **lifestyle diseases** such as **obesity**, **diabetes**, **hypertension** and **heart disease** because it helps to limit weight gain, lowers blood glucose levels and improves the functioning of the circulatory system.
- It helps to **balance** energy input and energy output. To maintain a **healthy body**, daily **energy input** from the food eaten should balance **energy output** as a result of daily activities, including exercise. If energy input exceeds output, a person **gains weight**. If energy output exceeds input, a person **loses weight**.

Effects of exercise on muscle toning

Muscle toning refers to the process of improving muscle shape, strength and firmness, and reducing body fat to give the body a lean and defined physique, and to improve overall physical health and appearance. **Exercise** plays a crucial role in muscle toning by stimulating muscle growth and improving muscle definition, overall strength and endurance whilst, at the same time, reducing excess body fat.

Figure 7.6 *Exercise improves muscle toning*

Hygiene

Personal hygiene practices

Personal hygiene refers to the practices carried out by individuals to maintain **cleanliness** of their bodies and clothing to preserve **good health** and **well-being**. It includes eliminating **body odours** and taking care of **genitalia**.

Eliminating body odours

Body odour or **BO** is caused mainly when a person **sweats** and the water evaporates leaving salts, urea and dead skin cells on the surface of the skin. **Bacteria** break down these substances and produce unpleasant smelling chemicals that cause BO, especially in the armpits, groin and genital areas. **Good personal hygiene practices** help **eliminate** BO and include the following.

- **Wash** the **body** daily using soap and water, especially the armpits, groin, genitals and feet.
- **Wash** the **hair** with shampoo at least once a week.
- **Dry** the **skin** and **hair** thoroughly after washing.
- Apply **antiperspirant** or **deodorant** to the clean, dry skin of the armpits and **foot powder** to the feet, if necessary.
- **Change** and **wash clothes** regularly, especially undergarments and socks, which should be changed daily.

Care of the genitalia

Keeping the **genitals** and surrounding areas clean is important to prevent **odours** and **infection**. **Good genital hygiene practices** include the following.

- Thoroughly **wash** the genitals daily using mild, unscented soap and water, especially during **menstruation** in females.
- **Wipe** female genitalia from front to back to prevent any faeces from reaching the vaginal area.
- Change female **sanitary products** regularly.
- Pull the male **foreskin** back and clean beneath it daily to remove **smegma** (secretions and dead cells).

Benefits of good personal hygiene

Two of the most important **benefits** of maintaining good personal hygiene are as follows.

- It promotes **social acceptance** by reducing or eliminating body odours, preventing bad breath, enhancing a person's appearance and promoting greater self confidence and self-esteem. These all help enhance social interactions, interpersonal relationships and professional endeavours.
- It significantly reduces the risk of **catching** and **spreading infections** such as skin, respiratory, gastrointestinal and sexually transmitted infections, by reducing the growth of microorganisms on the body. This then leads to **better health**.

Revision questions

1. **a** What are microbes?
   b Give THREE examples of groups of organisms that can be classified as microbes and distinguish among them.

2. Discuss THREE ways in which microbes are important in maintaining life on Earth and THREE ways in which they have harmful effects.

3. Define EACH of the following:
   a a disease
   b a communicable disease
   c a pathogen
   d a sexually transmitted infection.

4. Identify the causative agent and describe the signs/symptoms and treatment measures for EACH of the following sexually transmitted infections:
   a candidiasis
   b gonorrhoea
   c hepatitis B.

5. **a** Excluding treatment measures, outline FOUR measures that can be taken to control the spread of sexually transmitted infections in general.
   b Suggest FOUR measures, other than those outlined in **a**, that can be taken to control the spread of HIV/AIDS.

6. **a** Define the term 'immunity'.
   b Distinguish between innate immunity and acquired immunity.

7. Explain how the immune system is impacted by disease.

8. Define the terms 'antigen', 'antibody' and 'antitoxin'.

9. Kryssie suffered from chicken pox when she was a child and she then remained healthy several years later when there was an outbreak of the disease in her workplace. Explain fully why Kryssie remained healthy during the chicken pox outbreak.

10. Explain how Anton can be provided with immunity against polio without contracting the disease.

11. What is a non-communicable disease? Support your answer by reference to FOUR different types of non-communicable diseases.

12. Discuss the causes of EACH of the following and the effects that EACH has on the body:
    a hypertension
    b diabetes.

13. Identify TWO systems in the human body that benefit from a person taking regular exercise and outline the benefits of exercise to EACH system.

14. Discuss TWO important practices to maintain good personal hygiene and suggest TWO important benefits of practising good personal hygiene.

Drug use and abuse

Prescription drugs and non-prescription drugs

*A **drug** is any chemical substance that alters the functioning of the body physically and/or psychologically.*

Many **drugs** are used medically to **improve health**, whilst others are used for **recreational** purposes. Some drugs are **illegal** and **all** drugs can be **harmful** if used incorrectly, i.e. **misused** or **abused**.

Drugs can have **physiological effects** on the body because they affect the functioning of organs and systems, which affects a person's **physical health**, and they can have **psychological effects** because they affect the way a person's brain functions, which affects a person's **mental health**.

Drugs can be **classified** in various ways; one way is into **prescription drugs** and **non-prescription drugs**.

Prescription drugs

Prescription drugs are **legal** drugs that require a **prescription** from a medical practitioner to be dispensed. They include opioid pain relievers, sedatives, tranquillisers, antibiotics, steroids, diet pills, amphetamines and hormonal injections.

Non-prescription drugs

Non-prescription drugs can be obtained without a prescription. Some are **legal**, including **over-the-counter drugs**, e.g. cough medicines, decongestants, painkillers, nonsteroidal anti-inflammatory drugs (NSAIDs) and laxatives, **alcohol** and **tobacco**; the latter two can only be used legally by persons over a certain age, which differs from country to country. Other non-prescription drugs are **illegal**, e.g. **cocaine**, **methamphetamine**, **ecstasy** and **marijuana** (in most countries). Alcohol, tobacco and illegal drugs are also known as **recreational drugs** because they are usually used for **non-medical** purposes, e.g. **pleasure-seeking**.

Effects of drugs on the nervous system

Drugs can also be **classified** into **four** categories based on the effects they have on the **nervous system**: **stimulants**, **depressants**, **hallucinogens** and **narcotics**. Some can fall into more than one category.

- **Stimulants** or **'uppers' speed up** the body's functions and the functioning of the nervous system. They make a person feel more awake, alert, attentive and anxious, and produce feelings of pleasure and euphoria. They increase heart rate, blood pressure, energy levels and physical stamina. Examples include nicotine, caffeine, amphetamines, marijuana, methamphetamine, cocaine and ecstasy.

Figure 7.7 *Cocaine powder on dried coca leaves from which it is extracted*

- **Depressants** or **'downers' slow down** the body's functions and the functioning of the nervous system. They make a person feel calm, relaxed and drowsy or fatigued, and help reduce anxiety, tension and stress. They cause skeletal muscles to relax, impair coordination and lower heart rate, blood pressure and breathing rate. Examples include alcohol, sedatives, tranquillisers, marijuana and opioids, such as heroin.

- **Hallucinogens** alter a person's **perception of reality**. They can make a person see, hear, feel or experience things that are not there or are distorted from reality, and they lead to irrational or bizarre behaviour, aggression and/or paranoia, anxiety, mood changes and loss of contact with reality. Examples include ecstasy, LSD (acid), marijuana, mescaline and magic mushrooms.
- **Narcotics** or **opioids** give **relief** from moderate to severe **pain**. They can also cause relaxation, drowsiness, impaired judgment, confusion, nausea, slowed breathing and a feeling of euphoria. Examples include codeine, oxycodone, fentanyl, morphine and heroin.

Figure 7.8 *Marijuana flowers are dried, rolled into joints and smoked*

Social and economic effects of drug use and abuse

Drug abuse refers to the use of legal or illegal drugs for unintended purposes or in excessive amounts.

Drug use and **abuse** can have a devastating impact on **individuals**, **families** and **communities**. It harms the **user** physically and psychologically. It can shorten lives and lead to loss of self-worth and emotional stability, personal neglect, health issues, loss of earnings or job loss, and financial problems. Financial problems can cause the user to turn to crime or prostitution. Crime can lead to arrest and imprisonment, and prostitution exposes the user to infection with STIs. Use of intravenous drugs exposes the user to HIV/AIDS and hepatitis B and C, and babies born to addicts may have developmental problems or be addicts themselves. Higher suicide rates and anti-social behaviour are also associated with drug use.

Drug abuse upsets relationships with **family** and **friends**, and can lead to neglect or abuse of family members, especially children and the elderly, resulting in unstable and disturbing family environments. It may also cause children to lose one or both parents, and ultimately leads to dysfunctional families.

The cost to **society** of drug abuse is high. It can lead to reduced productivity and weakened economies. Automobile accidents resulting from drug use can cause serious injury and loss of life, and violent crimes associated with drug use cause injury, loss of life and communities to live in fear. More and more resources have to be used to treat and rehabilitate drug users and addicts, to fight drug-related crimes, and to apprehend, convict and imprison traffickers and pushers of illegal drugs. Ultimately, economies suffer, standards of living are reduced and human resources are lost due to drug use and abuse.

Pests and pest control

Pests, parasites, pathogens and vectors

A pest is any organism, usually a plant or animal, that has a harmful effect on humans, their food or their living conditions.

Pests can **damage** resources, e.g. crops and forests, and can **compete** with humans for resources. **Household pests** invade homes and can damage structures, deposit faecal pellets, eat human food, damage clothing and bite or sting humans. Many of these pests are also **vectors of disease**. Pests include **cockroaches**, **flies**, **ants**, **rats**, **mice**, **termites**, **bed bugs**, **clothes moths** and **mosquitoes**.

A parasite is an organism that lives in or on another living organism known as its host and gains benefit at the expense of the host.

Parasites obtain their **food** from their host and they usually harm their host to varying extents. Parasites include **tapeworms**, **hookworms**, **ticks**, **fleas**, **bed bugs** and **lice**.

*A **pathogen** is a parasitic microorganism that causes **disease** in its host.*

Pathogens damage their host's health. Pathogens include **viruses**, **bacteria**, **fungi** and **protozoans**.

*A **vector** is an organism that carries **pathogens** in or on its body and transmits the pathogens from one host to another.*

Vectors can carry pathogens **inside** their bodies, delivering them to the new host by biting the host, or on the **outside** of their bodies, delivering them by physical contact. Vectors are **not** usually harmed by the pathogens and include **mosquitoes**, **rats**, **houseflies**, **ticks** and **fleas**.

Figure 7.9 *Cockroaches are household pests*

Conditions that encourage breeding of household pests

The **improper disposal of waste** and **improper household hygiene** provide household pests with favourable environments in which to live and reproduce. Many of these pests are also **vectors**.

Improper disposal of waste

Waste produced by **domestic** and **industrial** activities must be properly treated and disposed of so that it does not provide a breeding ground for household pests. This waste includes **sewage** and **solid waste**.

- **Sewage** is **wastewater** from homes, schools, hospitals, industry and rainwater that runs into drains from streets. It can contain faeces, urine, detergents, organic matter and food particles.
- **Solid waste** such as **domestic refuse** includes paper and packaging, cans, glass and plastic items, textiles, food waste and garden waste.

Any **untreated** or **raw sewage** entering the environment or **solid waste** that is not stored in sealed containers, collected at least once a week and transported to a disposal facility promotes the **breeding** of household pests.

- **Organic matter**, **food particles** or **food waste** in raw sewage or unsealed refuse attracts pests such as cockroaches, flies, ants, rats and mice to feed and breed.
- **Standing water** in raw sewage or collecting in open containers provides a place for mosquitoes to lay their eggs.
- **Unpleasant odours** given off by raw sewage and unsealed refuse attract flies, cockroaches and rats to the waste, where they can then breed.
- **Unsealed refuse** provides pests with a place to **hide** from predators and a safe place to reproduce.

Figure 7.10 *Improper disposal of refuse provides ideal conditions for pests to breed*

Improper household hygiene

Household environments that are not kept properly **clean** and **tidy** can promote the **breeding** of household pests.

- **Uncovered** or **incorrectly stored food**, **food spills** and **food crumbs** left lying around attract pests such as flies, cockroaches, ants, rats and mice to feed and breed.
- Pools of **standing water** in sinks, uncovered drains or open containers, or caused by leaking taps, attract mosquitoes to lay their eggs.
- **Organic waste**, e.g. scraps of food, kitchen waste and pet faeces, provide flies with the ideal conditions to lay their eggs, and release unpleasant odours which attract cockroaches, rats and flies.
- **Unclean toilets**, **work surfaces**, **cupboards**, **tables** and **floors**, **unwashed dishes** and **uncovered drains** attract pests such as cockroaches, ants, flies and rats.
- **Clutter**, **debris** and any **unused items** left lying about create hiding places for pests such as cockroaches, rats, mice and mosquitoes, and safe places in which to reproduce.

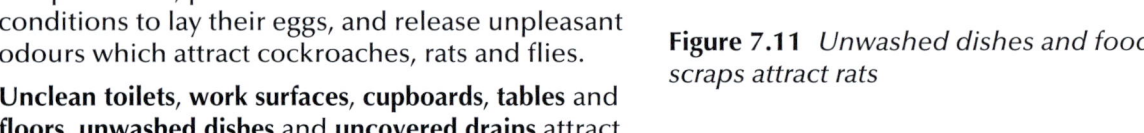

Figure 7.11 *Unwashed dishes and food scraps attract rats*

Methods of pest control

Pests can be controlled by **four** main methods: **biological**, **chemical**, **sanitary** and **mechanical control**.

- **Biological control** involves using **natural enemies** of the pests to kill them, e.g. predators or parasites. Care must be taken to ensure that the natural enemy does not become a pest itself.
- **Chemical control** involves using **chemicals**, known as **pesticides**, to kill pests, e.g. insecticides and rodenticides. These can be applied by spraying, dusting, fogging or as baits. These can be harmful to harmless and beneficial organisms, disrupt ecosystems and pollute land, water and the air.
- **Sanitary control** involves **removing conditions** that attract pests, e.g. treating all sewage, correctly storing and collecting refuse, and keeping household environments clean and tidy.
- **Mechanical control** involves **physically removing** or **excluding** pests from an area, e.g. using traps, sealing entry points and using fencing or other barriers such as nets and screens.

Figure 7.12 *Ladybird beetles can be used to control aphids*

Control of mosquitoes and houseflies

Controlling mosquitoes and **houseflies** is crucial to reducing the incidence of vector-borne diseases, some of which can be fatal. Understanding the **life cycles** of these vectors is essential to develop the most effective methods to control them.

The **life cycle** of a **mosquito** has **four distinct stages**.

- **Egg** – The adult female lays eggs in standing water and the eggs float on the surface of the water.
- **Larva** – The larva hatches from the egg. Larvae live in the water. They hang from the surface and breathe air through breathing tubes, **feed** on microscopic organisms in the water and **grow**.
- **Pupa** – The pupa develops from the larva. Pupae live in the water, where they hang from the surface and breathe air through breathing tubes. **Larval tissue re-organises** into adult tissue in the pupa.
- **Adult** or **imago** – The adult emerges from the pupa. Adults rest in cool, dark places around human residences during the day, and fly and feed on nectar and sugars from plants in the evenings. After mating, the female requires a **blood meal** to mature her eggs before she lays them. She usually obtains the blood from a human and can transmit any pathogens she is carrying whilst feeding.

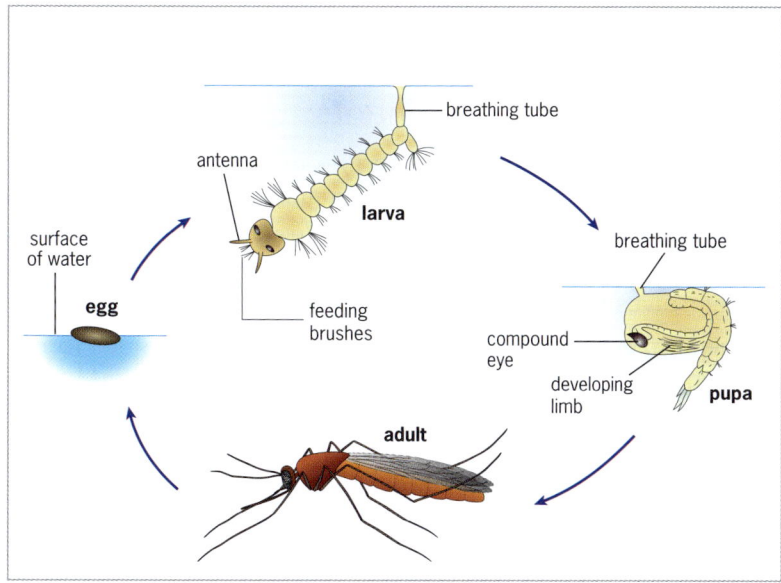

Figure 7.13 *The life cycle of a mosquito*

Note: The **life cycle** of a **housefly** has the **same four stages** as the mosquito. The female lays her **eggs** in decaying organic matter, and the **larvae** and **pupae** grow and develop in the same organic matter.

Table 7.2 *Methods used to control mosquitoes and houseflies*

Pest	Methods of control
Mosquitoes	To control mosquito **eggs**, **larvae** and **pupae**: • **Drain** all areas of standing water and remove all containers that collect water so that females have nowhere to lay their eggs. • Add **insecticides** to breeding areas to kill larvae and pupae. • Introduce **fish**, e.g. tilapia or mosquito fish, into breeding areas to feed on larvae and pupae. • Add mosquito dunks, containing a **bacterium** that produces toxins harmful to mosquito larvae, to areas of standing water. • Spray **oil**, **kerosene** or non-toxic **lecithins** onto still-water breeding areas to prevent females from laying eggs, and larvae and pupae from breathing.
	To control **adult** mosquitoes: • Spray with **insecticides** to kill adults. • Remove **dense vegetation** to reduce protection for adults during daylight hours. • Place **mosquito screens** over windows and doors to prevent adults entering buildings, and place **mosquito nets** impregnated with insecticide over beds at night.
Houseflies	• Spray adults with **insecticides** to kill them. • Use **fly traps** to kill adults. • **Dispose** of all human and animal waste properly so that females have nowhere to lay their eggs. • **Treat** all sewage so that adults are not attracted to it. • **Cover** food so that adults are not attracted to it and cannot land on it.

Food contaminants and the growth of microorganisms

Types of food contaminants

Food contaminants enter food supplies unintentionally and have the potential to harm the consumer. They include **pathogens**, **chemical contaminants** and **physical contaminants**.

- **Pathogens** that contaminate food include **viruses**, **bacteria**, **fungi** and **parasitic protozoans**. Consuming food containing pathogens causes **gastrointestinal infections** caused by the pathogens themselves and **food poisoning** caused by toxins produced by the pathogens.
- **Chemical contaminants** include pesticides; heavy metal ions, e.g. lead, mercury, cadmium and arsenic ions; industrial chemicals, e.g. polychlorinated biphenyls (PCBs) and dioxins; and chemicals from certain food packaging materials, e.g. plasticisers. **Food additives**, e.g. artificial colourings, flavour enhancers and preservatives, can also be considered chemical contaminants.
- **Physical contaminants** are **foreign objects** that accidentally get into food products during processing, packaging or preparation, e.g. pieces of broken glass, metal and plastic, splinters of wood, soil, fingernails, hair and dead insects or insect parts.

Conditions that promote the growth of microorganisms

All microorganisms require **moisture**, an **optimal temperature** and **nutrients** in order to grow. Some also need certain other conditions.

- Microorganisms require a certain amount of **moisture** because water is essential for enzyme activity and for chemical reactions to occur in their cells.
- Microorganisms require a certain **optimal (ideal) temperature** so that their **enzymes** can function as efficiently as possible (see page 110). For most, this is between about 20 °C and 40 °C.
- Microorganisms require a source of **food** to provide them with the **nutrients** they need to grow and reproduce.
- Many microorganisms require **oxygen** to carry out aerobic respiration to provide them with **energy**.
- Most microorganisms require a **pH** range of 6.5 to 7.5 for their enzymes to function efficiently.

Effects of microorganisms in food spoilage, production and processing

When provided with **moisture**, a suitable **temperature** and a source of **nutrients**, certain microorganisms cause **food** to **spoil**, while others are used in the **production** and **processing** of certain foods. These are explained on pages 65 to 66.

Procedures to retard and prevent the growth of bread mould

If **spores** from certain species of **mould** (multicellular fungi) land on bread, they can germinate and form thread-like **hyphae**, which grow over the surface of the bread by absorbing moisture and nutrients from the bread (see Figure 7.2 on page 66). The growth of bread mould can be **slowed** or **prevented** in various ways.

- Store the bread in a clean, dry, **air-tight container** to exclude oxygen and spores.
- Store the bread in a **cool temperature** such as a refrigerator to slow down enzyme activity.
- **Freeze** the bread in moisture-proof packaging to inhibit enzyme activity and remove moisture by converting it into ice crystals.
- Include a **preservative** in the dough when making bread to inhibit the growth of bread mould.

Food preservation

Principles used in food preservation

Food preservation refers to the process of treating and handling food in such a way as to stop or greatly slow down spoilage and prevent food-borne illness while maintaining nutritional value.

Food preservation aims to create conditions that are **unfavourable** for the growth of microorganisms, whilst maintaining the **quality** of the food. This can be achieved in several ways.

- Removing **moisture** from the food to prevent the growth of microorganisms by inhibiting enzyme activity and cellular reactions.
- Storing the food at **low temperatures** to slow down or stop the growth of microorganisms by slowing down or stopping enzyme activity.
- **Heating** the food to kill any microorganisms present and then sealing it in an **airtight** container to prevent microorganisms from re-entering and to exclude **oxygen**.
- Lowering the **pH** of the food to make conditions too **acidic** for microorganisms to grow due to their enzymes becoming denatured (see page 110).
- Adding chemicals known as **preservatives** to the food to inhibit the growth of microorganisms.

Methods used to preserve food

Table 7.3 *Methods of preserving food*

Method	Principles of the method
Salting	**Salt (sodium chloride)** is rubbed into the food or the food is placed into a concentrated salt solution known as **brine**. The salt **withdraws water** from the food by **osmosis** so microorganisms cannot survive and grow. It also enhances flavour, e.g. salt fish, salt pork and brined olives.
Adding sugar	The food is boiled in a concentrated **sugar** solution, sometimes to the point of crystallisation. The sugar **withdraws water** from the food by **osmosis** so microorganisms cannot survive and grow. It also adds sweetness, e.g. jams and crystallised or candied fruits.
Curing	A mixture of **salt**, **sodium nitrate** or **nitrite**, and sometimes **sugar**, is rubbed into the food, usually meat. The nitrates and nitrites have **antimicrobial properties** which inhibit the growth of microorganisms. The mixture also **withdraws water** from the food, enhances flavour and helps develop the pink colour of cured meats, e.g. ham, bacon, corned beef and salami.
Drying	The food is **dried** to **remove all water** present so microorganisms are unable to survive and grow. This can be done naturally or artificially. • When dried **naturally**, the food is placed on racks or trays in direct **sunlight** or in a **solar dryer**, which consists of a well-ventilated chamber with transparent walls that trap **heat** from the Sun. • When dried **artificially**, the food is placed in an **oven** at about 60 °C to 70 °C or in a **dehydrator** designed to dry food at between 35 °C and 60 °C. Examples include sun dried tomatoes, dried fruits and vegetables, grains, herbs and spices, and dried milk powders.
Pickling	The food is placed in **vinegar**, a dilute solution of **ethanoic (acetic) acid**. The vinegar **lowers** the **pH** of the food so it is too low (too acidic) for microorganisms to grow. It also adds a tangy flavour and crisp texture, e.g. pickled onions and cucumbers.

Method	Principles of the method
Heating	The food is **heated** to **kill** any microorganisms and **sealed** in airtight containers to prevent microorganisms re-entering. • **Canning** involves heating food in sealed jars or cans in a boiling water bath at **100 °C**, a steam bath or a pressure canner for a specified time, e.g. canned fruits, vegetables, fish and meat. • **Pasteurisation** involves heating the food to **72 °C** for 15 to 25 seconds and cooling it rapidly, e.g. pasteurised milk. • **Ultra-high temperature treatment (UHT)** involves heating food to temperatures higher than **135 °C** for 1 to 2 seconds and cooling it rapidly, e.g. UHT milk, soups and baby foods.
Refrigeration	The food is stored at **low temperatures** to **slow down** or **stop** the growth of microorganisms. • Placing the food in a **refrigerator** keeps it at about **4 °C**. This **slows** the growth of microorganisms. Energy efficient, **inverter type refrigerators** are preferable because they adjust their power output to match the cooling demands. • **Freezing** the food at temperatures of **−18 °C** and below **stops** the growth of microorganisms, e.g. frozen meat, fish and vegetables.
Treating with other preservatives	Various other **chemical preservatives** can be added to food to **inhibit** the growth of microorganisms. **Sulfur dioxide** and **sulfites** are used to preserve dried fruits, fruit juices, soft drinks, fish, wine and beer; **benzoates** are used to preserve fruit purees, jams, fruit juices and soft drinks; and **sorbates** are used to preserve cheese, baked goods and wine.

a a simple solar dryer

b herbs being dried in a solar dryer greenhouse

Figure 7.14 *Solar dryers*

Revision questions

15 **a** What is a drug?

b Distinguish between prescription drugs and non-prescription drugs. Support your answer with FOUR examples of drugs in EACH category.

16 Construct a table that classifies drugs into FOUR categories based on their effects on the nervous system, outlines the effects that drugs in EACH category have on the user and gives THREE examples of drugs in EACH category.

17 Describe the effects of drug use and abuse on:

  a the user   **b** society as a whole.

18 Differentiate between:
   a a pest and a vector   b a parasite and a pathogen.

19 Identify THREE ways that EACH of the following encourage household pests to breed:
   a improper waste disposal   b improper hygiene within the household.

20 Outline FOUR different methods of pest control.

21 Suggest the most appropriate methods to control the different stages in the life cycle of a mosquito.

22 Food contaminants can be divided into THREE categories. Identify these and write a short explanation of EACH.

23 List the THREE conditions that are required to promote the growth of all microorganisms.

24 Suggest THREE procedures that can be used to retard the growth of bread mould.

25 Define the term 'food preservation' and outline EACH of the following:
   a the aims of preserving food   b the principles involved in preserving food.

26 Explain how EACH of the following methods is used to preserve food:
   a pickling   b heating   c refrigeration
   d curing   e adding sugar.

27 What is a solar dryer?

Exam-style questions – Chapters 1 to 7

1 a) Figure 1 shows an unspecialised plant cell.

  i) Identify the structures labelled W and X. **(2 marks)**

  ii) State ONE function of W and ONE function of X. **(2 marks)**

  iii) Explain why Y is found in plant cells but not in animal cells. **(3 marks)**

  iv) In what way do the properties of Z and the cell membrane differ? **(2 marks)**

b) Diffusion and osmosis play important roles in living organisms, and diffusion can also be observed in the environment.

  i) Distinguish between 'diffusion' and 'osmosis'. **(2 marks)**

  ii) Explain how the process of diffusion contributed to the presence of volcanic ash on the island of Barbados several hours after the eruption of La Soufrière in St Vincent in 2021. **(2 marks)**

  iii) Jessica was advised to soak her wilted romaine lettuce leaves in water before making a salad with them. On following the advice, she noticed that the leaves became crisp and firm. Explain what happened to cause the changes that Jessica observed in her lettuce leaves. **(2 marks)**

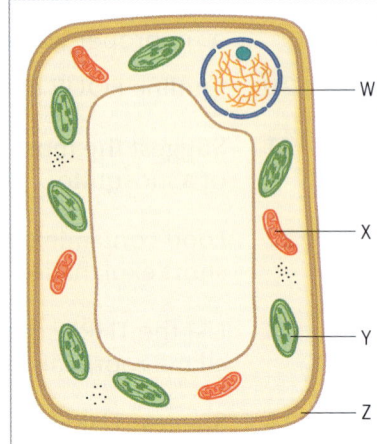

Figure 1 *An unspecialised plant cell*

Total 15 marks

2 a) Figure 2 shows a longitudinal section through a flower found on Farmer Ryan's sweet potato plants.

  i) Name the parts labelled L and M. **(2 marks)**

  ii) Identify the process by which pollen grains are transferred from L to the stigmas of other sweet potato flowers. **(1 mark)**

  iii) Suggest, with a reason, the most likely agent to transfer the pollen grains from L to other flowers. **(2 marks)**

  iv) Farmer Ryan always grows his sweet potato plants from cuttings that he takes from his mature plants a few weeks before he harvests the tubers. Suggest TWO advantages and TWO disadvantages of this method of propagation. **(4 marks)**

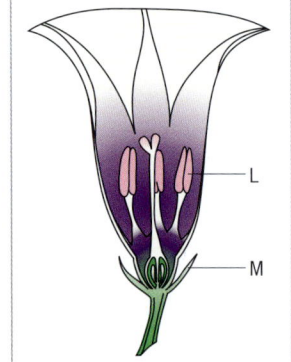

Figure 2 *Longitudinal section through a sweet potato flower*

b) Figure 3 shows some stages in the germination of a seed of an annual plant.

i) Identify ONE condition that must be present for the seed to germinate. **(1 mark)**

ii) At which stage, D, E or F, do you think that the overall mass will be increasing?
Give a reason for your answer. **(3 marks)**

iii) After reaching maturity and reproducing, the overall mass of an annual plant usually decreases before it dies. Explain the reason for this decrease in mass. **(2 marks)**

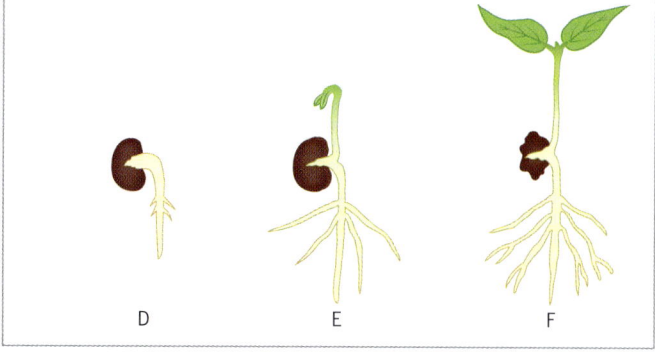

Figure 3 *Stages in the germination of a seed of an annual plant*

Total 15 marks

3 a) 'Soil is one of the world's most important natural resources because it is essential to the survival of all terrestrial organisms.'

i) Write an argument in support of the above statement. **(3 marks)**

ii) What is a fertile soil? **(1 mark)**

iii) Identify TWO properties that a fertile soil must possess. **(2 marks)**

b) In many regions of the world, including the Caribbean, farmers clear away natural vegetation to plant crops and they find that the soils in these regions become prone to erosion.

i) Explain why removal of natural vegetation can lead to soil erosion. **(2 marks)**

ii) The government of a Caribbean island on which there has been large-scale loss of agricultural soils has asked for assistance. Outline THREE measures that you would recommend to the government to help reverse the loss and prevent further problems. **(3 marks)**

iii) Describe TWO methods that farmers in regions that have experienced large-scale soil loss could implement to grow their crops that do not rely on the presence of fertile soil. **(4 marks)**

Total 15 marks

4 a) Figure 4 shows the male and female reproductive systems.

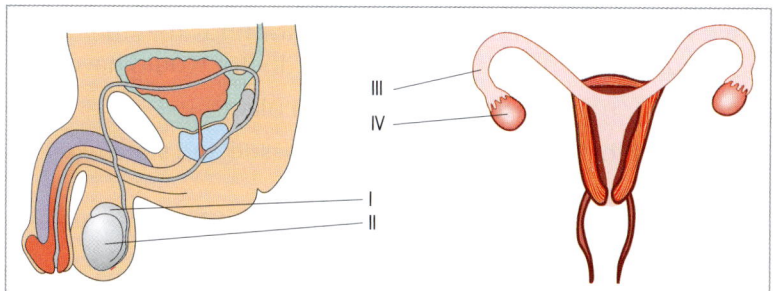

Figure 4 *Male and female reproductive systems*

i) Complete Table 1 below by writing the name and ONE function of EACH of the structures labelled I and III.

Exam-style questions – Chapters 1 to 7   85

Table 1 *Name and function of parts of the human reproductive system*

	Name of structure	Function
I		
III		

(4 marks)

  ii) Suggest TWO reasons why structures labelled II and IV in Figure 4 may be described as having similar functions. **(2 marks)**

  iii) Some types of sexually transmitted infections (STIs) can cause sterility if not treated. Use Figure 4 to help you explain ONE way this could happen in the case of a male. **(2 marks)**

b) After the birth of her baby, Jamal, Vanita is advised to breast feed him for a minimum of 6 months.

  i) Give TWO reasons why Vanita should follow this advice, if possible. **(2 marks)**

  ii) Suggest ONE other step that Vanita should take to ensure that Jamal grows and develops healthily. **(1 mark)**

c) Overpopulation is becoming a problem encountered in many developing countries.

  i) Discuss the TWO MAIN problems that could arise if these countries do not work towards controlling the growth of their populations. **(2 marks)**

  ii) Outline TWO effects that teenage pregnancy can have on human population growth. **(2 marks)**

Total 15 marks

5 a) Figure 5 shows the internal structure of Jo's heart.

  i) Identify the parts labelled R and S. **(2 marks)**

  ii) Explain why the wall of S is thicker than the wall of R. **(2 marks)**

  iii) Jo's friend, Ayanna, suffers from mitral valve prolapse in which the two flaps of her mitral valve do not close properly and she tires easily when she exercises. Suggest an explanation for Ayanna's tiredness. **(3 marks)**

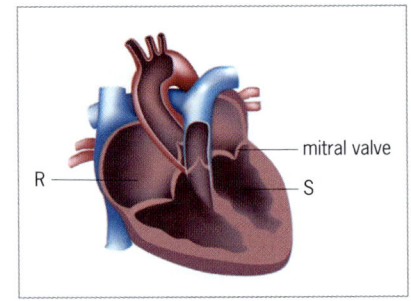

Figure 5 *The internal structure of Jo's heart*

b) Marissa's blood type is AB-negative.

  i) Explain why Marissa's blood type is AB-negative. **(2 marks)**

  ii) During her pregnancy with her first child, Marissa's doctor finds that the baby's blood type is AB-positive. Discuss the concerns that Marissa's doctor would have. **(3 marks)**

c) Vascular tissue is responsible for transporting substances in plants.

  i) Name the TWO tissue types that make up this vascular tissue and state what EACH tissue transports. **(2 marks)**

  ii) Suggest ONE major difference between transport in plants and transport in humans. **(1 mark)**

Total 15 marks

6 a) The process of excretion is important to all living organisms.

  i) What is meant by 'excretion'? **(1 mark)**

  ii) Name ONE excretory organ, other than the kidneys, found in the human body and identify ONE substance that it excretes. **(2 marks)**

b) Figure 6 is a diagram of a nephron found in a human kidney.
   i) Identify the structures labelled G and J. (**2 marks**)
   ii) Describe the process taking place in EACH of the structures labelled Y and Z. (**4 marks**)
   iii) Explain the likely cause of protein being found in a person's urine. (**2 marks**)
c) Table 2 shows the composition of Andrew's urine 1 hour after he drank 2 large glasses of water and again 1 hour after he played a 90 minute game of football on a hot day without drinking.

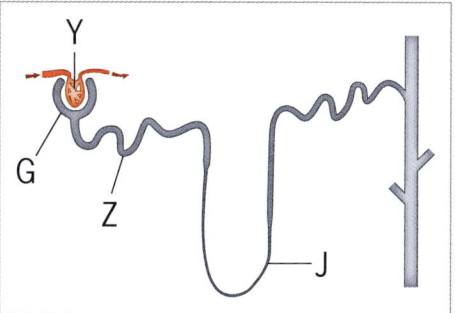

Figure 6 *Diagram of a nephron*

Table 2 *Composition of Andrew's urine*

Component	% in Andrew's urine after drinking	% in Andrew's urine after playing football
Urea	2.0	6.0
Salt	0.3	1.0
Water	95	90

   i) Which activity caused Andrew's urine to have the HIGHEST concentration? (**1 mark**)
   ii) Name the hormone released by Andrew's pituitary gland when there is not enough water in his body fluids. (**1 mark**)
   iii) Explain how the hormone named in **c) ii)** functioned to prevent Andrew losing too much water in his urine when his body fluids became too concentrated. (**2 marks**)

Total 15 marks

7 a) Ariel's ears are one of the five sense organs found in her body.
   i) Identify TWO different stimuli detected by Ariel's ears. (**2 marks**)
   ii) State the function of EACH of the following parts of Ariel's ear: the pinna, the ear drum and the Eustachian tube. (**3 marks**)
   iii) Would Ariel be able to hear a sound with a frequency of 25 000 Hz? Give a reason for your answer. (**1 mark**)
b) Cleavdon starts to find it increasingly difficult to focus on aircraft in the sky.
   i) Explain the possible cause of Cleavdon's difficulties. (**2 marks**)
   ii) What action must Cleavdon take to help him to focus on the aircraft? (**2 marks**)
c) The endocrine system is composed of endocrine glands that secrete hormones into the blood.
   i) What is the overall role of the endocrine system in the human body? (**1 mark**)
   ii) Which hormone is Merva MOST likely to release in large quantities immediately before she sits her CSEC Integrated Science examination? (**1 mark**)
   iii) Outline the effects that the hormone identified in **c) ii)** has on Merva's body. (**3 marks**)

Total 15 marks

8 a) The human nervous system is made up of nerve cells or neurones.
   i) Distinguish among the THREE types of neurones found in the nervous system. (**3 marks**)
   ii) Talia suffered a fall during a show jumping competition which resulted in her left arm becoming paralysed. Suggest an explanation for Talia's paralysis. (**3 marks**)
b) Figure 7 shows a human brain.
   i) Identify the structures labelled D and E. (**2 marks**)

Exam-style questions – Chapters 1 to 7

ii) Give ONE function of the structure labelled F. **(1 mark)**

c) Gwinette sees her 3-year-old son, Otis, fall down and hurt himself. She immediately runs to his assistance.

   i) Is Gwinette's action voluntary or involuntary? Support your answer with a reason. **(2 marks)**

   ii) Describe the events occurring in Gwinette's nervous system to bring about her response. **(4 marks)**

Total 15 marks

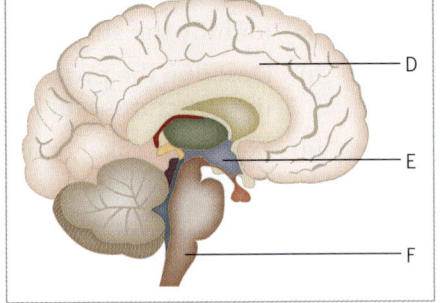

Figure 7 A human brain

9 a) Diseases can be classified as communicable or non-communicable.

   i) Define the term 'disease'. **(1 mark)**

   ii) Explain why hypertension and diabetes are both classified as non-communicable diseases. **(2 marks)**

   iii) Elwyn suffers from hypertension. Outline TWO reasons why it is important for Elwyn to exercise regularly. **(2 marks)**

   iv) Suggest why practising good personal hygiene is important to control the spread of skin infections. **(1 mark)**

b) HIV/AIDS and syphilis are both sexually transmitted infections (STIs), however it is much easier to control the spread of syphilis than HIV/AIDS.

   i) Suggest the MAIN factor that makes the spread of syphilis relatively easy to control. **(1 mark)**

   ii) Identify TWO factors that make it difficult to control the spread of HIV/AIDS. **(2 marks)**

   iii) To date, the only STI whose spread can be controlled by vaccination is hepatitis B. Outline how the hepatitis B vaccine can protect a person against contracting the disease. **(3 marks)**

c) Based on their effects on the nervous system, certain drugs can be described as being hallucinogens.

   i) What is a drug? **(1 mark)**

   ii) Give ONE example of a drug that can be described as being a hallucinogen and explain why. **(2 marks)**

Total 15 marks

10 a) Mr. Scruffy always has a pile of refuse outside his home and his neighbour, Mr. Clean, is worried that it will attract household pests.

   i) Distinguish between a pest and a parasite. **(2 marks)**

   ii) Name TWO household pests that might be attracted to Mr. Scruffy's refuse. **(2 marks)**

   iii) Discuss why Mr. Clean is so worried that the refuse will attract pests. **(3 marks)**

b) Onika bought a jar of pickled onions and decided to keep it in the refrigerator after opening.

   i) Discuss, citing examples, how microorganisms are used in the production and processing of food. **(2 marks)**

   ii) Define the term 'food preservation'. **(1 mark)**

   iii) Identify and explain the method of food preservation used to make the pickled onions. **(3 marks)**

   iv) Explain how the refrigerator assisted in preserving the onions after after Onika had opened the jar. **(2 marks)**

Total 15 marks

Module 2 – Energy

8 Conservation of energy

The Sun supplies us with energy to do the work of the many processes occurring on Earth. Motion of the winds, oceans and rivers; changes in our weather; the production of food by photosynthesis; the light that we see and the sounds that we hear, are all linked to several energy transformations.

Energy, work and change

Energy is the ability to do **work** (the ability to produce change).

Work is done when the point of application of a **force moves** and is measured as the product of the **force** and the **distance** moved in the **direction of the force**.

1 joule is the work done (or energy used) when a **force** of 1 newton moves an object through a **distance** of 1 m in the direction of the force.

Figure 8.1 shows FOUR forces (F, f, mg, R) acting on a block pushed through a distance, d. Forces are discussed in more detail in Chapter 16.

- Work done (energy used) by F is **positive** since F and d are **similarly** directed: $W = F \times d$
- Work done (energy used) by f is **negative** since f and d are **oppositely** directed: $W = -f \times d$
- Work done (energy used) by mg is **zero** since mg and d are **perpendicular** to each other: $W = 0$
- Work done (energy used) by R is **zero** since R and d are **perpendicular** to each other: $W = 0$

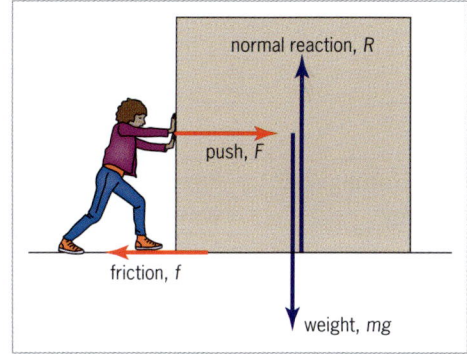

Figure 8.1 *Forces can do work*

Energy can produce different types of change

- Energy stored in a compressed spring causes a change of position (motion) when released.
- Energy absorbed or released from water in the form of heat causes a change of temperature.
- Energy absorbed by melting ice or released by freezing water causes a change of state.
- Energy released from the spark plug of a car causes a chemical change as the gasoline combusts to form water vapour and oxides of carbon.

Some types of energy

Table 8.1 *Some types of energy*

Energy	Description
Gravitational potential energy	Energy **stored** by a body due to its **position** in a **gravitational** force field.
	It increases as the body rises above the Earth's surface.
Elastic potential energy	Energy **stored** by a body due to its position in an **elastic** force field.
	It increases as the body is stretched or compressed.
Nuclear potential energy	Energy **stored** by an atomic **nucleus** due to its **physical state**. It is released (often as gamma radiation) if the nucleus is unstable.

Energy	Description
Chemical potential energy	Energy **stored** by a body due to its **chemical state**. It is usually released when the chemicals react.
Electrical energy	Energy carried by moving **charged particles**, for example electrons or ions.
Kinetic energy	Energy possessed by a body **due to its motion**.
Thermal energy	Energy possessed by a body **due to the motion of its atoms, molecules or ions**.
Heat energy	Thermal energy transferred from **hotter to cooler** bodies.
Sound energy	Energy transported as **particle vibrations** of a **longitudinal** wave (see page 92). **Speed:** Greater in solids than in liquids and greater in liquids than in gases, since vibrations **transmit** easier by the **closer packing of particles**. Sound waves **cannot pass through a vacuum**, so communication in space is done by **radio waves**. **Pitch:** Increases with **frequency** of the vibration. **Loudness:** Increases with **amplitude** (max. displacement) of the vibration. Table 8.1 *Pitch and loudness* |	Loudness	Pitch
---	---	---
A	High	Low
B	Low	High
C	High	High
D	Low	Low	
Electromagnetic energy	Energy transported as electric and magnetic field vibrations forming a **transverse wave** (see page 92).

A closer look at electromagnetic wave energy

Electromagnetic waves are the **only waves** that can **travel through the vacuum of outer space**. This is how we receive our **energy from the Sun** to do things on Earth.

Table 8.2 *Wavelengths, production, detection, properties and uses of electromagnetic waves*

Type of electro-magnetic wave	Relative wavelength	Production, detection, properties and uses
Radio (includes microwaves – the shortest of the radio waves)		Emitted and detected by **radio transmitters** and **receivers**, respectively. • Can better **deviate around obstacles** than can other electromagnetic waves and so are used for **radio, TV and cell phone communication**. • They are also used to **heat food**.
Infrared (IR)		All bodies emit IR. • Detected as **warmth** when incident on our skin. • Detected by thermal **sensors** and prints on certain **photographic** film.
Light (ROYGBIV)		Bodies above **1100 °C** emit light. • Is comprised of seven colours (red, orange, yellow, green, blue, indigo, violet) which **form white light** when **combined**. • Detected by light **sensors** and prints on certain **photographic** film.

Type of electro-magnetic wave	Relative wavelength	Production, detection, properties and uses
Ultraviolet (UV)	∿∿∿	Emitted by very hot bodies, such as the Sun, lightning and lamp filaments. • Produces **fluorescence** of certain materials (see pages 143 and 144). • Used to **cut** and **weld metals**. • Harms living tissue and can cause **sunburn**. • Detected by UV **sensors** and prints on certain **photographic** film.
X and gamma	∿∿∿∿∿∿	X-rays can be produced by **high-voltage machines**. **Gamma rays** are emitted by the **atomic nuclei** of **unstable atoms**. • Both can be detected by specialised **sensors** and can pass through **flesh and metals** to form **images** on certain photographic film. • Both are **ionising**, **high frequency** radiations which are **hazardous to our health** and can produce **cancerous tumours**. • Both can also be used as **therapy** in **destroying cancerous tumours**. • Both cause **fluorescence**.

Transformation, transfer and transport of energy

*Energy is **transformed** as it changes from **one form to another**.*

*Energy is **transferred** as it passes from **one body to another**.*

*Energy is **transported** between bodies as it transfers **between bodies in different locations**.*

Figure 8.2 shows how energy can be transformed, transported and transferred.

Figure 8.3 shows how solar energy can be transported through space and transferred to Earth.

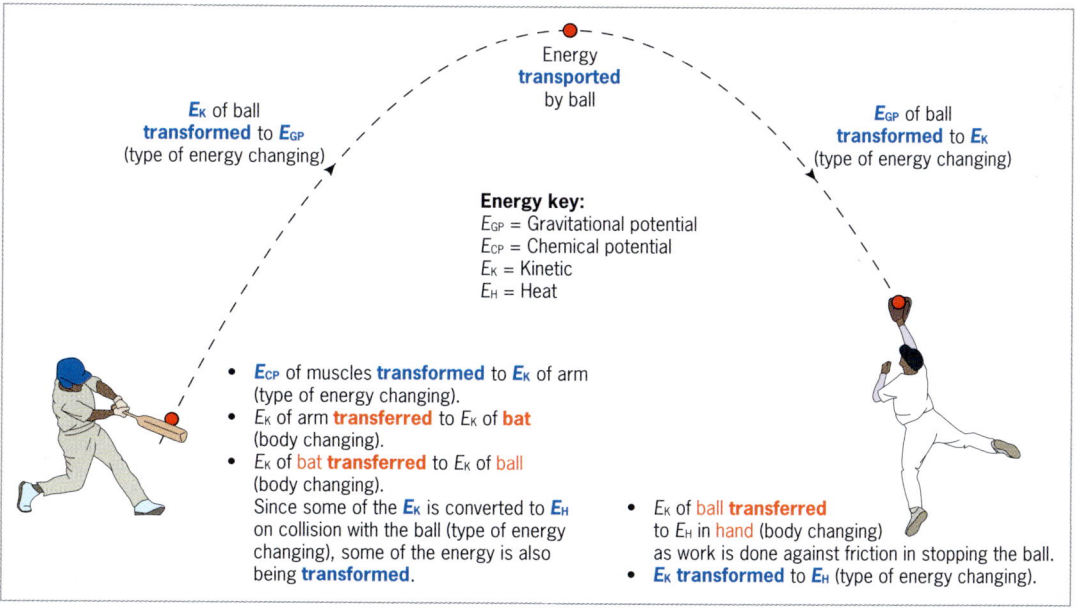

Figure 8.2 *Transformation, transfer and transport of energy*

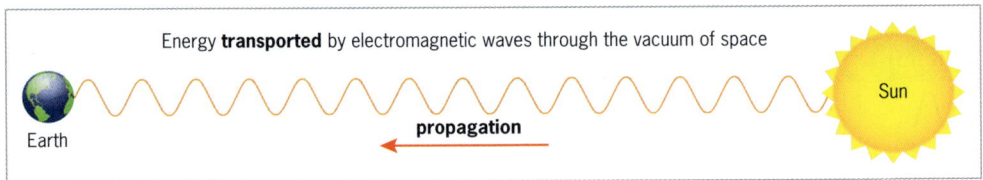

Figure 8.3 *Electromagnetic waves transporting energy between bodies*

Wave energy can propagate as transverse or longitudinal vibrations

Transverse waves have vibrations that are **perpendicular** to the direction of travel of the wave.

They are characterised by **crests** and **troughs**.

Examples are **water** waves, **electromagnetic** waves and a **rope vibrated** as shown in Figure 8.4.

Longitudinal waves have vibrations that are **parallel** to the direction of travel of the wave.

They are characterised by a series of regions of **high pressure** (compressions, C) and **low pressure** (rarefactions, R).

Examples are **sound waves** (Figure 8.5) and a **spring vibrated** as shown in Figure 8.6.

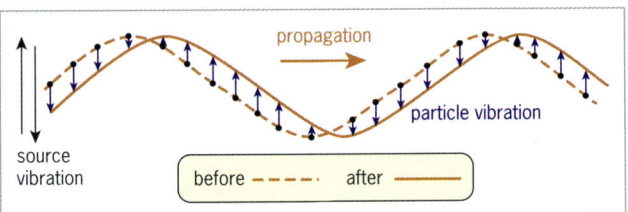

Figure 8.4 *Transverse wave along a rope*

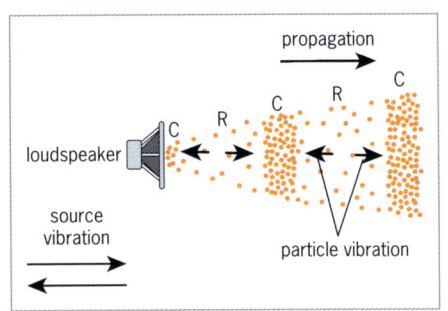

Figure 8.5 *Sound waves in air*

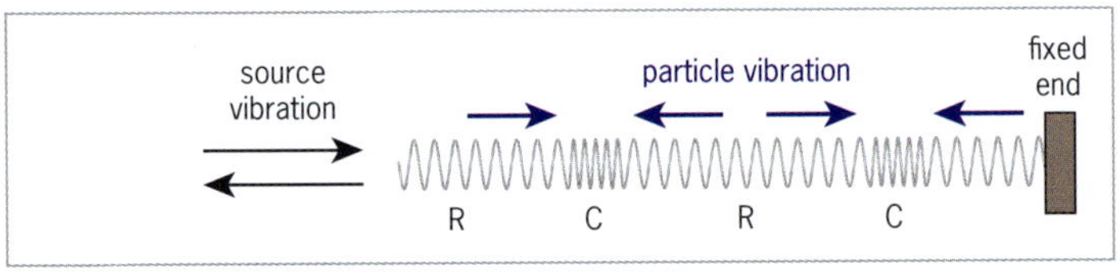

Figure 8.6 *Longitudinal wave along a spring*

Wave energy can be focused by reflection

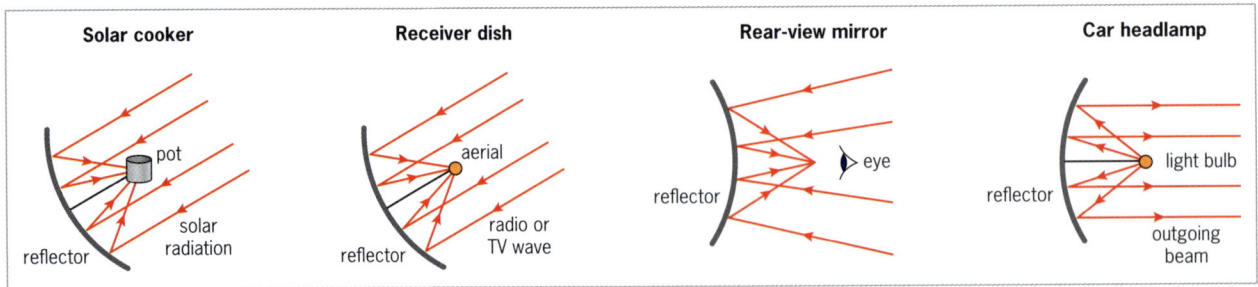

Figure 8.7 *Curved reflectors*

Law of conservation of energy

*The **law of conservation of energy** states that energy cannot be created or destroyed but can be transformed from one type to another.*

Several examples of energy transformations are outlined on pages 93 to 95.

Note: When heat energy is produced, it is usually absorbed by the surroundings.

Examples of energy transformations

Coconut falling to the ground

gravitational potential energy (due to its height in the gravitational field) → kinetic energy (due to motion) → heat energy (due to friction on striking ground) / sound energy (due to vibration on striking ground)

Charging a battery

electrical energy (from charger) → chemical potential energy (stored in chemicals of the battery)

Filament lamp in use

electrical energy (from power socket) → heat energy (due to electrical friction as electrons move against the resistance of the filament) / light energy (due to electrical friction as electrons move against the resistance of the filament)

Photovoltaic (PV) electrical generation (solar panels)

light energy (from Sun – 'solar energy') → electrical energy (generated by PV panel)

Photosynthesis

light energy (from Sun – 'solar energy') → chemical potential energy (stored in carbohydrate molecules of green plants)

Solar powered calculator

light energy (from Sun – 'solar energy') → electrical energy (generated by PV panel) → light energy (generated as electrons transfer through the panel) / heat energy (small amount due to electrical resistance)

Electrical motor lifting a load

electrical energy (from power socket or battery pack) → gravitational potential energy (as the load rises) / heat energy (due to friction within the motor as it spins)

8 Conservation of energy   93

Hydro-electric power station (using an elevated reservoir or river dam)

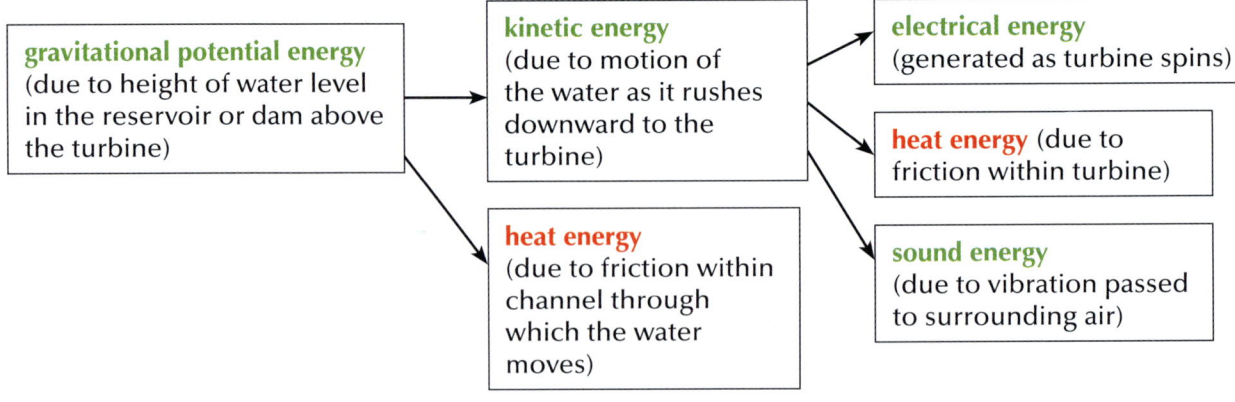

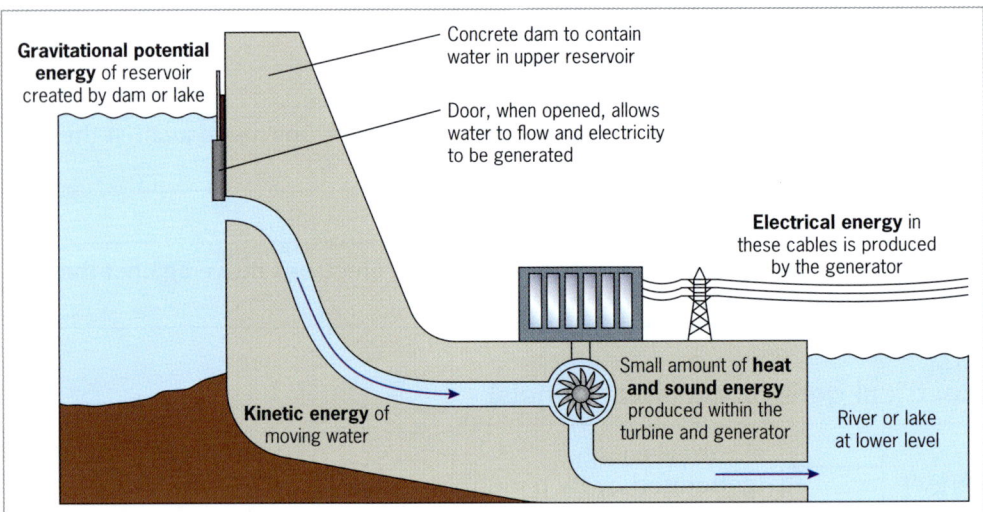

Figure 8.8 *Hydro-electric power*

Nuclear-electric power station

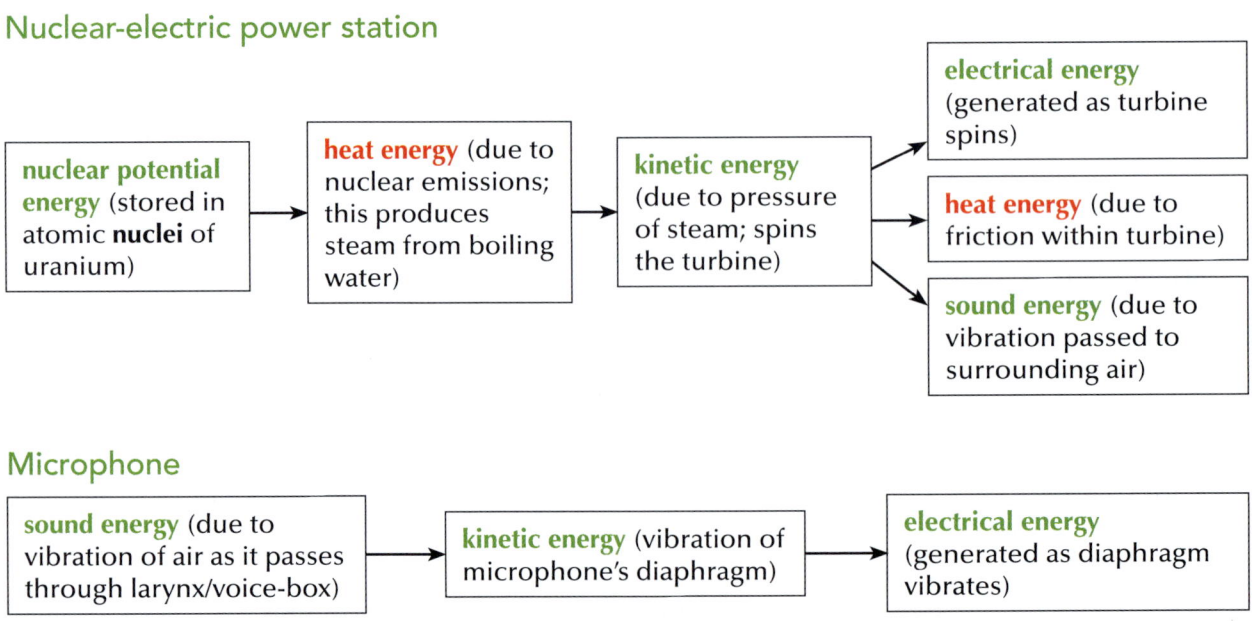

Microphone

sound energy (due to vibration of air as it passes through larynx/voice-box) → kinetic energy (vibration of microphone's diaphragm) → electrical energy (generated as diaphragm vibrates)

Loudspeaker

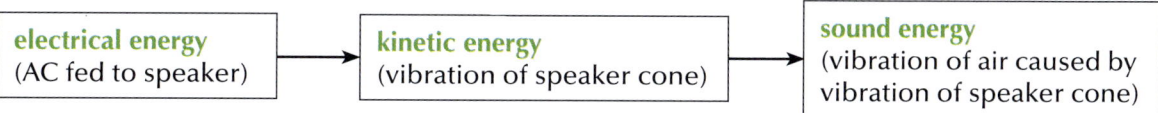

Note: The energy changes in the microphone and speaker are the reverse of each other.

Microwave oven warming food

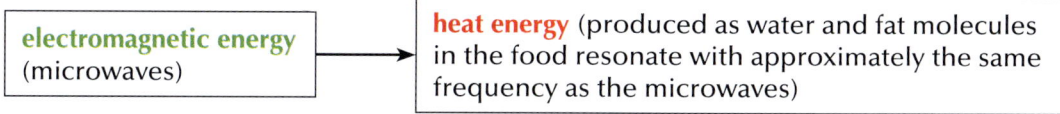

Vehicle braking and coming to rest

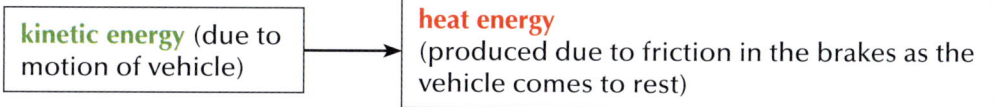

Block sliding down rough incline

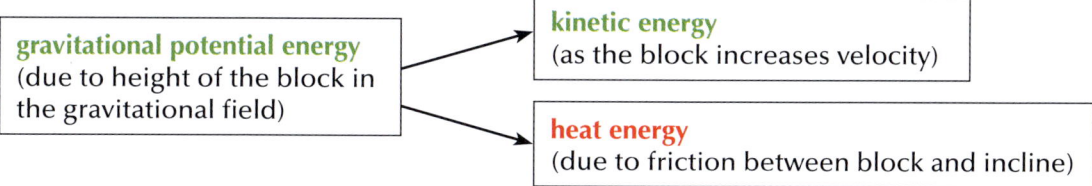

Swinging simple pendulum

Figure 8.9 shows a pendulum at different points of its swing.

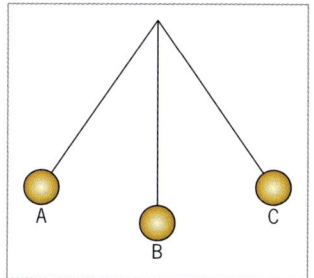

At A and C: $E_K = 0$ since the ball is momentarily at rest.

E_{GP} is then maximum since the ball is now at its maximum height for that swing.

At B: E_K is maximum since the ball moves fastest.

E_{GP} is then minimum since the ball is at its lowest point.

As the ball swings, a **small amount** of **heat** is **continuously lost** due to **friction**.

For a '**simple** pendulum', this is ignored.

Figure 8.9 *Simple pendulum*

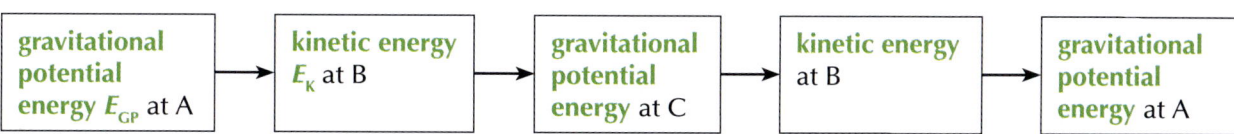

Interconversion of mass and energy

Energy can convert to mass and mass can convert to energy.

8 Conservation of energy

Nuclear fission is the **splitting** of a large atomic nucleus into smaller atomic nuclei, resulting in a **large output of energy** and a **small loss in mass**.

Nuclear fission of uranium or plutonium produces electricity in nuclear reactors.

Nuclear fusion is the **joining** of two very **small atomic nuclei** to form a larger atomic nucleus, resulting in a **large output of energy** and a **small loss in mass**.

Nuclear fusion in the very hot core of the Sun provides us with energy.

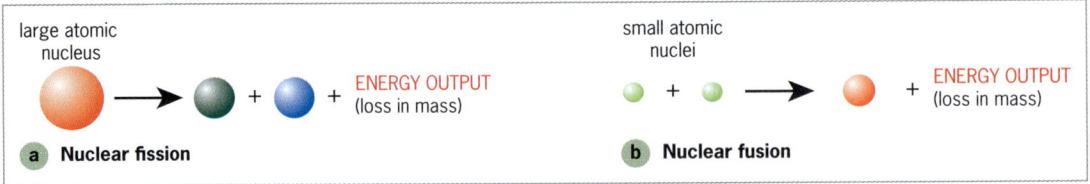

Figure 8.10 *Nuclear fission and nuclear fusion*

Simple nuclear equations

Fission: uranium → barium + krypton + energy   … compare with Figure 8.10a
Fusion: hydrogen + hydrogen → helium   … compare with Figure 8.10b

Energy supplies for use in space

Satellites and spacecraft require energy sources to (1) **operate devices** and (2) **provide propulsion**.
- **Nuclear fission** devices can be **small**, **portable** and **reliable** and can deliver **large amounts of energy**.
- **Solar cells** are **lightweight**, **do not produce harmful emissions** and use **free energy input**.

The internal combustion engine (ICE)

An internal combustion engine is a **heat engine** that burns **fossil fuel internally** within cylinders. The hot gases produced move pistons, engaging a mechanical system to propel a vehicle or other device.

Harmful exhaust contaminants of the internal combustion engine

The following **FIVE** harmful contaminants are produced by burning fossil fuels: **carbon monoxide, oxides of nitrogen and sulfur, volatile hydrocarbons, particulate matter** and **heavy metal ions**.
See Table 10.3, page 126 to learn more about these harmful emissions.

Industrial methods of increasing efficiency and reducing pollution caused by the ICE

- **Government regulations** ensure that air quality meets specific standards.
- **Improved quality of engines and fuels** produce higher fuel efficiency and less pollution.
- **Streamlined aerodynamic** shapes of vehicles increase efficiency by reducing air friction.
- **'Idle off' systems** save fuel by switching off the engine when the car is stationary.
- **Catalytic converters** fitted to exhaust pipes convert pollutants to harmless gases.
- **Solar cells**, **LED** lighting and **energy efficient systems** reduce fuel consumption.
- **Modern cooling systems** reuse heat energy from the engine.
- **Modern suspension systems** generate electricity by absorbing energy from vibrations.
- **Planting trees** helps remove the greenhouse gas, carbon dioxide (see Table 10.2, page 126).
- **Regenerative braking systems** recharge batteries of electric vehicles.
- The **switch to electric vehicles** reduces pollution and provides better energy conversion.

Methods of individuals to increase efficiency and reduce pollution by the ICE

- Ensure correct **tyre pressure maintenance**.
- Keep **windows closed** to reduce friction and so increase fuel efficiency.
- **Regularly change engine oil**, **oil filters** and **air filters**.
- Ensure engine is **well tuned**.
- Avoid unnecessary **idling**.
- Avoid unnecessary **braking** by keeping your distance from the vehicle in front.
- Avoid unnecessary **speeding** to improve fuel efficiency.
- **Combine chores** to make a single trip.
- **Carpool**.

Revision questions

1. **a** Define: **i** energy   **ii** work.
   b State the SI unit of work and energy.

2. **a** Define: **i** kinetic energy   **ii** elastic potential energy   **iii** gravitational potential energy   **iv** heat energy   **v** chemical potential energy.
   b What type of energy can pass through a vacuum?

3. **a** What can be said of the pitch and loudness of a musical note which has a low frequency and a large amplitude?
   b Does sound travel faster through air or through metal? Give a reason for your answer.
   c Can sound pass through a vacuum?

4. State the type of electromagnetic energy that:
   a has the largest wavelength
   b contributes to body warmth
   c causes sunburn
   d is emitted by an atomic nucleus
   e destroys cancerous cells
   f is used to weld and cut metals
   g ionises body cells to produce cancers.

5. Distinguish between the different modes of vibration that transverse and longitudinal waves use to propagate their energy, giving ONE everyday example of each.

6. Draw a labelled diagram to show how radio waves can be focused by a satellite receiver dish.

7. **a** State the law of conservation of energy.
   b Name a process that converts mass to energy.
   c Using **arrow diagrams**, show the MAIN energy transformations occurring during the following:
      i a book falling from a shelf
      ii a vehicle braking and coming to rest
      iii a battery being charged
      iv a filament lamp plugged into a power outlet.

8. **a** Define: **i** nuclear fission   **ii** nuclear fusion.
   b Which of the processes in part **a** occurs in:
      i the core of the Sun?
      ii a nuclear power plant?

9   **a** State TWO energy sources useful to spacecraft.
     b Give TWO reasons spacecraft need to carry energy sources.

10   **a** What is an internal combustion engine (ICE)?
     b State FIVE harmful exhaust contaminants produced by such an engine.

Photosynthesis and energy transfer in the environment

Photosynthesis and energy conversion

Photosynthesis is the process by which green plants convert carbon dioxide and water into glucose by using energy from sunlight, which is absorbed by chlorophyll in chloroplasts. Oxygen is produced as a by-product.

Photosynthesis occurs in any plant structure that contains **chlorophyll**, i.e. is green; however, it mainly occurs in **leaves**. Chlorophyll molecules in the **chloroplasts** of leaf cells absorb **energy** from sunlight and use it to bring about the reaction between **carbon dioxide** and **water** to produce **glucose** and **oxygen**.

Photosynthesis can be summarised by the following **equations**:

Word equation

$$\text{carbon dioxide} + \text{water} \xrightarrow{\text{energy from sunlight absorbed by chlorophyll}} \text{glucose} + \text{oxygen}$$

Chemical equation

$$6CO_2 + 6H_2O \xrightarrow{\text{energy from sunlight absorbed by chlorophyll}} C_6H_{12}O_6 + 6O_2$$

Photosynthesis is known as a **photochemical reaction** because it is initiated by the absorption of **light energy**. This light energy is **converted** into **chemical energy**, which is stored within glucose molecules.

Substrates and conditions for photosynthesis

Photosynthesis requires the following **two** raw materials, also known as **substrates**.
- **Carbon dioxide**, which diffuses into the leaves from the air through **stomata** (see Figure 9.16, page 119).
- **Water**, which is absorbed from the soil by the roots.

In addition to the these substrates, photosynthesis requires several **conditions**.
- **Energy from sunlight**, which is absorbed by the chlorophyll in chloroplasts of leaf cells.
- **Chlorophyll**, the green pigment which is present in chloroplasts.
- **Enzymes**, which are present in chloroplasts.
- A **suitable temperature** of between about 5 °C and 40 °C, so that the enzymes can function.

Certain **mineral ions** are also indirectly required, including magnesium (Mg^{2+}), iron (Fe^{3+}), nitrate (NO_3^-) and potassium (K^+) ions.

The products of photosynthesis and their importance

Photosynthesis produces **two** products: **glucose** and **oxygen**.
- Some of the **glucose** is used by plant cells in **respiration**, during which the stored **energy** is released and used by the plants in life processes. The rest can be converted to various other **organic compounds** (carbon-containing compounds), including the following.

- **Sucrose** and **starch**, which are stored by plants for later use.
- **Cellulose**, which is used to make cell walls in growing plant parts.
- **Amino acids** and **proteins**, which plants use for growth.
- **Lipids**, which plants store, mainly in seeds for use during germination.
* The **oxygen** is used by leaf cells in **aerobic respiration**. Any unused oxygen diffuses out of the leaves into the air and is used by other living organisms.

An introduction to the environment

Certain **terms** are used when discussing interactions between living organisms and their environment.

* **Environment** refers to the combination of factors that surround and act on an organism.
* **Habitat** is the place where a particular organism lives.
* A **species** is a group of organisms of common ancestry that closely resemble each other and are normally capable of interbreeding to produce fertile offspring.
* A **population** is composed of all the members of a particular species living together in a particular habitat.
* A **community** is composed of all the populations of different species living together in a particular area.
* An **ecosystem** is a community of living organisms interacting with each other and with their physical (non-living) environment, e.g. a pond, a coral reef, a mangrove swamp and a forest.

Transfer of energy in the environment

During **photosynthesis**, **energy** from sunlight is incorporated as **chemical energy** into the **organic food molecules** made by plants. Because **green plants** make their own food, they are known as **producers**. These organic food molecules and the **energy** they contain are then passed on to other organisms, known as **consumers**, through **food chains** and **food webs**.

Food chains

*A **food chain** is a linear diagram showing the flow of food and energy from one organism to the next in an ecosystem.*

Any **food chain** always begins with a producer and includes the following.
* A **producer**, i.e. a green plant.
* A **primary consumer** that eats the producer.
* A **secondary consumer** that eats the primary consumer.
* A **tertiary consumer** that eats the secondary consumer.

Consumers can also be classified according to what they **consume**.
* **Herbivores** consume plants or plant material only, e.g. cows, grasshoppers, snails, slugs, parrot fish and sea urchins.
* **Carnivores** consume animals or animal material only, e.g. lizards, toads, spiders, centipedes, eagles, octopuses and sharks.
* **Omnivores** consume both plants and animals, or plant and animal material, e.g. hummingbirds, crickets, mice, humans and crayfish.

Organisms in a **food chain** are linked by **arrows** showing the direction in which the **food** and **energy** flow, and the **position** of an organism is known as its **trophic level**.

***Trophic level** refers to the position or level that an organism occupies in a food chain.*

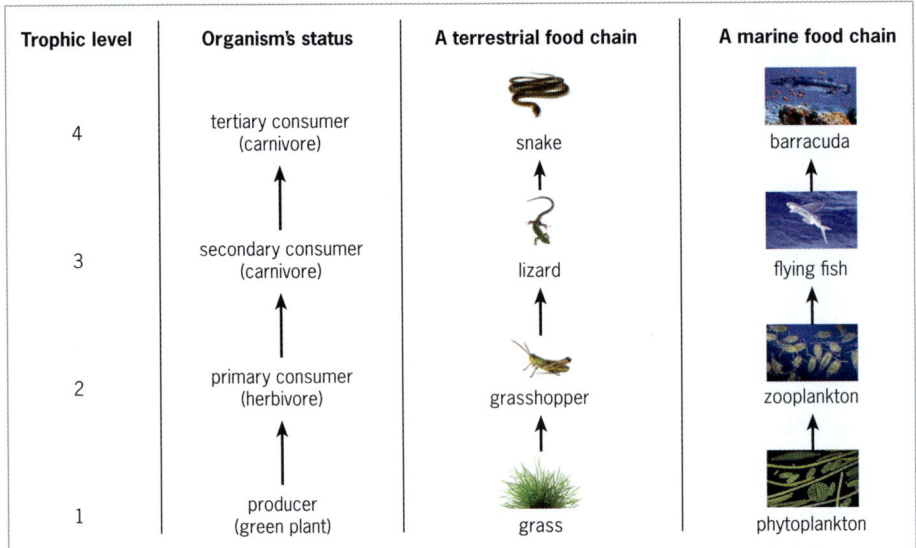

Figure 8.11 *Examples of food chains*

Food chains can also be written horizontally, as in the following example from a **mangrove swamp**:

green algae ⟶ mangrove crab ⟶ mudskipper ⟶ egret

Food webs

Any ecosystem usually has more than one producer and most consumers have more than one source of food. Consequently, food chains are interconnected to form **food webs**.

*A **food web** is a diagram that links food chains together to show all the pathways along which food and energy flow between organisms in an ecosystem.*

Organisms that are not consumed eventually **die**, and **decomposers**, which are mainly bacteria and fungi, obtain their food from the bodies of these dead organisms. This causes the dead organisms to **decompose**, which releases **mineral nutrients** back into the environment.

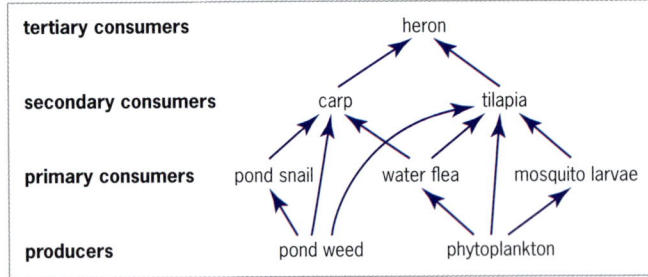

Figure 8.12 *An example of a food web from a freshwater lake*

Energy transfer in food chains

Not all the **energy** incorporated into the organic food molecules made by green **plants** during photosynthesis is passed along a food chain; some is **used** and some is **lost** at each trophic level. Plants use some of the food in **respiration**. This releases energy which the plants **use** in life processes or **lose** as heat energy. The rest of the food is used by the plants for **growth** or is **stored**.

When plants are eaten by **herbivores**, some of the organic matter containing energy is **lost** in **faeces** or **excretory products**, e.g. urea, and some is used in **respiration**, during which the stored energy is released and **used** by the herbivores or is **lost** as heat energy. The remaining food containing energy is used to **build body tissues** or is **stored**, and is then passed on to the next trophic level when herbivores are consumed. This then continues at each trophic level.

When organisms that have not been consumed **die**, these dead organisms, together with faeces and excretory products, are **decomposed** by **decomposers**, and the stored energy is released by the decomposers during respiration and **used** or **lost**. In general, only about **1** to **2%** of the energy from sunlight is absorbed by plants and used in photosynthesis, and only about **10%** of the energy from one trophic level is transferred to the next level.

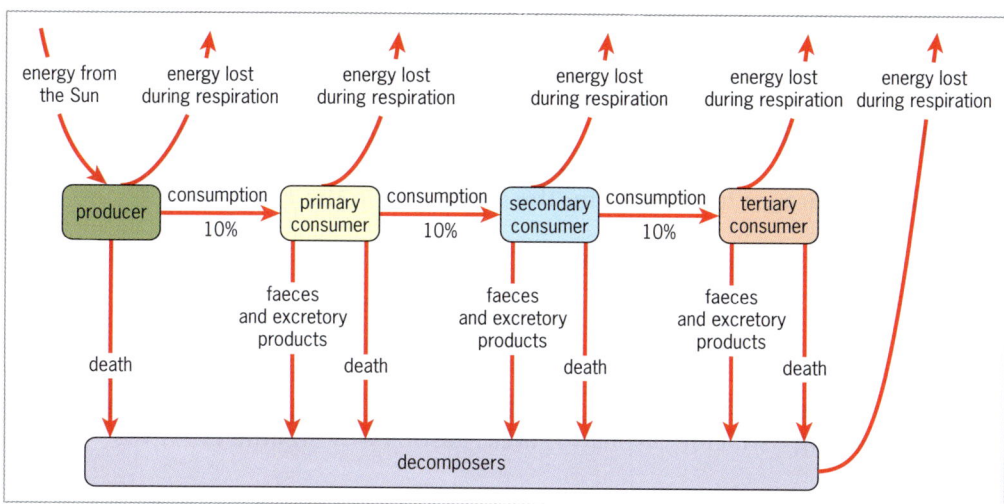

Figure 8.13 *Energy flow through a food chain*

Ecological pyramids

*An **ecological pyramid** is a graphical representation in the form of a pyramid that shows the relationship between organisms at different trophic levels in a food chain.*

Because the total amount of **energy decreases** at successive trophic levels in a food chain, there are usually **fewer** organisms with a **lower** total **biomass** at successive levels. **Energy, number of organisms** and **biomass** at successive levels can be represented by **ecological pyramids** in which the **width** of each bar is proportional to the quantity of the factor being measured. There are **three** types of ecological pyramids: a **pyramid of energy**, a **pyramid of numbers** and a **pyramid of biomass**.

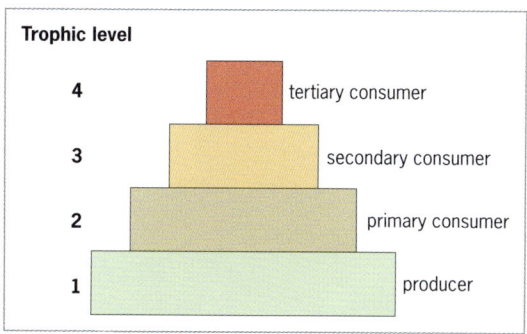

Figure 8.14 *A pyramid of energy, biomass or numbers*

Ecological balance and environmental sustainability

An **ecologically balanced** ecosystem is one in which species coexist with each other and with their environment in a **stable** and **sustainable** way. **Maintaining** ecological balance is essential to ensure the **stability** and **sustainability** of ecosystems for all forms of life. It helps **preserve biodiversity**, which ensures the continuous existence of species and availability of natural resources, and it maintains the **health** of ecosystems. If ecological balance is **disrupted** by natural or human-induced factors it can lead to habitat loss, the loss of biodiversity and the extinction of species, and it has a negative impact on human life, e.g. creating shortages of food and other natural resources, and sometimes leading to the outbreak of disease.

Revision questions

11 a Define the term 'photosynthesis' and give a word and chemical equation to summarise the process.

   b Identify the substrates, conditions and products of photosynthesis.

   c Explain the energy interconversion that happens during photosynthesis.

12 Identify FOUR ways that a green plant can make use of the products of photosynthesis.

13 a Provide a definition for EACH of the following: a habitat, a population, an ecosystem, a food chain, a food web and trophic level.

   b Some aphids were observed on the underside of the leaves of tomato plants in a garden and ladybird beetles were seen feeding on the aphids. A toad was also seen eating a dragonfly, and dragonflies are known to eat ladybird beetles. Use this information to draw a food chain for the organisms in the garden.

14 From the organisms in named in question **13 b**, identify:

   a a carnivore   b a producer   c a herbivore

   d a primary consumer   e a secondary consumer.

15 When Jared eats a barracuda he only gets about 10% of the energy that the barracuda obtained from the flying fish it ate. Explain THREE reasons why so little energy is passed on to Jared.

16 What does an ecological pyramid show?

17 Explain why it is important that ecosystems remain ecologically balanced.

9 Energy in life processes

Living organisms need a variety of **nutrients** to provide them with the **energy** and important chemicals they need to carry out the different life processes. These nutrients are contained in the **food** that they make or obtain. After humans consume their food, it must be broken down by **digestion** into a form that is useful for the body's activities. To obtain **energy** from the digested food, it must then be **respired** by body cells. Most living organisms respire **aerobically**, and the process of **breathing** is responsible for taking in the oxygen they need for aerobic respiration and getting rid of the carbon dioxide produced.

The human diet

A balanced diet

The **food** an animal eats is called its **diet**. The human diet must contain the following **seven** components.

- **Carbohydrates**, **proteins** and **lipids** (**fats and oils**), also known as **macronutrients**.
- **Vitamins** and **minerals**, also known as **micronutrients**.
- **Dietary fibre** or **roughage** and **water**.

*A **balanced diet** is a diet that contains carbohydrates, proteins, lipids, vitamins, minerals, dietary fibre and water in the **correct proportions** to maintain **growth** and **good health**.*

Humans must consume a **balanced diet** each day to supply the body with enough **energy** for daily activities, the correct materials for **growth**, **repair**, **development** and the **manufacture** of biologically important molecules, and to keep the body in a **healthy state**.

A **balanced diet** should contain a **variety** of foods selected from each of the **six Caribbean food groups** shown in Figure 9.1. Each group contains foods which supply similar nutrients in similar proportions. The **size** of each sector indicates the **relative amount** of each group that should be eaten daily.

Figure 9.1 The six Caribbean food groups

Table 9.1 *Nutrients supplied by the six Caribbean food groups*

Food group	Nutrients supplied
Staple foods	Carbohydrates (mainly starch), vitamins, minerals, fibre.
Legumes and nuts	Proteins, carbohydrates (mainly starch), vitamins, minerals, fibre.
Foods from animals	Proteins, lipids, vitamins, minerals.
Fruits	Carbohydrates (mainly sugars), vitamins, minerals, fibre.
Non-starchy vegetables	Vitamins, minerals, fibre.
Fats and oils	Lipids, vitamins.

Carbohydrates, proteins and lipids

Carbohydrates, **proteins** and **lipids** are required in relatively large quantities. They are **organic** compounds, i.e. their molecules contain carbon, hydrogen and oxygen atoms. They supply the body with the **energy** it needs, and the materials it needs for **growth**, **repair**, **development** and the **manufacture** of biologically important molecules.

- **Carbohydrates** include **sugars** and **starch**, and can be classified into **three** groups based on the chemical structure of their molecules: **monosaccharides**, **disaccharides** and **polysaccharides**.
    - **Monosaccharides** are the simplest carbohydrate molecules. Many have the formula $C_6H_{12}O_6$. They include glucose, fructose and galactose.
    - **Disaccharides** are formed by chemically joining two monosaccharide molecules together by **condensation** (see Figure 9.2). They have the formula $C_{12}H_{22}O_{11}$. They include maltose, sucrose and lactose.
    - **Polysaccharides** are formed by joining many monosaccharide molecules into straight or branched chains. They include starch, cellulose and glycogen (animal starch).

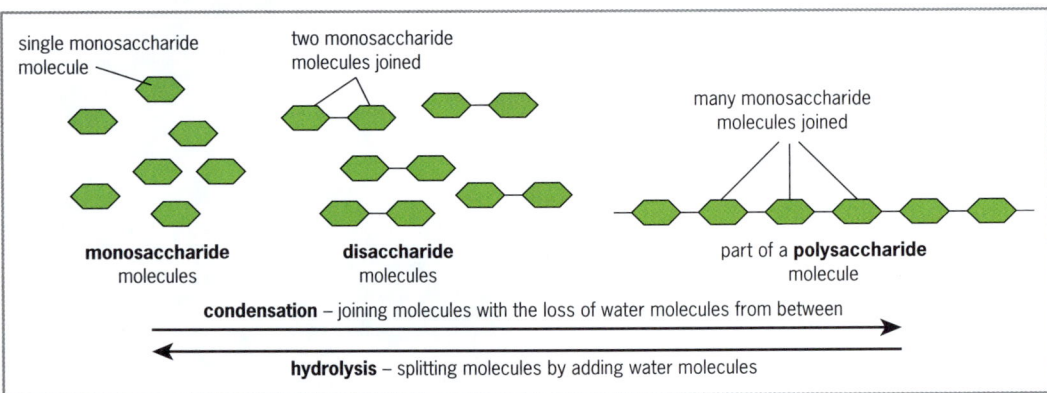

Figure 9.2 *The three types of carbohydrates*

- **Proteins** are large molecules composed of hundreds or thousands of small molecules known as **amino acids**, which are joined together in long chains by **peptide links**.

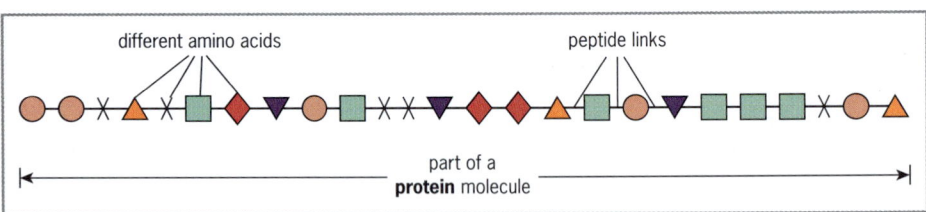

Figure 9.3 *Part of a protein molecule*

- **Lipids** are **fats** and **oils**. Their molecules are composed of **four** smaller molecules joined together: three **fatty acid** molecules and one **glycerol** molecule.

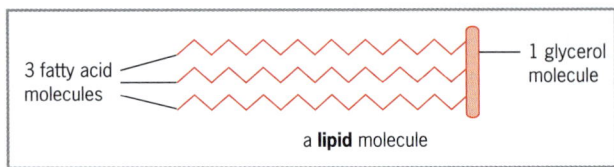

Figure 9.4 *A lipid molecule*

Table 9.2 *Sources and functions of carbohydrates, proteins and lipids in the human body*

Macronutrient	Sources	Functions
Carbohydrates	**Sugars**: fruits, cakes, sweets, jams, sugar-sweetened beverages. **Starch**: yams, potatoes, rice, pasta, bread.	• To provide **energy** (16 kJ g^{-1}); energy is easily released when respired. • For **storage**: glycogen (animal starch) granules are stored in the cytoplasm of many cells.
Proteins	Fish, lean meat, milk, cheese, eggs, peas, beans, nuts.	• To make **new cells** for growth and to repair damaged tissues. • To make **enzymes** to catalyse (speed up) chemical reactions in the body. • To make **hormones** to control various processes in the body. • To make **antibodies** to fight disease. • To provide **energy** (17 kJ g^{-1}); used only when stored carbohydrates and lipids have been used up.
Lipids	Butter, vegetable oils, margarine, nuts, fatty meats.	• To make **cell membranes** of newly formed cells. • To provide **energy** (38 kJ g^{-1}); used after carbohydrates because their metabolism is more complex and takes longer. • For **storage**: fat is stored under the skin and around organs. • For **insulation**: fat under the skin acts as an insulator.

Vitamins and minerals

Vitamins and **minerals** are required in small amounts for healthy growth and development.

Table 9.3 *Some important vitamins required by the human body*

Vitamin	Sources	Functions
A	Liver, cod liver oil, yellow and orange vegetables, and fruits, e.g. carrots and pumpkin, and green leafy vegetables, e.g. spinach.	• To help keep the skin, cornea and mucous membranes healthy. • To help vision in dim light (night vision). • To strengthen the immune system.
B$_{12}$	Animal products, e.g. meat, liver, poultry, fish, dairy products and eggs.	• To help regulate the formation of red blood cells in red bone marrow. • To aid in respiration to produce energy. • To help the nervous system function properly.
C	West Indian cherries, citrus fruits and raw green vegetables, e.g. kale.	• To keep tissues healthy, especially the skin and connective tissue. • To strengthen the immune system. • To help the body absorb iron in the ileum.
D	Oily fish, eggs and cod liver oil. Made in the body by the action of sunlight on the skin.	• To promote the absorption of calcium and phosphorus in the small intestine. • To help build and maintain strong bones and teeth. • To strengthen the immune system.

Table 9.4 *Some important minerals required by the human body*

Mineral	Sources	Functions
Calcium (Ca)	Dairy products, e.g. milk, cheese and yoghurt, and green vegetables, e.g. broccoli.	• To build and maintain healthy bones and teeth. • To help blood to clot at cuts.
Iron (Fe)	Red meat, liver, eggs, beans, nuts and dark green leafy vegetables, e.g. spinach.	• To make haemoglobin, the red pigment in red blood cells, which transports oxygen around the body for use in respiration.
Iodine (I)	Iodised table salt, milk, eggs and sea foods, e.g. fish, shellfish and seaweed.	• Used by the thyroid gland to make the hormone thyroxine (see Table 6.2, page 63).

The importance of dietary fibre and water to health

- **Dietary fibre** is food that **cannot be digested**. It consists mainly of the cellulose of plant cell walls, lignin in the walls of plant xylem vessels, husks of brown rice and bran of whole-grain cereals. **Dietary fibre** adds **bulk** to the food, which keeps it moving through the digestive system. This helps prevent **constipation** and reduces the risk of **colorectal (bowel) cancer**. It helps make a person feel **full**, which reduces overall food intake and helps **reduce obesity**. It also **lowers cholesterol**, reducing the risk of **cardiovascular disease** and **reduces** the risk of a person developing **type 2 diabetes**.

Figure 9.5 High-fibre foods

- **Water** makes up about **65%** of the human body. Consumption of **water** is essential to keep body cells **hydrated** so chemical reactions can occur, to **dissolve** and **transport** nutrients and other useful substances to body cells and to **remove** their waste (see page 38). It also aids in **digestion** of food, **cools** the body when it evaporates from sweat, helps **lubricate** joints, keeps skin **hydrated** and **healthy**, and plays a crucial role in the functioning of the kidneys in **excreting** waste.

Factors affecting dietary needs

The amount of **energy** required daily from the diet depends on a person's **age**, **occupation** or **level of physical activity** and **sex**. If energy input exceeds energy output, the person will **gain weight**. If energy output exceeds energy input, the person will **lose weight**.

- **Effect of age** – As **age increases** up to about 18 years, daily **energy** requirements **increase**. They then remain fairly **constant** throughout adulthood up to **old age**, when **less** energy is required daily. Children and adolescents also need a higher proportion of **protein** in their diets than adults.
- **Effect of occupation** or **level of physical activity** – As **activity increases**, daily **energy** requirements **increase**, e.g. a manual labourer requires more energy than a person working in an office, and a sportsperson requires more energy than someone who never plays any sport.
- **Effect of sex** – Daily **energy** requirements are generally **higher** in **males** than in females of the same age and occupation. However, women require more iron than men up to menopause, after which women require more vitamin D and calcium due to their higher risk of developing osteoporosis.

Food additives and their effects on health

Food additives are **chemicals** added to food to prevent it from spoiling or to improve its colour, flavour or texture. They include **artificial colourings**, **artificial sweeteners**, **salt** (sodium chloride), **monosodium glutamate** or **MSG** and **preservatives** (see Table 7.3, page 82). Some people are **sensitive** to certain additives and can experience **allergic reactions**, e.g. skin rashes, itching, swelling, rhinitis, sinusitis and difficulty breathing. Additives can also cause **asthma**, **digestive problems**, e.g. diarrhoea and abdominal pain, and **hyperactivity** in children. Some may also increase a person's risk of developing **cancer**.

An unbalanced diet

An **unbalanced diet** does not contain the correct proportions of nutrients. It may be **lacking** in certain nutrients or certain nutrients may be in **excess**. Both can lead to several serious health conditions.

- **Protein-energy malnutrition (PEM)** refers to a group of related disorders, including **kwashiorkor**, caused by a severe shortage of **protein** in the diet, and **marasmus**, caused by a severe shortage of **protein** and **energy-rich foods**, e.g. carbohydrates. Common symptoms include weight loss, fatigue and irritability. PEM mainly affects young children in developing countries.
- **Deficiency diseases** are caused by a shortage or lack of a particular vitamin or mineral in the diet. **Night blindness** is caused by a deficiency of vitamin A, **scurvy** by a deficiency of vitamin C, **rickets** by a deficiency of vitamin D or calcium, **anaemia** by a deficiency of iron and **goitre** by a deficiency of iodine.
- **Obesity** is generally caused by the excessive consumption of energy-rich foods high in **carbohydrates**, especially sugar, and/or **fat**, especially animal fat, and a **lack of physical activity**. It is characterised by an excessive accumulation and storage of **fat** in the body.
- **Type 2 diabetes** (see page 71) can be caused by consuming a diet high in **refined carbohydrates**, e.g. white bread, flour and rice, **sugars** and **unhealthy fats**, e.g. animal fats. It can also be caused by being overweight or obese.

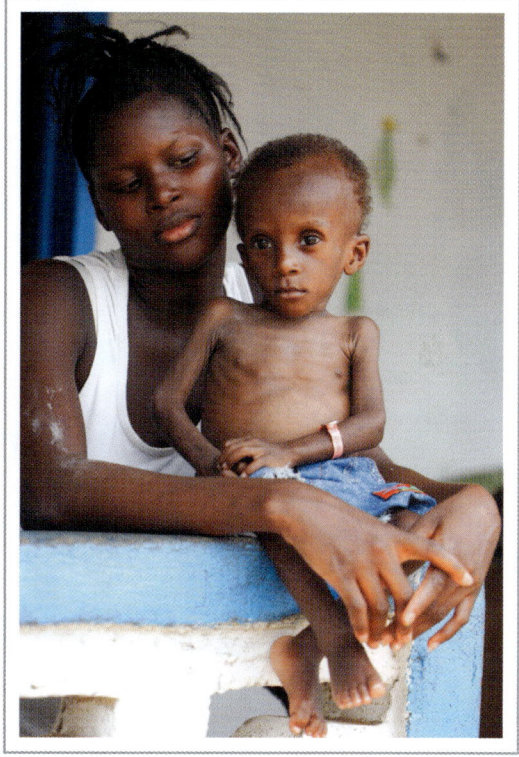

Figure 9.6 *A child with marasmus*

Digestion in humans

*Digestion is the process by which food is **broken down** into a form that is useful for body activities.*

During **digestion**, large food molecules, i.e. polysaccharides, disaccharides, proteins and lipids, are broken down into **simple**, **soluble food molecules**, namely monosaccharides, amino acids, fatty acids and glycerol, to make them **useful** for body activities.

Digestion occurs in the **alimentary canal**, which is a tube 8 to 9 metres long with muscular walls that runs from the **mouth** to the **anus**. The alimentary canal and its various associated organs, including the liver, gall bladder and pancreas, make up the **digestive system** shown in Figure 9.7 on page 108.

Digestion involves **two** processes: **mechanical digestion** and **chemical digestion**.

- During **mechanical digestion**, large **pieces** of food are broken down into smaller pieces. It begins in the **mouth**, where food is chewed by the **teeth**, and continues in the **stomach**, where contractions of the stomach walls churn the food, which helps to break it down.

- During **chemical digestion**, large, usually insoluble **food molecules** are broken down into small, soluble food molecules by **enzymes** (see pages 109 to 111). It begins in the **mouth** and is completed in the **small intestine**.

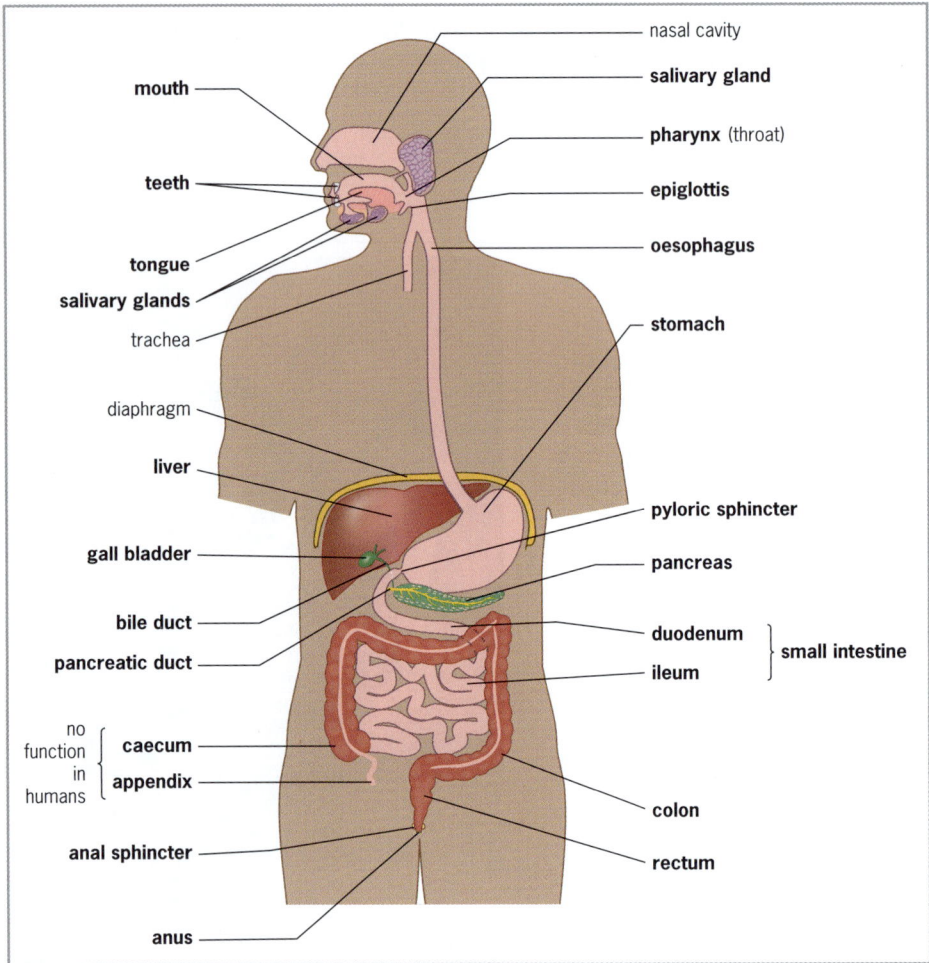

Figure 9.7 *The human digestive system*

The teeth and mechanical digestion

When food is **chewed** in the mouth, the teeth help to break up **large pieces** of food into **smaller pieces**. This is important in digestion for **two** reasons.

- It gives the pieces of food a **larger surface area** for digestive enzymes to act on, making chemical digestion quicker and easier.
- It makes food easier to **swallow**.

Types of teeth

Humans have **four** different types of teeth: **incisors**, **canines**, **premolars** and **molars**. An adult has **8** incisors (i), **4** canines (c), **8** premolars (pm) and **12** molars (m). The **dental formula** of an adult gives the number of teeth of each kind in one half of the upper and lower jaw as follows:

$$i\ \frac{2}{2} \quad c\ \frac{1}{1} \quad pm\ \frac{2}{2} \quad m\ \frac{3}{3}$$

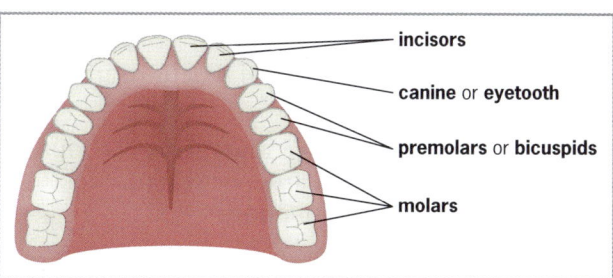

Figure 9.8 *Teeth of the upper jaw of an adult human*

Table 9.5 *The different types of teeth in humans*

Type	Position	Shape		Functions
Incisor	At the front of the jaw.	Chisel-shaped with sharp, thin edges.	crown / root	To cut food. To bite off pieces of food.
Canine (eye tooth)	Next to the incisors.	Cone-shaped and pointed.		To grip food. To tear off pieces of food.
Premolar	At the side of the jaw next to the canines.	Have a fairly broad surface with two pointed cusps.	cusp / root	To crush and grind food.
Molar	At the back of the jaw next to the premolars.	Have a broad surface with four or five pointed cusps.		To crush and grind food.

Importance of tooth care

Caring for teeth and gums is **important** to help prevent **tooth decay** and **gum disease**.

- **Brush** teeth and gums in the proper way for about 2 minutes, at least twice a day.
- Use a **fluoride tooth paste** and a good quality **tooth brush** when brushing.
- Use **dental floss** and an **interdental brush** once a day to clean between the teeth.
- Use an **antibacterial mouthwash** after brushing and flossing to kill any remaining bacteria.
- Eat plenty of **tooth-healthy foods**, e.g. fresh fruits and raw vegetables, and drink water or unsweetened and non-carbonated beverages. **Avoid** eating sugary and starchy foods and drinking sugar-sweetened and carbonated beverages, especially between meals and before going to bed.
- Visit a **dentist** regularly for a check-up and have teeth **professionally cleaned** twice a year.

Enzymes

***Enzymes** are **biological catalysts** that are produced by all living cells. They speed up chemical reactions occurring in living organisms without being changed themselves.*

Enzymes are **protein molecules** that cells produce from **amino acids**. Enzymes **speed up reactions** by **lowering** the energy required to initiate the reactions, known as the **activation energy**. Without enzymes, chemical reactions would occur too slowly to maintain life.

Effects of temperature and pH on enzymes

Enzymes are **specific**, meaning that each type of enzyme catalyses only **one** type of reaction, and they are affected by **temperature** and **pH**.

- Enzymes work best at a particular **temperature**, known as the **optimum temperature**. This is about **37 °C** for human enzymes. High temperatures **denature** enzymes, i.e. they are inactivated. Most enzymes start to be **denatured** at about 40 °C to 45 °C and can no longer function.

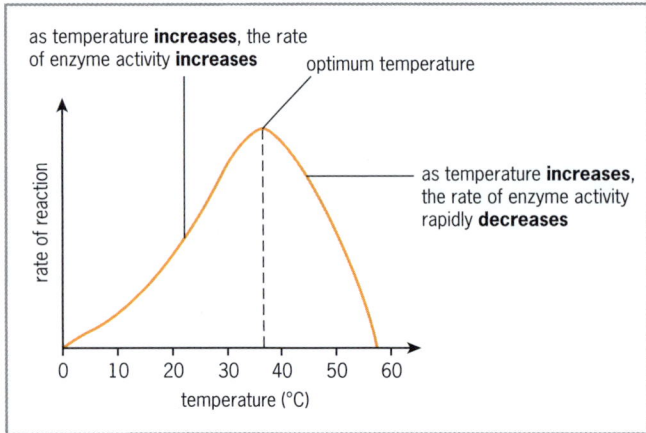

Figure 9.9 *The effect of temperature on the rate of a reaction catalysed by enzymes*

- Enzymes work best at a particular **pH** known as the **optimum pH**. This is about **pH 7** for most enzymes. Extremes of acidity or alkalinity **denature** most enzymes.

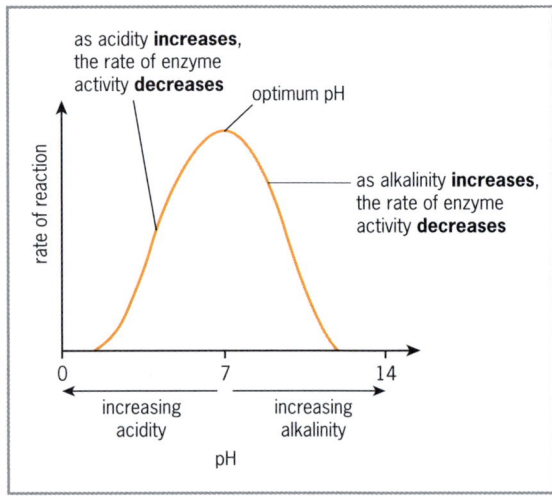

Figure 9.10 *The effect of pH on the rate of a reaction catalysed by enzymes*

Chemical digestion

Digestive enzymes catalyse the breakdown of food molecules during **chemical digestion**. There are **three** categories of **digestive enzymes** and several different enzymes may belong to each category. These enzymes are present in the **digestive juices** produced by different regions of the digestive system.

Table 9.6 *The three categories of digestive enzymes*

Category of digestive enzyme	Food molecules digested	Products of digestion
Carbohydrases	Polysaccharides and disaccharides	Monosaccharides
Proteases	Proteins	Amino acids
Lipases	Lipids	Fatty acids and glycerol

Table 9.7 *A summary of the enzymes involved in chemical digestion*

Organ (digestive juice)	Enzyme(s) present	Function of the enzyme
Mouth (saliva)	• Salivary amylase	• Begins to digest **starch** into **maltose** (a disaccharide).
Stomach (gastric juice)	• Rennin	• Produced in infants to clot the soluble **proteins** in milk so the proteins are retained in the stomach for pepsin to act on.
	• Pepsin	• Begins to digest **proteins** into **peptides** (shorter chains of amino acids).
Small intestine (duodenum and ileum) (pancreatic juice and intestinal juice)	• Pancreatic amylase	• Continues to digest **starch** into **maltose**.
	• Trypsin	• Continues to digest **proteins** into **peptides**.
	• Pancreatic lipase	• Digests **lipids** into **fatty acids** and **glycerol**.
	• Maltase	• Digests **maltose** into **glucose** (a monosaccharide).
	• Sucrase	• Digests **sucrose** (a disaccharide) into **glucose** and **fructose** (monosaccharides).
	• Lactase	• Digests **lactose** (a disaccharide found in milk) into **glucose** and **galactose** (monosaccharides).
	• Peptidase or erepsin	• Digests **peptides** into **amino acids**.

The role of bile

Bile is produced in the **liver**, stored in the **gall bladder** and enters the small intestine via the **bile duct**. It contains **organic bile salts** which **emulsify lipids**, i.e. they break large lipid droplets into smaller droplets. This increases their **surface area** for the pancreatic lipase to work on, which speeds up their digestion.

Absorption

Absorption is the process by which soluble food molecules, produced in digestion, move into the body fluids.

Absorption occurs in the **small intestine** and **colon**.
- Monosaccharides, amino acids, fatty acids, glycerol, vitamins, minerals and water are absorbed through the lining of the **small intestine**, mainly the **ileum**, and into **blood** flowing through the **blood capillaries** and **lymph** flowing through the **lacteals** (**lymph capillaries**) in its walls.
- Water and mineral salts can be absorbed from any undigested food in the **colon**. As the undigested waste moves along the colon to the rectum, it becomes progressively more solid as water is absorbed.

Egestion

*Egestion is the process by which undigested food material is **removed** from the body.*

The almost solid material entering the rectum is called **faeces** and consists of undigested dietary fibre, dead bacteria and intestinal cells, mucus and bile pigments. Faeces is stored in the rectum and **egested** at intervals through the **anus** when the **anal sphincter** relaxes.

Assimilation

*Assimilation is the process by which the body **uses** the soluble food molecules absorbed after digestion.*

- Any non-glucose **monosaccharides** are converted to **glucose** by the liver. The **glucose** is then used by all body cells in **respiration** to release **energy**. Excess glucose is converted to **glycogen** (animal starch) by **liver** and **muscle** cells and stored, or is converted to **fat** by cells in **adipose (fat) tissue** and stored under the skin and around organs.
- **Amino acids** are used by body cells to make **proteins** for cell growth and repair. They are also used by cells to make **enzymes**, **hormones** and **antibodies**. Excess amino acids are converted into glucose, glycogen or fat, and **urea**. The urea is excreted by the kidneys.
- **Fatty acids** and **glycerol** are used to make **cell membranes** of newly forming cells. They can also be used by body cells in **respiration** and any excess are converted to **fat** and stored in **adipose tissue**.

Revision questions

1. Identify the SIX Caribbean food groups and give TWO examples of foods belonging to EACH group.

2. Give TWO functions of EACH of the following macronutrients supplied by the six Caribbean food groups: carbohydrates, proteins and lipids.

3. Identify THREE foods that provide a good source of EACH of the following and outline the importance of EACH in a balanced diet:

    a vitamin B_{12} b vitamin D   c iron   d calcium.

4. Discuss the importance of dietary fibre and water to health.

5. a Distinguish between a balanced diet and an unbalanced diet.

    b Identify THREE factors that can affect a person's dietary needs, explain EACH of these effects and discuss THREE possible consequences of having an unbalanced diet.

6. Suggest why it is advisable to avoid eating foods that contain food additives.

7. a What happens during digestion?

    b Why are teeth important in digestion?

8. Keenan has FOUR types of teeth in his mouth. Identify these and state the function of EACH.

9. Melissa develops a cavity in one of her teeth. Suggest FOUR things she should do to prevent cavities forming in her other teeth.

10. a What are enzymes and why are they important in the digestive process?

    b Explain the effect that temperature has on enzyme activity.

11 Identify the enzymes that are active in the different regions of the alimentary canal, name the region where EACH is found and state the function of EACH enzyme named.

12 Distinguish among absorption, egestion and assimilation.

13 How does the body make use of the amino acids produced by the digestion of proteins?

Respiration

Respiration is the process by which energy is released from food by all living cells.

Respiration provides cells with a constant supply of **energy**. It is catalysed by **enzymes** and occurs **slowly** in a large number of **stages**. During respiration, about 60% of the energy is released as **heat energy**, which helps organisms to maintain a constant body temperature, e.g. **37 °C** in humans. The rest of the energy released is **used** by cells in the following ways.

- To **manufacture** complex, biologically important molecules, e.g. proteins and DNA.
- For **cell growth** and **repair**.
- For **cell division**.
- In **active transport** to move molecules and ions into and out of the cells through their membranes.
- For **special functions** in specialised cells, e.g. contraction of muscle cells and transmission of impulses in nerve cells.

There are **two** types of respiration: **aerobic respiration** and **anaerobic respiration**.

Aerobic respiration

Aerobic respiration is the process by which energy is released from food by living cells using oxygen.

Aerobic respiration occurs in most cells. It **uses oxygen** and takes place in the **mitochondria**. The amount of energy produced depends on the type of molecules respired, called **respiratory substrates**.

Table 9.8 *Amount of energy produced by different respiratory substrates*

Respiratory substrate	Energy produced (kJ g^{-1})
Carbohydrates	16
Lipids	38
Proteins	17

The main respiratory substrate is **glucose**. The glucose molecules are **completely** broken down and **all** their stored energy is released. Aerobic respiration always produces **carbon dioxide, water** and **energy**.

$$\text{glucose} + \text{oxygen} \xrightarrow{\text{enzymes in mitochondria}} \text{carbon dioxide} + \text{water} + \text{energy}$$

or

$$C_6H_{12}O_6 + 6O_2 \xrightarrow{\text{enzymes in mitochondria}} 6CO_2 + 6H_2O + \text{energy}$$

Anaerobic respiration

Anaerobic respiration is the process by which energy is released from food by living cells without the use of oxygen.

Anaerobic respiration occurs in some cells. It takes place **without oxygen** in the **cytoplasm**. The main respiratory substrate is **glucose** and the products **vary**. It produces considerably **less energy** per molecule of glucose than aerobic respiration because the glucose molecules are only **partially** broken down and at least one product still contains energy, e.g. ethanol, lactic acid or methane. Yeast cells, muscle cells and certain bacteria can carry out anaerobic respiration.

Relevance of anaerobic respiration

Making bread and alcoholic beverages

Yeast cells carry out anaerobic respiration, known as **fermentation**, which produces **ethanol, carbon dioxide** and **energy**.

$$\text{glucose} \xrightarrow{\text{enzymes in cytoplasm}} \text{ethanol + carbon dioxide + energy}$$

or

$$C_6H_{12}O_6 \xrightarrow{\text{enzymes in cytoplasm}} 2C_2H_5OH + 2CO_2 + \text{energy}$$

When **making bread**, yeast ferments sugars present in the dough. The **carbon dioxide** produced forms **bubbles** which cause the dough to rise. When baked, heat from the oven causes the bubbles to expand, kills the yeast and evaporates the ethanol.

When making **alcoholic beverages**, e.g. beer, wine, rum and other spirits, yeast ferments sugars in grains, fruits or molasses. Fermentation stops when the **ethanol** concentration reaches about 14–16% because it kills the yeast cells, so the ethanol content of beer and wine is always below about 16%. Spirits are made by **distillation** of the fermentation mixture to increase the ethanol concentration.

Anaerobic respiration in sports

During **intense physical activity**, the body's demands for energy can exceed the amount that can be supplied by aerobic respiration. As a result, muscle cells begin to respire **anaerobically**. This produces **lactic acid** and provides the muscle cells with a rapid source of **energy**.

$$\text{glucose} \xrightarrow{\text{enzymes in cytoplasm}} \text{lactic acid + energy}$$

$$C_6H_{12}O_6 \xrightarrow{\text{enzymes in cytoplasm}} 2C_3H_6O_3 + \text{energy}$$

Lactic acid builds up in the muscle cells and begins to harm them, causing **fatigue** and even collapse as they stop contracting. The muscle cells are said to have built up an **oxygen debt**. This debt must be **repaid** directly after exercise by resting and breathing deeply to remove the lactic acid by respiring it **aerobically**.

Regular **anaerobic training** involving short bursts of high-intensity exercises improves the **efficiency** of anaerobic respiration and allows athletes to perform at higher intensities for longer periods before becoming fatigued. This improves overall performance.

Breathing

Breathing refers to the movements that cause air to be moved into and out of the lungs.

Breathing must not be confused with **respiration**. Breathing is a **physical process** involving movements, whereas respiration is a **chemical process** involving chemical reactions. The **respiratory system** is responsible for **breathing** in humans.

Breathing is **essential** to humans for **two** reasons.

- It ensures that air containing **oxygen** is continually drawn into the body. This ensures that humans have a **continual supply** of oxygen to meet the demands of aerobic respiration.
- It ensures that the **carbon dioxide** produced in aerobic respiration is **continually removed** from the body so that it does not build up and poison cells.

The structure of the human respiratory system

Humans have **two** lungs composed of thousands of air passages called **bronchioles** and millions of pocket-shaped air sacs called **alveoli** (singular: **alveolus**), each surrounded by a **network** of **blood capillaries**. Each lung is surrounded by two **pleural membranes** which have **pleural fluid** between. A single **bronchus** leads into each lung from the **trachea**, which leads downwards from the **larynx**, **pharynx** or **throat**, **mouth** and **nasal cavities**.

Each lung receives blood from the heart via a **pulmonary artery** and blood is carried back to the heart via a **pulmonary vein**. The lungs are situated inside the **chest cavity** or **thorax** and are surrounded by the **ribs**, which form the **rib cage**. The ribs have **intercostal muscles** between them and a dome-shaped sheet of muscle, the **diaphragm**, stretches across the floor of the thorax.

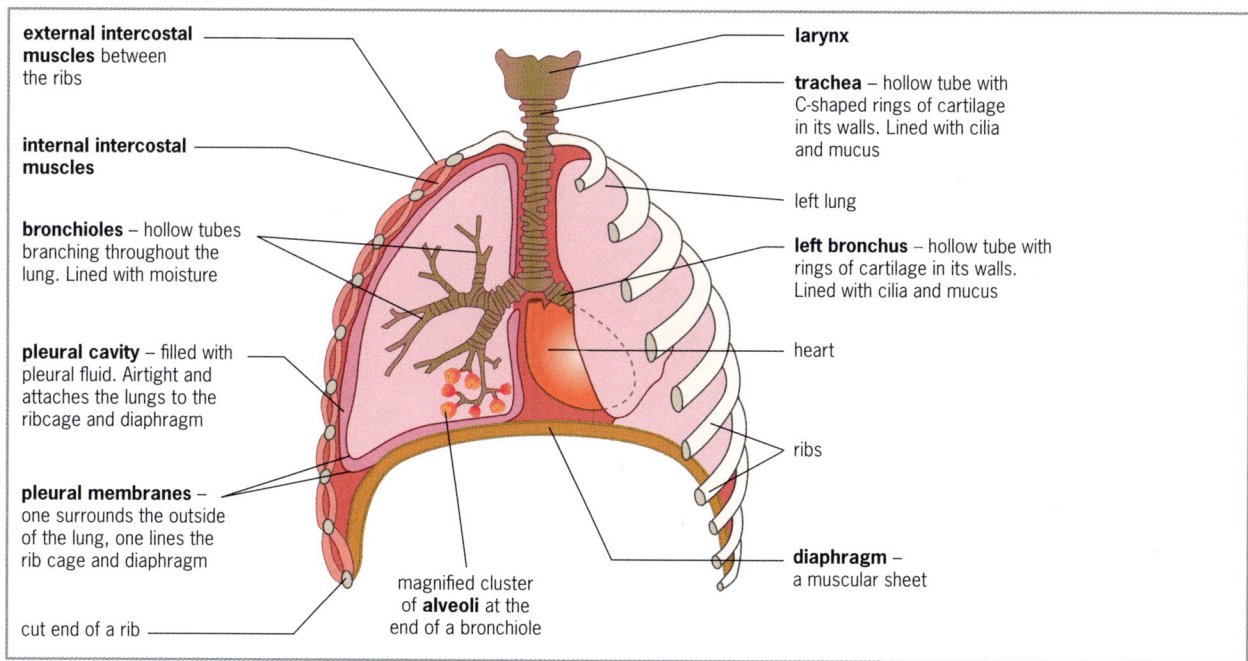

Figure 9.11 *The human thorax showing the structure of a lung*

9 Energy in life processes

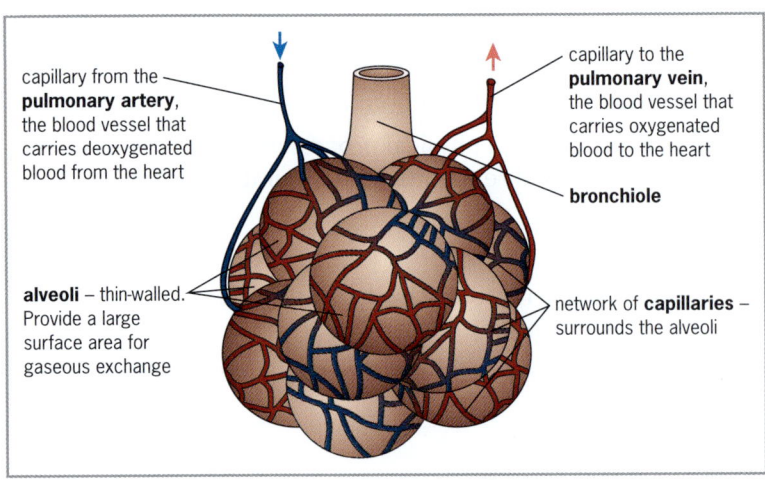

Figure 9.12 *Surface view of a cluster of alveoli showing the blood supply*

The mechanism of breathing

Breathing is brought about by the **intercostal muscles** and the **diaphragm muscles** that bring about **movements** of the rib cage and diaphragm, which cause the **volume** inside the chest cavity to change. This causes the volume inside the lungs to change because they are attached to the rib cage and diaphragm by the airtight pleural cavity.

Table 9.9 *The mechanism of breathing summarised*

	Features	Inhalation (inspiration)	Exhalation (expiration)
1	External intercostal muscles	Contract	Relax
	Internal intercostal muscles	Relax	Contract
	Ribs and sternum	Move upwards and outwards	Move downwards and inwards
2	Diaphragm muscles	Contract	Relax
	Diaphragm	Moves downwards or flattens	Domes upwards
3	Volume inside thorax and lungs	Increases	Decreases
	Pressure inside thorax and lungs	Decreases	Increases
4	Movement of air	Air moves into the lungs due to the decrease in pressure	Air moves out of the lunges due to the increase in pressure

As air enters the lungs during **inhalation**, it passes through the bronchi and bronchioles and into the alveoli, where **gaseous exchange** occurs between the air and the blood passing through the capillaries surrounding the alveoli. Some **oxygen** in the inhaled air diffuses into the blood whilst **carbon dioxide** diffuses out of the blood into the air in the alveoli, and is **exhaled** (see Figure 9.14, page 118).

Table 9.10 *The composition of inhaled and exhaled air compared*

Component	Inhaled air	Exhaled air	Reason for the differences
Oxygen (O_2)	21%	16%	Oxygen diffuses into the blood from the inhaled air and is used by body cells in respiration.
Carbon dioxide (CO_2)	0.04%	4%	Carbon dioxide is produced by body cells during respiration and excreted by the lungs.
Nitrogen (N_2)	78%	78%	Nitrogen gas is not used by body cells.
Water vapour (H_2O)	Variable	Saturated	Moisture from the respiratory system evaporates into the air being exhaled.

The role of kinetic energy in breathing

Air moves into and out of the lungs mainly due to the **kinetic energy** of its molecules. The molecules in air possess **large amounts** of kinetic energy (see page 241), move around freely and rapidly, and always move from areas of **high pressure** to areas of **low pressure**. During **inhalation**, the pressure inside the lungs **decreases**, which causes the air molecules to move into the lungs. During **exhalation**, the pressure inside the lungs **increases**, which causes the air molecules to move out of the lungs.

Cardiopulmonary resuscitation (CPR)

CPR is an emergency procedure performed on a person whose heart has stopped beating (cardiac arrest) and/or who has stopped breathing (respiratory arrest). During CPR, the rescuer performs **chest compressions** to maintain circulation and deliver oxygen to vital organs. **Rescue breathing** or **mouth-to-mouth resuscitation** can also be performed to deliver oxygen to the victim's lungs by persons who are **trained** to carry out CPR. The following **steps** are used to perform chest compressions with rescue breathing.

- Lay the victim on his or her back and kneel next to the victim's shoulders. Place the heel of one hand on the breastbone. Place the other hand on top of the first and interlock the fingers.
- With straight elbows, push straight downwards on the victim's chest so that it is compressed by approximately **5 cm**. Release the compression so the chest returns to its original position.
- Perform **30** of these **chest compressions** at a rate equivalent to **100** to **120 per minute.**
- Tilt the victim's head backwards and lift the chin to open the airways. Open the victim's mouth, remove any debris and pinch his or her nose.
- Inhale and seal your lips over the victim's open mouth and breathe out into the victim's mouth for **1 second**. If the victim's chest rises, breathe into the mouth a second time for **1 second**.
- Continue to alternate **30 chest compressions** at a rate of **100** to **120 per minute** with **two rescue breaths** until the victim shows signs of recovery or medical help arrives.

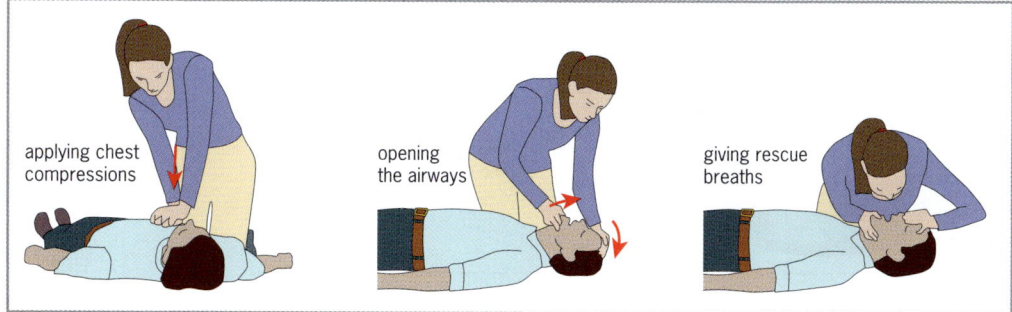

Figure 9.13 *Performing chest compressions and rescue breathing*

Gaseous exchange and smoking

Gaseous exchange and its importance

Gaseous exchange is the process by which oxygen diffuses into an organism and carbon dioxide diffuses out of an organism through a respiratory (gaseous exchange) surface.

- Gaseous exchange is the means by which living organisms obtain the **oxygen** they need to sustain aerobic respiration from their environment, which ensures cells are provided with **energy**.
- Gaseous exchange is the means by which living organisms get rid of the **carbon dioxide** produced during aerobic respiration, which ensures that carbon dioxide does not build up and poison cells.

Breathing movements speed up the supply of oxygen to **respiratory surfaces** and the removal of carbon dioxide from these surfaces in organisms that breathe, e.g. humans and fish.

Respiratory surfaces

Respiratory surfaces or **gaseous exchange surfaces** are surfaces through which gases are exchanged, and they have several **adaptations** to make the exchange of gases through them as **efficient** as possible.

- They have a **large surface area** so that large quantities of gases can be exchanged.
- They are very **thin** so that gases can diffuse through them rapidly.
- They have a **rich blood supply** (if the organism has blood) to quickly transport gases between the surface and the body cells.
- They are **moist** so that gases can dissolve before they diffuse through the surface.

The respiratory surface and gaseous exchange in humans

In humans, **gaseous exchange** occurs between the air in the **alveoli** of the lungs and the blood passing through the capillaries surrounding the alveoli. The **walls** of the **alveoli** form the **respiratory surface**.

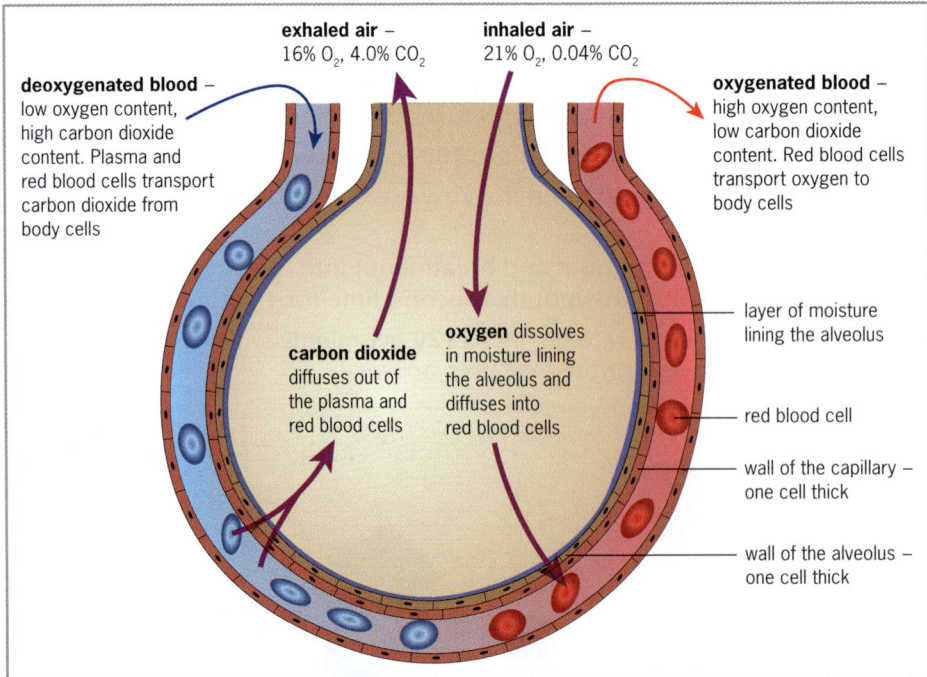

Figure 9.14 *Gaseous exchange in an alveolus*

The respiratory surface and gaseous exchange in a fish

Gaseous exchange occurs in the **gills** of fish. A bony fish usually has four gills at each side of its pharynx. Each gill has two rows of long, thin, finger-like projections called **gill filaments**. Each filament has a wall **one cell thick** and a **network of capillaries** inside. The **walls** of the **filaments** form the **respiratory surface**.

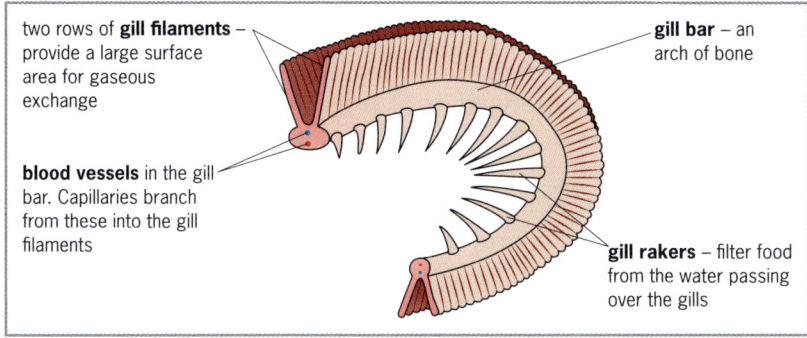

Figure 9.15 *The gill of a fish showing the gill filaments*

The fish takes water into its mouth and the water passes between the gill filaments. **Oxygen**, dissolved in the water, diffuses through the walls of the filaments and into the blood, and **carbon dioxide** diffuses from the blood into the water passing out of the gills.

The respiratory surface and gaseous exchange in a flowering plant

Gaseous exchange occurs in the **leaves**, **stems** and **roots** of plants by **direct diffusion** between the **air spaces** around all the cells and the **cells** themselves. Gases diffuse between the **atmosphere** and the air spaces in leaves through the **stomata** (singular: **stoma**), and in stems and roots through **lenticels** (see page 51). The **walls** and **membranes** of all the **cells** inside the leaves, stems and roots form the **respiratory surface**. Movement of gases into and out of **leaves** depends on the time of day.

- During the **night**, only **respiration** occurs. **Oxygen** diffuses **into** the leaves and **carbon dioxide** diffuses **out**.
- During the **day**, the rate of **photosynthesis** is greater than the rate of respiration. **Carbon dioxide** diffuses **into** the leaves and **oxygen** diffuses **out**.

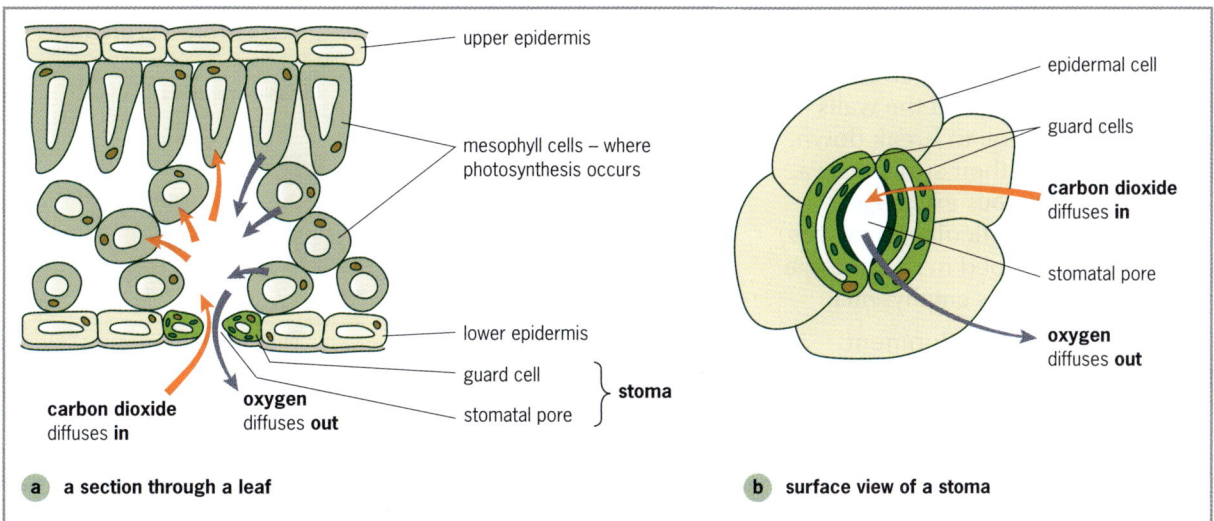

Figure 9.16 *Gaseous exchange in a leaf during the day*

9 Energy in life processes

Respiratory surfaces summarised

Table 9.11 *A summary of the features of the gaseous exchange surfaces of different organisms*

Adaptation of the gaseous exchange surface	Organism		
	Human	Fish	Flowering plant
Large surface area	Each alveolus is pocket-shaped and a human has two lungs, each with over 350 million alveoli.	Most fish have eight gills, each with a large number of long, thin filaments arranged in two rows.	Leaves are broad, thin and numerous, stems and roots have a branching structure, and all structures are made up of a very large number of cells.
Thin	The walls of the alveoli are only one cell thick.	The walls of the filaments are only one cell thick.	Cell walls and membranes are extremely thin.
Rich blood supply	A network of capillaries surrounds each alveolus.	A network of capillaries is present inside each filament.	No blood supply; direct diffusion occurs between the air and the cells.
Moist	The walls of the alveoli are lined with a thin layer of moisture.	A fish lives in water containing dissolved oxygen.	All the cells are covered with a thin layer of moisture.

The effects of smoking on gaseous exchange

Smoking usually refers to **inhaling smoke** produced by burning plant material in **cigarettes**, **cigars**, **pipes** and **hookahs**, whilst **vaping** refers to **inhaling vapour** produced by **electronic cigarettes** (**e-cigarettes**). The most commonly **smoked** plant materials are **tobacco** and **marijuana**. Smoke from **tobacco** contains over 7000 different chemicals, including **nicotine**, **carbon monoxide** and **tar**. Smoking affects the **gaseous exchange** process in various ways.

- It reduces the amount of **oxygen** carried by the blood, which reduces **respiration** in body cells and the smoker's ability to exercise.
- It causes increased **mucus** production in the airways and paralyses the **cilia** lining the airways, which stops them from beating and removing the mucus. The excess mucus builds up and makes it harder for **oxygen** in inhaled air to reach the alveoli and **carbon dioxide** to be removed.
- It irritates and inflames the walls of the **bronchi** and **bronchioles**, which makes it harder for **oxygen** in inhaled air to reach the alveoli and **carbon dioxide** to be removed, and leads to **chronic bronchitis**.
- It causes the walls of the **alveoli** to become less elastic and the walls between the alveoli to break down, which decreases their surface area. This reduces **gaseous exchange**, makes exhaling difficult and causes air to remain trapped in the lungs, a condition known as **emphysema**.
- It can lead to the development of **lung cancer** in which tumours replace normal, healthy lung tissue and this reduces **gaseous exchange**.

Note: Chronic bronchitis and emphysema are two types of **chronic obstructive pulmonary disease** or **COPD**.

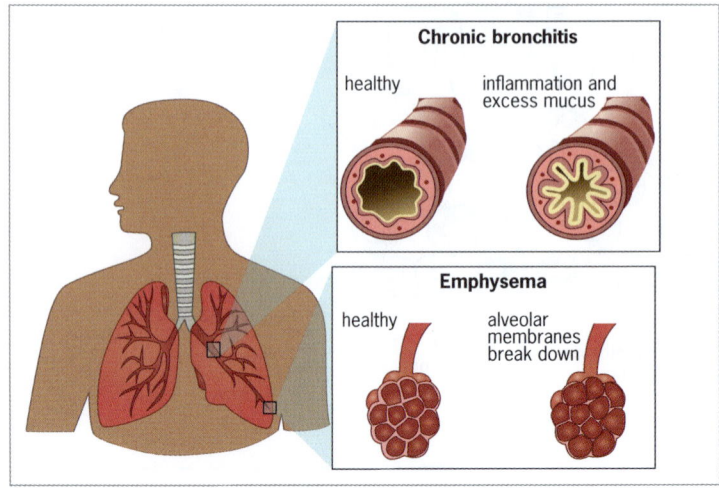

Figure 9.17 *Chronic obstructive pulmonary disease (COPD)*

Second-hand smoke and smoke-free environments

Second-hand smoke consists of smoke exhaled by smokers and smoke from the lit ends of cigarettes. It has a higher concentration of some **cancer-causing chemicals** or **carcinogens** and smaller particles than mainstream smoke, and it poses considerable **health risks** to non-smokers, e.g. it increases their risk of developing asthma and other respiratory disorders, heart disease, stroke and cancer.

Smoke-free environments, where smoking is not allowed, protect individuals against exposure to second-hand smoke. They help reduce air pollution and contribute to improved air quality. This reduces the harmful effects of air pollutants (see page 248), creates healthier environments and lowers healthcare costs associated with smoke-related illnesses. They also aim to reduce smoking in general and to help people to stop smoking all together.

Revision questions

14
  a Define the term 'respiration'.
  b What is the main use that the body makes of the energy released in respiration?

15
  a Construct a table to give FOUR differences between aerobic and anaerobic respiration.
  b Write a word and a chemical equation to summarise the process of aerobic respiration.

16 Explain how Kendra makes use of anaerobic respiration when making bread, and write a word and a chemical equation to summarise the reaction occurring.

17 What is meant by the term 'breathing' and why is breathing important to humans?

18 Describe the structure of a human lung.

19 Explain the mechanism by which air is expelled from the lungs. Your answer must include an explanation of the role of kinetic energy in the process.

20 Discuss the differences in composition of inhaled air and exhaled air.

21
  a What is CPR?
  b Outline the steps that a trained rescuer would take when performing CPR.

22 What is gaseous exchange and why is it important to living organisms?

23 Identify the respiratory surfaces in a bony fish and a human, and give a detailed explanation of FOUR features that the surfaces have in common.

24 Explain how smoking affects the gaseous exchange process.

25 What are smoke-free environments and why are they important?

10 Fossil fuels and alternative sources of energy

Plants absorb light energy and convert it to **chemical energy**, which is then **passed on to animals** as they eat the plants. After hundreds of millions of years, energy stored in the **fossils of plants and animals** have become fuels which are now widely used by humans.

However, burning fossil fuels produces several pollutants that negatively affect our health and the environment, and this has led to the search for alternative sources of energy.

Fossil fuels

*Fossil fuels are buried combustible deposits of **decayed plant and animal** matter that have been converted to **crude oil**, **natural gas** and **coal** by subjection to **heat** and **pressure** in the Earth's crust for **millions of years**.*

Figure 10.1 shows how crude oil and natural gas may be mined.

NOTE: Charcoal is not a fossil fuel since it is readily formed by heating materials such as wood in a low oxygen environment. It is therefore a renewable form of energy (see page 127).

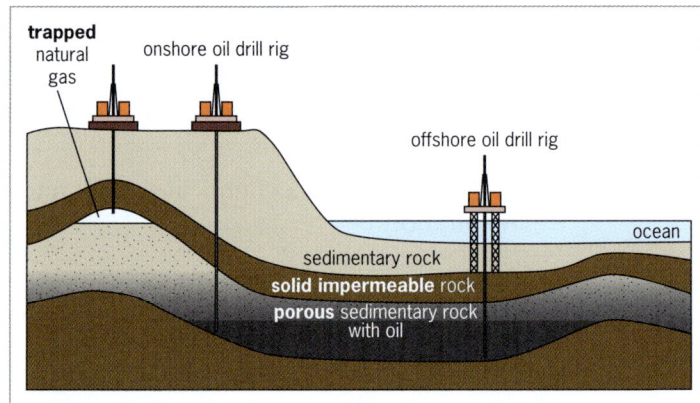

Figure 10.1 Mining crude oil and natural gas

Table 10.1 Types of fossil fuels

Fossil fuel	Description	Formation
Crude oil	A yellow-to-black mixture, mainly of **solid** and gaseous hydrocarbons dissolved in **liquid hydrocarbons**. Hydrocarbons are compounds consisting entirely of hydrogen and carbon.	Produced as the remains of microscopic **marine animals** and **plants** fell to the **ocean floor** where they were covered by mud and subjected to intense **heat** and **pressure**. After millions of years these fossilised organisms converted to **crude oil** (petroleum). Crude oil is generally found **within layers of sandstone.**
Natural gas	A mixture of **hydrocarbon gases** (mainly **methane**).	Formed together with coal and crude oil. It is generally found **trapped** under **impermeable rock**.
Coal	A black or dark brown rock consisting mainly of **carbon**.	Produced as the remains of plants in **swampy, forested areas** were covered by dirt and rock and subjected to intense **heat** and **pressure**.

Refining crude oil by fractional distillation

Crude oil is a **mixture of hydrocarbons** of **different sized molecules**. The **larger the molecule**, the **higher is its boiling point** (or condensing point). The crude oil is **heated** to over **400 °C** in a **furnace** where **most of it vaporises** and is then passed to a **fractionating tower**, as shown in Figure 10.2.

The **larger** hydrocarbon **molecules condense first** and **sink** to the bottom where they are drained into trays. The **remaining** gases rise, **condensing** into trays at the **levels** having the temperatures of their **boiling points**. The condensed oils are individually drained from the trays and stored separately. Hydrocarbon gases that are not cooled **to below their boiling points**, exit from the top of the tower.

Liquid petroleum gas (LPG) *is a mixture of hydrocarbon gases (mainly propane and butane) used as a source of energy.*

Liquid natural gas (LNG) *is a mixture of hydrocarbon gases (mainly methane) used as a source of energy.*

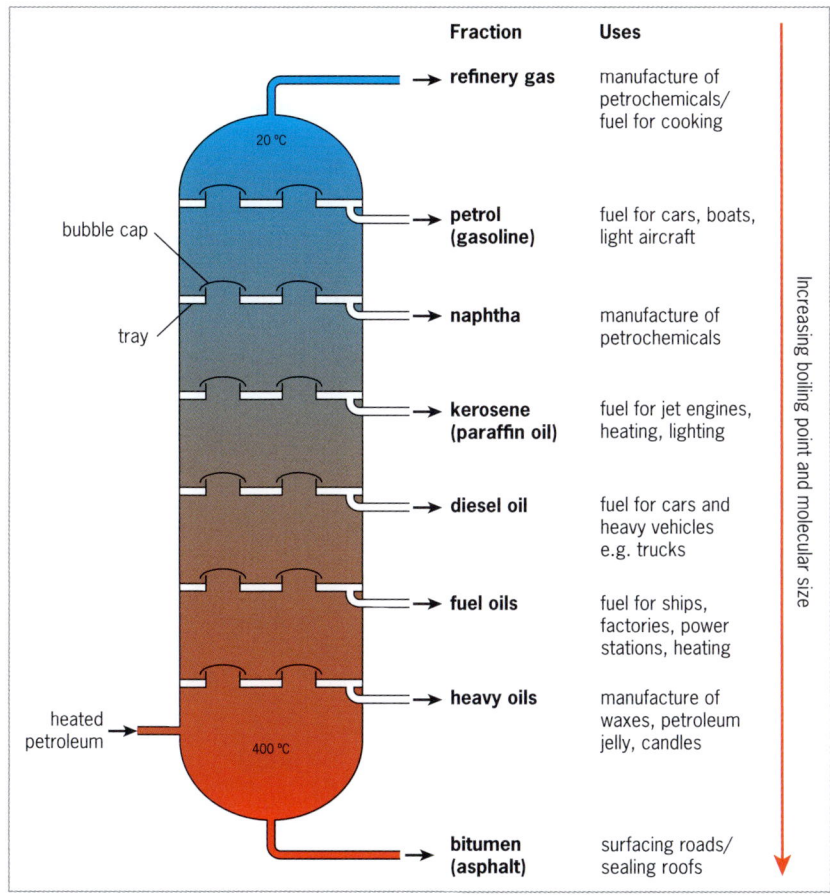

Figure 10.2 *Refining crude oil by fractional distillation*

Examples of interconversion of energy using fossil fuels

Gasoline-fuelled car accelerating on level road

Aircraft taking off

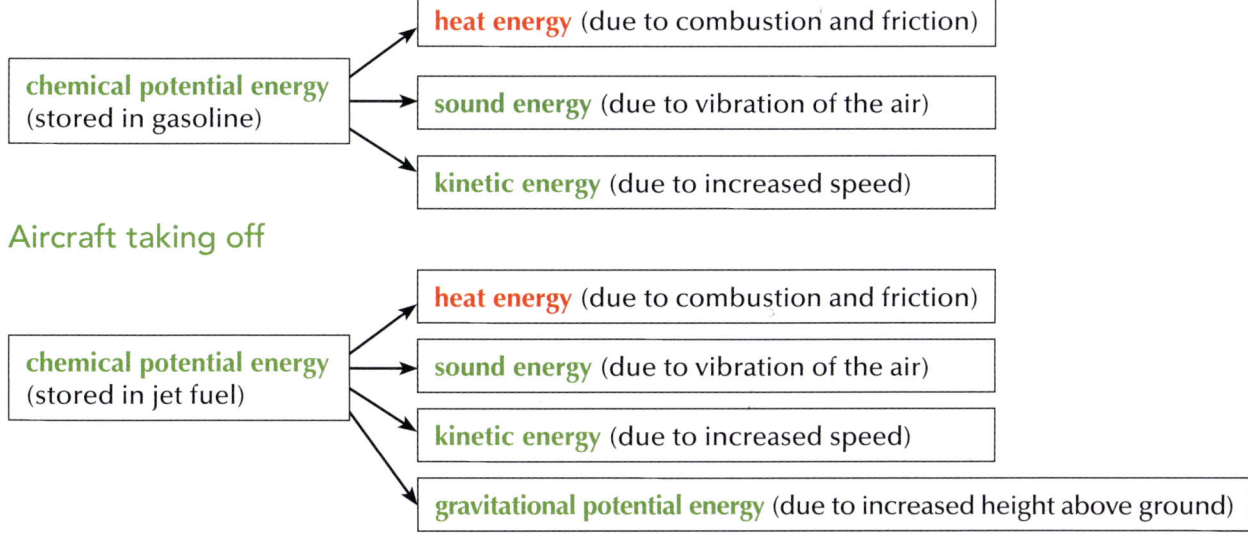

10 Fossil fuels and alternative sources of energy

Electrical generation using fossil fuels

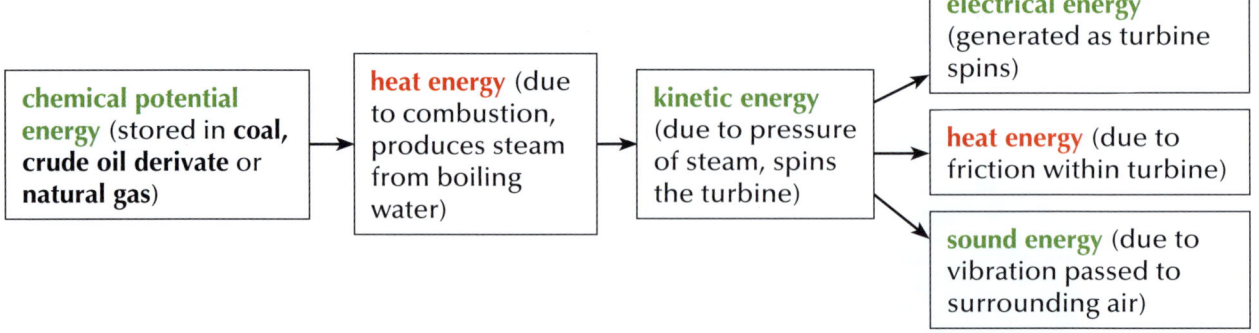

Figure 10.3 shows how fuel oil and diesel are typically used in the Caribbean for generating electricity.

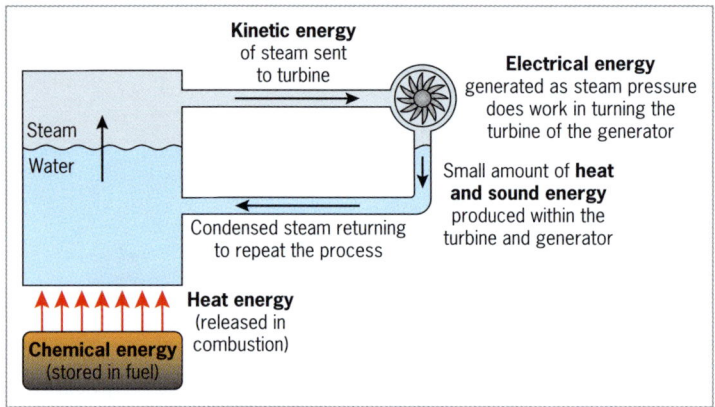

Figure 10.3 *Electricity from fuel oil or diesel*

Advantages of using fossil fuels

- **Relatively cheap**.
- **Polymorphic**, meaning they can be used as solid coal, liquid petroleum or as natural gas.
- **Have a high energy density**, so a small amount of fuel can produce a large amount of energy.
- **Great economic benefit** as they contribute heavily to natural economies.
- **Have shaped our infrastructure** as they are the basis of many items used in our daily lives.
- **Predictable**, unlike solar and wind energy.
- **Can be stored easily** to be used when required.
- **Can be efficiently extracted, transported** and **used** by well-established infrastructure.
- Are **non-renewable** (see page 127) but **large reserves** exist.

Disadvantages of using fossil fuels

- **Non-renewable**, so despite the large reserves, they will eventually be **exhausted**.
- **Negative environmental effects: acid rain, global warming** and **climate change**.

Acid rain

*Acid rain is the precipitation of highly acidic droplets produced as **oxides of sulfur and nitrogen**, created by burning fossil fuels, **dissolve in rainwater**.*

Figure 10.4 shows how acidic rain is formed and how it harms the environment.
- **Increases the bleaching of coral** reefs.
- **Damages soils** and **forests** affecting agriculture and animal farming.
- **Corrodes** metal objects, buildings and landscapes.
- **Dissolves metals** forming solutions that flow to **poison** life in rivers, lakes and oceans.

Figure 10.4 *Acid rain*

Global warming and the greenhouse effect

Greenhouse gases (Table 10.2) are a group of gases in the atmosphere which are responsible for **increasing atmospheric temperatures**. With the **extensive use of fossil fuels** and the increased **burning of trees** as forests are cleared, levels of greenhouse gases have risen. Figure 10.5 shows how Earth is warmed by the greenhouse effect.

The **glass** of a greenhouse **traps heat** in a **similar** way to **greenhouse gases**. It **allows high frequency** radiation **to enter** but **prevents** the **lower frequency** infrared radiation (IR) from **leaving** (see Figure 10.6).

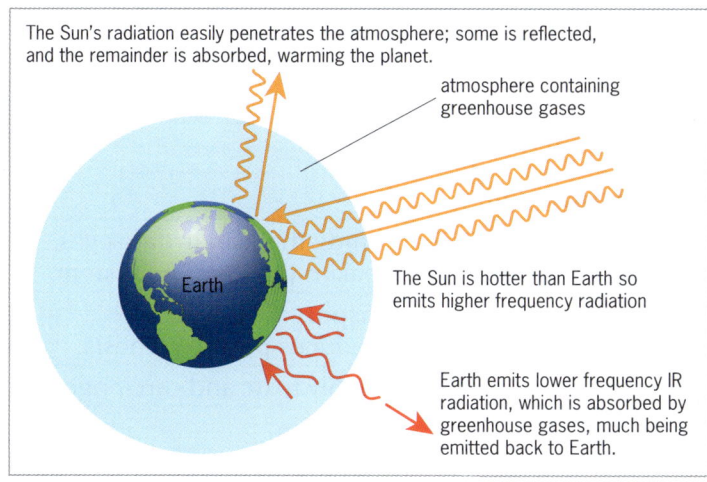

Figure 10.5 *The greenhouse effect*

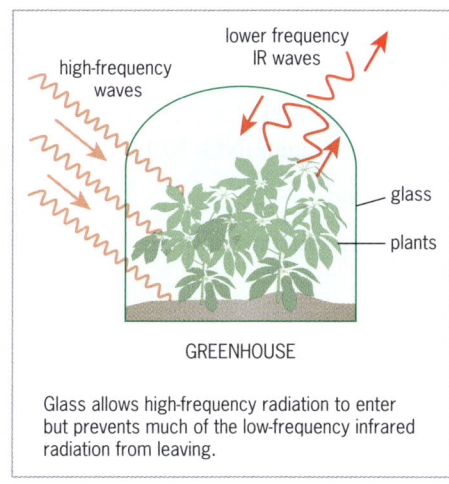

Figure 10.6 *The glass greenhouse*

10 Fossil fuels and alternative sources of energy

Table 10.2 *Generation of greenhouse gases*

Greenhouse gas	Generation of greenhouse gas
Carbon dioxide (CO_2)	Burning fossil fuels/deforestation as trees are burnt/respiration
Methane (CH_4)	Released by livestock manures/leakage from sites of fossil fuels
Nitrous oxide (N_2O)	Released by fertilisers/ released by some industrial processes
Ozone (O_3)	UV radiation/welding/lightning and spark plug discharges from engines
Water vapour (H_2O)	Evaporation/respiration/burning of fossil fuels
Chlorofluorocarbons (CFCs)	Leakage from refrigerants/aerosol propellants/industrial processes

Effects of climate change due to global warming

- **Animal migration** due to harsh conditions such as polar caps melting and higher temperatures allows invasive species to enter their space.
- **Rising sea levels** threaten the assets of coastal regions and islands, including their tourism industry.
- **Increased frequency of bad weather** (hurricanes, floods, etc.) will burden the national budget.
- **Increased frequency of wildfires** (bush fires) destroys ecosystems.
- **'Coral bleaching'** threatens the fishing industry, the production of sand on the beaches and tourism.
- **Forest shrinkage** releases carbon dioxide, disrupts ecosystems which sustain ecological balance, causes soil degradation which reduces agriculture and deprives the indigenous people of resources.
- **Decrease in agriculture** and **animal farming** due to scorching temperatures affects food supply.
- **Decrease in water supply** due to excessive evaporation threatens water security.
- **Increase in air conditioning usage** needed to counteract increased atmospheric temperatures.

Negative effects of fossil fuels

Pollution caused by fossil fuels is mainly due to (1) **electricity and heat production**, (2) **transport** and (3) the **manufacturing industry**.

Table 10.3 *Harmful emissions from fossil fuels*

Harmful emission	Negative effect
Carbon monoxide (CO)	Prevents haemoglobin in blood from transporting oxygen and causes eye irritation, headaches and nausea. Can result in death by suffocation and heart stress.
Sulfur dioxide (SO_2) **Oxides of nitrogen (NO, NO_2)**	Cause acid rain (see pages 124 and 125), eye irritation, respiratory irritation and disorders.
Nitrous oxide (N_2O)	Greenhouse gas (see page 125).
Volatile organic compounds (VOCs)	Mixed with nitrogen oxides, they create smog that can irritate eyes, cause lung cancer, harm the heart and weaken the immune system.
Particulate matter	Irritates the eyes and airways and can transport allergens. Falls on leaves, blocking sunlight and so reduces photosynthesis.
Heavy metals	Mercury, lead, arsenic and cadmium are highly toxic and carcinogenic.

General equation for the reaction occurring as fossil fuels are burnt:

fossil fuel + oxygen → carbon dioxide + water vapour + harmful gases + particulate matter

Table 10.4 *Other problems associated with using fossil fuels*

Problem	Negative effect
Oil spills	Cause damage to our environment, plants and animals and are costly to correct.
Leaked natural gas	Methane (a greenhouse gas) leaks during extraction, processing and distributing.
Healthcare costs	Healthcare costs to treat illnesses associated with fossil fuels are rising.
Competition	Unlike fossil fuels, new technologies in renewables are becoming cheaper.

Tetraethyl lead is a **toxic** substance previously added to gasoline, but its addition is **now banned**.

Carbon dioxide is a greenhouse gas but generally is **not considered a pollutant** to our health.

Non-renewable, renewable and alternative sources of energy

Non-renewable energy sources are those which are not readily replenished by natural processes.
Examples are fossil fuels and radioactive materials.

Renewable energy sources are those that are readily replenished by natural processes (sustainable).
Examples are solar, biofuels, wind, hydro-electric, tidal, wave and geothermal energy.

Zero energy buildings (ZEB) or *net zero-energy (NZE)* buildings are those whose annual energy *production* from renewable sources is *equal* to their annual energy *consumption*.

Alternative energy sources are those which are not fossil fuels.
Note: Although **radioactive** material is an **alternative** energy source, it is **non-renewable**.

Advantages of most alternative sources of energy

- **Renewable** and so **unlimited in supply** and have **free raw energy input**.
- Produce zero or **minimum environmental pollution**.
- Require **minimum operational costs**, although initial plant costs are generally high.

Various alternative sources of energy

Solar

Table 10.5 *General advantages and disadvantages of various types of solar systems*

General advantages of solar systems	General disadvantages of solar systems
Low operational costs.	**Except** for **solar dryers**, start-up costs are high.
Durable with low maintenance costs.	**Except** for **PV systems**, the energy cannot be stored in cells.
Limitless supply of free sunshine.	Large production requires much space.
Clean (non-polluting) energy source.	Poor performance on cloudy days/zero production at night.

Photovoltaic (PV) systems

These **convert solar radiation into electrical energy** see Figure 10.7 – they are not solar water heaters. In the northern hemisphere they should be on a south-facing roof to better absorb the rays.

An added advantage of these systems is that the energy can be **returned to the grid** for reimbursement. It can also be **stored in batteries** so off-grid regions of rural areas can obtain electricity.

Figure 10.7 *PV panels*

Water heater systems

These **heat water directly** with solar energy and are relatively **easy to install**. In the **northern hemisphere** the panels should be on a **south-facing** roof to better absorb the rays (see page 157, Figure 12.9). **Back-up electrical heating** should be used for **cloudy days**.

However, the **conversion to electricity** is of **low efficiency**, particularly if the panels are **not kept cool**.

Solar cookers

These use mirrors to reflect solar energy to a furnace or pot which should be painted **matt black** (see page 92, Figure 8.7). They take **longer to cook** food than conventional cookers would.

Solar dryers

Direct solar dryers are racks where items to be dried are exposed to the Sun in the open air.

Indirect solar dryers dry items in a partially enclosed container, as shown in Figure 10.8.

Biofuels

These provide **energy from plants and animals**. Although they produce carbon dioxide when burnt, the plants from which they are formed removed carbon dioxide from the air when they were alive, cancelling the effect on environmental pollution. Various biofuels are outlined as follows.

Wood

Advantages: Can be burnt directly; requires no maintenance; is cheap and its ash is a useful fertiliser.

Disadvantages: Pollutes the air; increases global warming by releasing CO_2; requires a large storage area; can lead to deforestation causing soil erosion and the inability to remove CO_2 from the air.

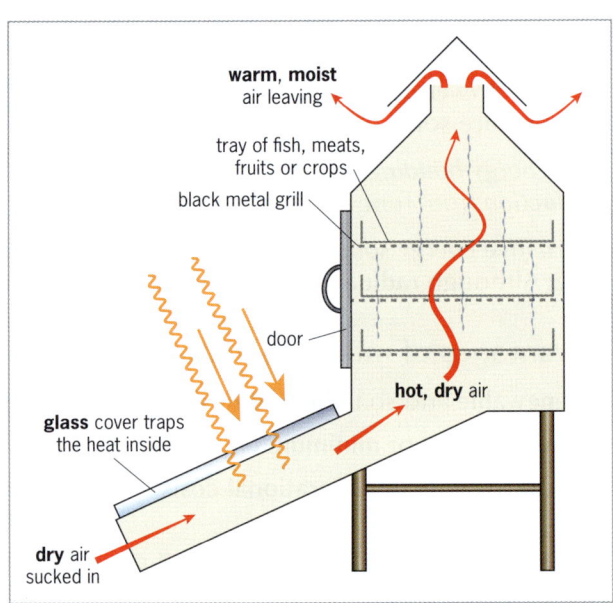

Figure 10.8 *An indirect solar dryer*

Biogas

Biogas is a mixture of gases (mainly methane) obtained from the **decay of plant and animal wastes** in the **absence of oxygen**. A biogas generator is shown in Figure 10.9. It is used for heating boilers, cooking and for driving electrical generators on farms.

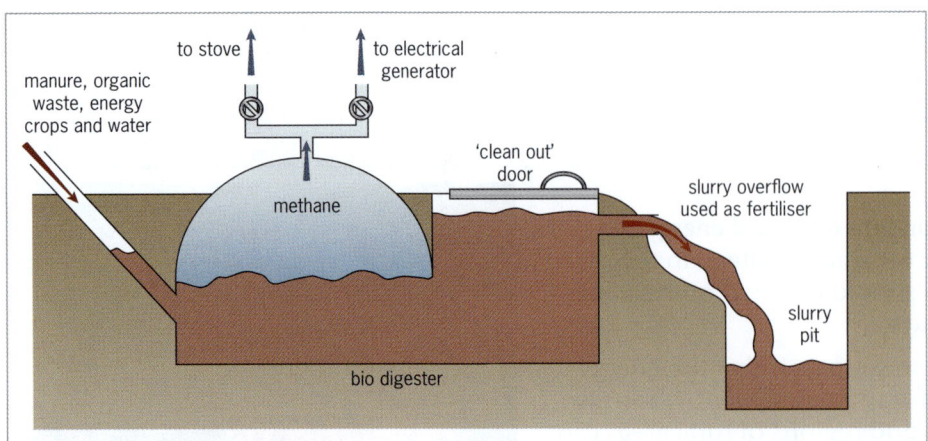

Figure 10.9 *Biogas generator*

Advantages: Cheaper than natural gas and, unlike natural gas, is a renewable energy source. Removes animal manure, wastes from slaughterhouses and sludge from treatment plants. Produces organic fertiliser that is safer and better than manufactured fertilisers. Healthier alternative to burning wood since it does not contaminate the environment with smoke.

Disadvantages: Biogas corrodes machinery and so its use is restricted to heating, lighting and cooking. The process of production is inefficient, producing a great deal of waste. It has a foul odour and so it is unsuitable for generation in residential areas.

Biodiesel

Produced from **vegetable oils** or **animal oils/fats** by a process known as **transesterification**.

Advantages: Has a lower viscosity and is less polluting than petroleum diesel. Is a great way to dispose of used cooking oil.

Disadvantages: Is more costly than petroleum diesel, destroys rubber hoses and clogs fuel filters.

Gasohol

A **mixture of gasoline and ethanol** produced by the **fermentation of crops** such as sugar cane.

Advantages: Gasohol is partially renewable (unlike gasoline), is cheaper, is less polluting and produces better engine performance than regular gasoline.

Disadvantages: Agricultural land is used up to produce ethanol and so leaves less land to grow food. It yields less energy content per litre than gasoline and some engines require modification to use it.

Wind

Figure 10.10 shows an onshore wind farm on the brim of a mountainous region. Wind farms can also be placed offshore.

Advantages: Produces no air pollution or global warming and reduces the use of fossil fuels as it captures strong Caribbean breezes. Harnessing wind energy requires low maintenance (onshore farms only), uses free energy input, can be stored in batteries and provides job opportunities of several skills.

Disadvantages: High construction plant costs; fluctuation in wind speed; vulnerability to storms; noise pollution and the hazard of birds flying through its propellers.

Figure 10.10 *Onshore wind farm*

Hydro-electric

Figure 10.11 shows water collected by a dam being released to turn the turbines of generators.

Advantages: Hydro-electric plants are non-polluting with low greenhouse gas emissions, have low maintenance and operational costs, and use free energy input. Unlike solar and wind generation, electricity can be produced when required by controlling when the sluice gates open. The plants provide opportunities for work and reduce the dependence on fossil fuels. Water from them can be used for irrigation and water parks, and purified for drinking.

Figure 10.11 *Hydro-electric plant*

10 Fossil fuels and alternative sources of energy

Disadvantages: These plants have high construction costs. They displace communities previously living there and disturb the ecology downstream by supplying less water and sediment to those regions. They are vulnerable to drought and climate change. It is difficult to locate a site that (1) has adequate water supply, (2) does not significantly affect the ecology and (3) is close to the electrical grid.

Tidal

Tidal **barrage generators** collect water at **high tide** and release it at **low tide** to turn electrical generators (Figure 10.12).

Tidal stream generators are placed in fast-flowing rivers. They do not collect and release water.

Advantages: Tidal plants are non-polluting, use free energy input and are reliable since tides are predictable. The power output is strong and the plant protects the coast from bad weather.

Disadvantages: Tidal plants are costly to construct. They must be on coastal regions having significant difference in tide levels and they are a hazard to marine life. Timing of the high tide is usually different than when demand is highest and so a storage system of batteries is often required.

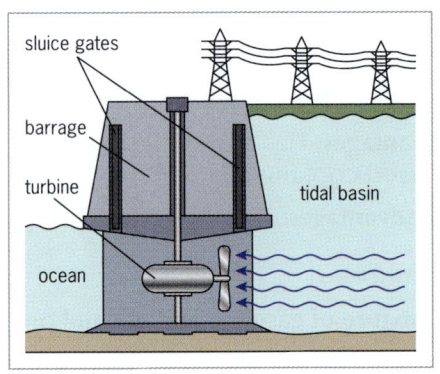

Figure 10.12 *Tidal barrage generator*

Wave

Buoys connected to the turbines of generators **rise and fall** with the waves to produce electricity.

Advantages: Same as for using tidal energy.

Disadvantages: Wave energy plants have high construction costs since the equipment must withstand harsh elements and rough seas. They destroy the landscape and seabed and disturb sea animals with noise pollution. They occupy space which may have been used by shipping vessels.

Geothermal energy

Geothermal energy is thermal energy generated in the Earth's crust and mantle (1) by **radioactivity** and (2) by **friction between tectonic plates**. Lava, hot water geysers or pressurised water injected through the geothermal reservoir as shown in Figure 10.13 can bring the energy to the surface.

Advantages: Geothermal plants are relatively clean and use free energy input. They use minimum space, can last for centuries and are very reliable (unlike solar and wind plants).

Disadvantages: The high-pressure injection can produce small earthquakes. Small amounts of greenhouse gases are emitted and small amounts of toxic elements are gathered by the water.

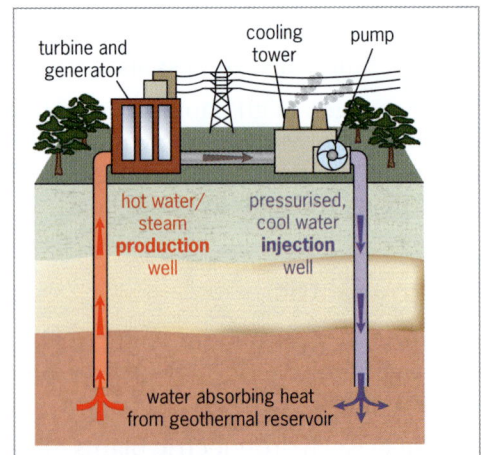

Figure 10.13 *Geothermal energy plant*

Nuclear

This is an **alternative** source of energy that is also **non-renewable**. Energy released by the nuclear fission of uranium or plutonium is used to produce steam to turn the turbines of electrical generators.

Advantages: Uranium is available in large quantities and is a reliable source, independent of weather conditions. Only small amounts are needed to produce enormous amounts of energy and so transport and storage are inexpensive. No greenhouse gases are emitted in the process. It provides a suitable backup to alternative energy sources such as wind or solar when weather conditions are unsuitable.

Disadvantages: Nuclear plants are expensive to construct. Nuclear radiation is extremely dangerous and so mining, transporting and using the material in the plant is hazardous. A critical malfunction at the plant can cause huge explosions and result in radioactive fallout (spreading of radioactive material by air masses – see page 180). Spent nuclear fuel is difficult to dispose of and plant closures are necessary as plants become contaminated with use.

Hydrogen (an energy carrier – not an energy source)

Hydrogen mainly exists as compounds on Earth. It is not generally considered a **source** of energy as we need to extract it from its compounds before we use it. However, scientists have recently discovered that free hydrogen is much more abundant from underground sites than previously thought.

White hydrogen is naturally occurring hydrogen gas which can be obtained from underground deposits. Its production is **free of carbon emissions**.

Grey hydrogen is produced by burning fossil fuels – generally from methane in a process known as **steam reforming** – and results in the **emission of greenhouse gases such as carbon dioxide**.

Blue hydrogen is also produced by steam reforming, but the **carbon dioxide emissions are captured**.

Green hydrogen (renewable hydrogen) is produced by the **electrolysis of water** into hydrogen and oxygen. It **does not emit greenhouse gases** since it is produced from a renewable source.

Advantages: Hydrogen is **abundant**, **readily available**, **cheap** and has a **high energy** density.

Disadvantages: Hydrogen can easily **leak from containers** since its molecules are very small. It is **expensive to transport** and is **explosive**.

Variables affecting the collection of solar and wind energy

Solar

Intensity is greater:

- **at low latitudes** (near equator) since the energy concentrates on a smaller area of Earth (see Figure 10.14)
- **at high altitudes** (tops of mountains) since it is absorbed by less atmosphere
- **during summer** since the Sun is then more directly overhead
- when there are **no tall objects or clouds** casting shadows
- when the air contains **less particulate matter** to absorb the energy.

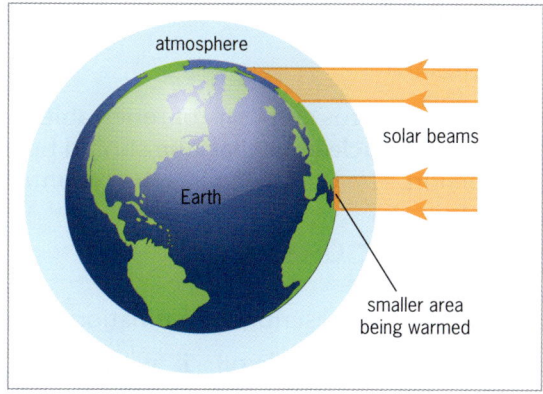

Figure 10.14 *Greater intensity near equator*

Wind

Strong and **consistent** winds deliver more kinetic energy to the turbines of generators. Wind **speed is greater** over the **oceans** and at **high altitudes** where there is **less friction**.

Appraisal of the extent to which alternative sources of energy can be used in the Caribbean

Economic benefits – more money in pocket

- The availability of **intense insolation**, **strong and consistent winds**, **geothermal reservoirs** and **water catchment reservoirs** in a few mountainous regions, offer potential for obtaining **cheaper** energy.
- **New jobs** available in research, engineering, manufacturing, installation, marketing and sales.
- **Stable energy costs** since the raw energy input is **free** and **never depleted**.
- **Less reliance on fossil fuels** as some renewables can be **stored in batteries** for when the grid is down.
- Solar heaters and PV systems **increase property value** and unproductive land **can now be productive**.

Environmental benefits – safer and cleaner surroundings

- **Improved air quality** results since there would be less pollutants from burnt fossil fuels.
- **Reduction in oil spills** destroying the environment.
- **Reduction** in **global warming** and **climate change** since there would be less carbon dioxide emission.

Social benefits – happier, healthier and less stressed

Due to the economic and environmental benefits outlined above, individuals and **families can feel more secure** and **capable of running their homes** and are often **happier**, **healthier** and **less stressed**.

General problems of implementing the switch to alternative sources of energy

Renewable energy plants are often **expensive to construct**, and large-scale generation **usually requires much space**. Some can **disturb wildlife** and **damage the ecology** and/or produce **visual or noise pollution**. Solar and wind energy can have **intermittent outputs**.

Banks are hesitant to grant loans as many lack understanding of the new technologies. There is a **lack of accountability and transparency of the borrowers** and there is **no firm credit history of the installers**.

Conclusion

Owing to the pros and cons of renewable energy plants, a **mixture of various types is preferred**. A **fossil fuel** or **nuclear fuel back-up** should be used should any of the renewable sources fall short in production. The Caribbean has the potential to be a significant producer of alternative energy.

Revision questions

1
  a Describe the formation of EACH of the following fossil fuels:
     i crude oil   ii natural gas.
  b Why is charcoal not considered a fossil fuel?
  c Distinguish between LPG and LNG.

2
  a What component of the crude oil mixture:
     i forms a sludge at the bottom of the fractionating column?
     ii rises through the top of the tower as a gas?
  b List FOUR other components of the mixture.
  c Give ONE use of EACH component mentioned in part **a**.
  d What can be said of the relative molecular size of the components of crude oil having high melting points?

3 **a** Draw an **arrow diagram** to show the energy transfer occurring in EACH of the following.
  i An aircraft rising from a runway.
  ii Electrical generation using fossil fuels.
  b Write a general **word equation** indicating the reaction occurring as fossil fuels are burnt.

4 **a** Which fossil fuel emission causes the haemoglobin in blood to become incapable of transporting oxygen?
  b Other than the emission mentioned in part **a**, name THREE fossil fuel emissions that cause eye irritation, heart problems and respiratory problems.

5 **a** State TWO ways that UNBURNT fossil fuels can be harmful to the environment.
  b A main contributor to fossil fuel pollution is the production of electricity. List TWO other main contributors to the pollution caused by burning fossil fuels.
  c List THREE advantages of using fossil fuels.

6 List TWO ways that EACH of the following greenhouse gases are produced: carbon dioxide; ozone; water vapour; methane; chlorofluorocarbons.

7 **a** Distinguish between renewable and non-renewable energy sources.
  b What is meant by an alternative energy source?
  c Name the non-renewable energy source that is also an alternative energy source.
  d What is meant by a net zero-energy (NZE) building?

8 Briefly outline how EACH of the following are produced:
  a biogas   **b** biodiesel   **c** gasohol.

9 **a** List THREE GENERAL advantages of most alternative sources of energy.
  b List TWO disadvantages of EACH of the following alternative sources of energy:
   i solar water heater   **ii** wood   **iii** biogas   **iv** biodiesel
   vi wind   **vii** tidal   **viii** geothermal.

10 State TWO advantages and TWO disadvantages of having hydroelectric power plants in the Caribbean.

11 **a** What is the function of a photovoltaic (PV) system?
  b In which cardinal direction should the roof carrying a solar panel face in the northern hemisphere?

12 List THREE variables that can alter the intensity of solar energy at a given location.

13 List THREE advantages and THREE disadvantages of using nuclear energy to produce electricity.

14 **a** Why is hydrogen not considered a renewable energy source?
  b Distinguish among 'white hydrogen', 'green hydrogen' and 'grey hydrogen'.
  c List TWO advantages and TWO disadvantages of obtaining energy from hydrogen.

15 List TWO environmental benefits and TWO economic benefits of using renewable energy in the Caribbean.

11 Electricity and lighting

Electricity is important in providing us with devices necessary to run our modern lifestyles. It is essential to our infrastructure, is clean, can be stored in batteries and is easily transported by electrical wires.

Electrical components and circuits

Conductors, semiconductors and insulators

Conductors of electricity are materials containing electrical charges that can flow freely.

Examples are **metals**, **graphite** (a non-metal), **ionic solutions** and **molten ionic substances**.

In metals and graphite, the free charges are **electrons** (negative charges).

In ionic solutions and molten ionic substances, the free charges are **positive and negative ions**.

Insulators (poor conductors) are materials through which electrical charges cannot flow freely.

Semiconductors are materials with conductivity between that of good conductors and insulators.

Table 11.1 *Examples of conductors, semiconductors and insulators*

Good conductors	Semiconductors	Insulators
Silver, gold, copper, aluminium, graphite, electrolytes such as blood, sweat, salt water (brine), molten table salt	Silicon, germanium	Glass, wood, plastic, rubber, cork, paper

Table 11.2 *Uses of conductors, semiconductors and insulators*

Material	Uses
Conductors	**Electrical wires** are generally made of **copper** and **aluminium**.
	Electrical contacts which repeatedly connect and disconnect are made of **graphite**.
	Electrolytes (ionic solutions) in **batteries**.
Semi-conductors	**Electronic components** such as sensors and LEDs are made of silicon and germanium.
	PV cells for solar panels are typically made from silicon.
Insulators	**Electrical wire insulation** (to prevent electrical shock) is made of rubber and plastic.

Electrical circuits

An *electric circuit* is a complete conducting path or loop through which electric charges can flow.

Charges cannot flow if this path is broken. Electrical circuits are described using circuit diagrams.

Table 11.3 *Electric circuit symbols*

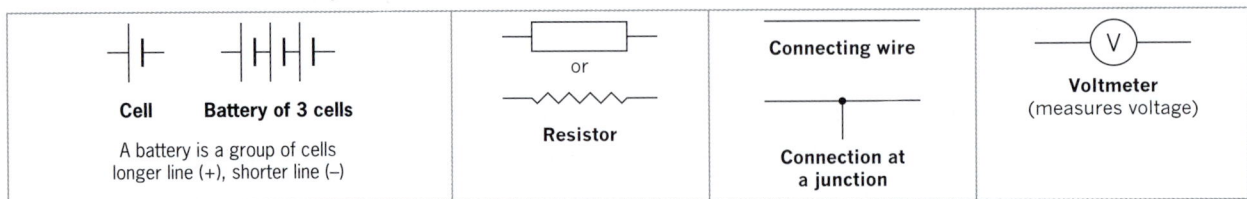

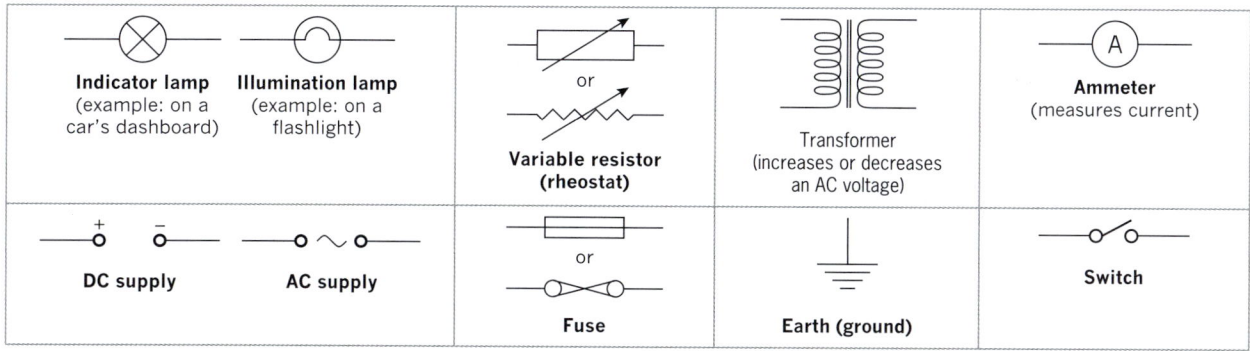

Important electrical quantities

Table 11.4 *Important electrical quantities and their SI units*

Quantity	Definition	SI unit
Voltage (V)	The electrical **energy per unit charge** used in **pushing** electrons against the resistance of the circuit.	volt (V)
Current (I)	The **rate of flow** of **electrical charge** through a circuit.	ampere (A)
Resistance (R)	The **opposition** to the **flow of electrical current** through a material.	ohm (Ω)
Power (P)	The **rate** of **using energy** (or the rate of doing work).	watt (W)

A **battery** is analogous to a water pump, raising water to the top of a hill to achieve a **potential difference**, and then releasing it to flow as a **current** through the **resistance** of rocks or bushes.

Note: Current flows through a circuit from the positive terminal of a battery to its negative terminal.

Potential difference (pd) is the **voltage** (energy per unit charge) required to push electric charges through a section of the resistance of a circuit.

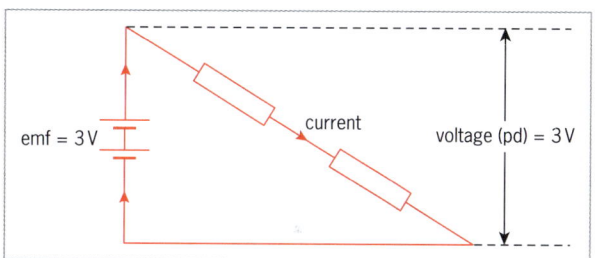

Figure 11.1 *The emf of a battery provides a voltage (pd) across the resistors of a circuit*

Electromotive force (emf) of a cell or battery is the **total voltage (pd)** it provides across all the components of a circuit.

Useful equations

Ohm's law states that the current through a metallic conductor is proportional to the voltage across it, providing the temperature is constant.

Voltage: voltage = current × resistance $V = IR$ ... equation (i)

Resistance: From equation (i) it follows that: $R = \dfrac{V}{I}$ ... equation (ii)

By calculating the **GRADIENT** (or **SLOPE**) of a graph of several **voltages** against their corresponding **currents** we can obtain the resistance of a resistor (see question 6, page 145).

Power: power = voltage × current $P = VI$ ... equation (iii)

From **Table 11.4**: power = $\dfrac{\text{energy}}{\text{time}}$ $P = \dfrac{E}{t}$ ... equation (iv)

Energy: From equation (iv) it follows that: $E = Pt$ ... equation (v)

11 Electricity and lighting

Voltage and power rating of appliances

The voltage and power rating of an appliance indicate the **voltage** to be used on the appliance under **normal operation** and the **power it will consume at that voltage**.

> **Example 1**
>
> A device is rated at 1800 W, 120 V. Calculate the following.
>
> a The current flowing when in normal operation.
> b The energy consumed in 5 minutes.
> c The resistance of the device.
>
> ---
>
> **NOTE:** Use the format shown in part **a** for all your equation solving.
>
> a $P = VI$ 　　Lay out the equation as you have learnt it.
>
> 　　$\dfrac{P}{V} = I$ 　　Make the subject of the equation the quantity you are solving for.
>
> 　　$\dfrac{1800 \text{ W}}{120 \text{ V}} = I$ 　　Substitute values with correct SI units.
>
> 　　$15 \text{ A} = I$ 　　Represent final answer with correct SI unit.
>
> b $P = \dfrac{E}{t}$ 　　$\therefore E = Pt$ 　　$E = 1800 \text{ W} \times (5 \times 60 \text{ s})$ 　　$E = 540\,000 \text{ J}$
>
> c $V = IR$ 　　$\therefore R = \dfrac{V}{I}$ 　　$R = \dfrac{120 \text{ V}}{15 \text{ A}}$ 　　$R = 8 \; \Omega$

Current in series and parallel circuits

Ammeters have **negligible resistance** and are **connected in series** with components.

Series circuit: Figure 11.2 shows that when components are connected in series (one in front of the next), **the current is the same at each point**.

Parallel circuit: Figure 11.3 shows that when components are connected in parallel (different branches), the **current outside the branches is equal to the sum of the currents within the branches.**

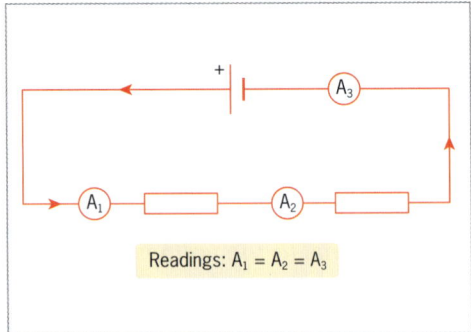

Figure 11.2 Current in series circuits　　Figure 11.3 Current in parallel circuits

Voltage in series and parallel circuits

Voltmeters have very **high resistance** and are **connected in parallel** with components.

Series circuit: Figure 11.4 shows that voltage from the battery is split amongst the components in proportion to their resistance. **The sum of the voltages across the resistors is equal to the emf of the battery.**

Parallel circuit: Figure 11.5 shows that **voltage across components in parallel is always the same.**

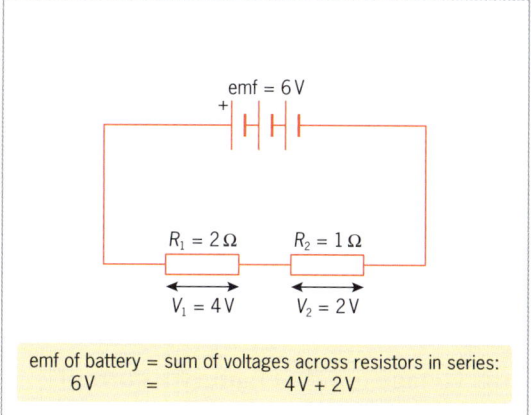

Figure 11.4 *Voltage in series circuits*

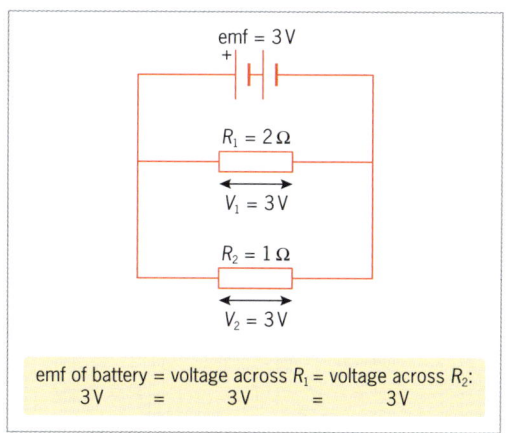

Figure 11.5 *Voltage in parallel circuits*

Example 2

Figure 11.6 shows an electrical circuit. Calculate the following.
a The current flowing.
b The power used.
c The energy used in 5 minutes.

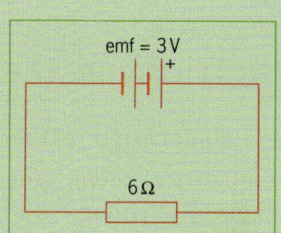

Figure 11.6 *Example 2*

a $V = IR \therefore I = \dfrac{V}{R}$ $I = \dfrac{3\,V}{6\,\Omega}$ $I = 0.5\,A$

b $P = VI$ $P = 3\,V \times 0.5\,A$ $P = 1.5\,W$

c $P = \dfrac{E}{t} \therefore E = Pt$ $E = 1.5\,W \times (5 \times 60\,s)$ $E = 450\,J$

Example 3

Figure 11.7 shows an electrical circuit. Calculate the following.

a The total resistance.
b The pd (voltage) across:
  i the variable resistor (rheostat)   ii the lamp.
c The power used by:
  i the variable resistor   ii the lamp.
d The power delivered by the battery.

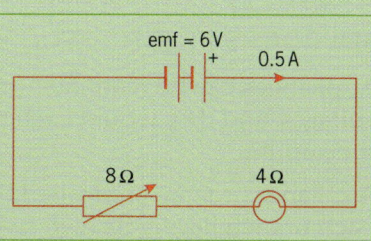

Figure 11.7 *Example 3*

Notes:
- **Lamps** have **resistance**.
- **Current is the same at all points in a series circuit**, so 0.5 A flows through both resistors.
- **A portion of the voltage** from the battery is **across the variable resistor**, the remainder is **across the lamp**.
- The power **delivered by the battery** is equal to the **total power used by the resistors**.

11 Electricity and lighting

a Total resistance = 8 Ω + 4 Ω = 12 Ω
b i V = IR   V = 0.5 A × 8 Ω   V = 4 V   Using: current through rheostat × resistance of rheostat.
   ii V = IR   V = 0.5 A × 4 Ω   V = 2 V   Using: current through lamp × resistance of lamp.
c i P = VI   P = 4 V × 0.5 A   P = 2 W   Using: voltage across rheostat × current through rheostat.
   ii P = VI   P = 2 V × 0.5 A   P = 1 W   Using: voltage across lamp × current through lamp.
d P = VI   P = 6 V × 0.5 A   P = 3 W   Using: voltage (emf) of battery × current through battery.

Alternatively, the power provided by the battery is the same as the power consumed by both resistors, and this is: 2 W + 1 W = 3 W

Electricity in the home

*Alternating current (AC) repeatedly **reverses direction**.*

The electrical **power grid delivers AC** to our power outlet sockets. The voltage at the sockets in the Caribbean is usually between **110 V** and **120 V** but some territories also receive 240 V; the frequency is usually **50 Hz**.

*Direct current (DC) flows in **one direction** only.*

Batteries produce DC. The cylindrical cells (known as dry cells) we use in our TV remote controls are typically 1.5 V. The end with the 'bump' is the positive pole and the flat end is the negative pole. The total emf of cells in series is the sum of the emfs.

Figure 11.8 *Power outlet socket*   Figure 11.9 *A 1.5 V dry cell*

Advantages of parallel connection of appliances in domestic wiring

- Appliances can be manufactured to **operate on standard voltages**, and they would each receive that voltage from the supply.
- Appliances can be controlled by **individual switches**, each branch having its own switch.
- Appliances can **draw different currents** from the same voltage and so operate at **different powers**.

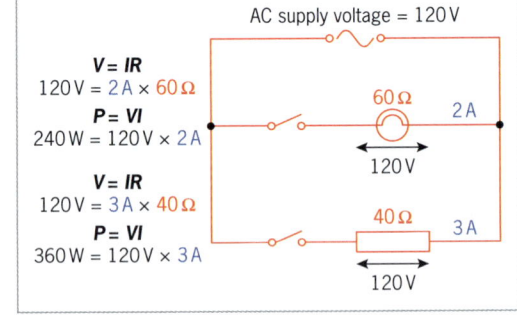

Figure 11.10 *Appliances in parallel*

Measuring electrical energy consumption using the kilowatt-hour

We use the **kilowatt-hour** instead of the **joule** to describe the energy consumption in our homes since if expressed in joules, the numbers would be very large.

energy = power × time

1 J = 1 W × 1 s

1 kW h = 1 kW × 1 h = 1000 W × (60 × 60 s) = 3 600 000 J

Example 4

Calculate the total weekly energy consumed in kW h, given the following appliance usage.
- One 250 W refrigerator, always in use (24 hours, 7 days per week).
- One 1200 W electric oven used for 3 hours each day from Monday to Friday inclusive.
- Twelve 60 W filament lamps used 8 hours each night.
- One 1400 W electric kettle used for 15 minutes (1/4 h) every morning.

Appliance	Power (kW)	Time (h)	Energy (kW h)
One 250 W refrigerator	1 × 0.250	24 × 7	42.00
One 1200 W electric oven	1 × 1.2	5 × 3	18.00
Twelve 60 W filament lamps	12 × 0.060	7 × 8	40.32
One 1400 W electric kettle	1 × 1.4	7 × 1/4	2.45
Total (kW h)			102.77

Note: Power must be in **kW (not W)** and time must be in **h (not s)** since energy is to be in kW h.

Electricity meters

Energy used within a period is found from the difference between the previous and current readings.

Digital meters

- Interpreted as one direct reading (see Figure 11.11).

Analogue meters

- The dials are read from left to right.
- The hands of successive dials rotate alternately clockwise and anticlockwise.
- The digits selected are those to which the hands point or have just passed.
- The reading shown in Figure 11.12 is 05358.

Figure 11.11 *Digital electric meter*

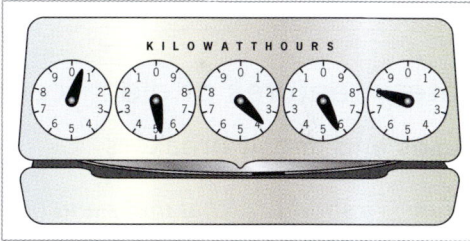

Figure 11.12 *Analogue electric meter*

Electricity bills

- **Energy charge** – **Fixed rate** (for total usage) or a **blocked rate** (varies with usage).
- **Fixed charge** – Applied to cover administrative costs and/or meter rentals.
- **Fuel adjustment charge** – Applied per kW h to cover fluctuating fuel costs.

> **Example 5**
>
> A meter reading was 27 538 kW h on 31st May and 27 798 on 30th June. The charge for the first 200 kW h is 50 cents per kW h and additional usage is charged at 25 cents per kW h. There is a fuel adjustment charge of 2 cents per kW h and an administrative fee of $12. Calculate the bill for the period.
>
> Energy used during the month = 27 798 kW h − 27 538 kW h = 260 kW h
>
> **Energy charges:** 200 kW h × $\frac{\$0.50}{\text{kW h}}$ = $100.00
>
> 60 kW h × $\frac{\$0.25}{\text{kW h}}$ = $15.00
>
> **Fuel adjustment:** 260 kW h × $\frac{\$0.02}{\text{kW h}}$ = $5.20
>
> **Admin. fee:** = $12.00
>
> **TOTAL CHARGE** $132.20

Average power consumption of commonly used appliances

Appliances that produce **heat** generally use more power than others.

Table 11.5 *Average power consumption of commonly used appliances*

Appliance (heating type)	Power (W)
Microwave	1000
Electric kettle	1200
Electric oven	2000
Electric clothes dryer	5400

Appliance	Power (W)
Laptop	60
65-inch LED TV	100
Refrigerator	400
CRT TV	500

Safety features of electrical devices

Fuses and circuit breakers

A **fuse** is a short wire that melts and breaks a circuit when its current exceeds a certain value (Figure 11.13).

A **circuit breaker** is an electromagnetic switch that breaks a circuit when its current exceeds a certain value (Figure 11.14).

The **current rating** of a fuse or circuit breaker is the current which is **slightly greater** than that which should flow in the circuit **during normal operation**.

Figure 11.13 *Typical fuse*

Figure 11.14 *Circuit breaker*

> **Example 6**
>
> An air fryer rated at 1500 W is to be operated on a 120 V electrical mains supply.
> a Calculate the current (amperage) it takes.
> b State, with reasons, which of the fuses listed below should be used to protect the device: 2 A, 7 A, 14 A, 18 A.
> c Determine the energy used by the air fryer in 20 minutes.
>
> ---
>
> a $P = VI$ $\therefore I = \dfrac{P}{V}$ $I = \dfrac{1500\ W}{120\ V}$ $I = 12.5\ A$
>
> b The 14 A fuse should be used since its current rating is just above the normal operating current.
> - The 2 A and 7 A fuses will 'blow' as soon as the device is switched on.
> - The 18 A fuse will allow a larger current to flow than should be taken by the device, which can result in overheating or destruction of the device, or may even cause an electrical fire.
>
> c $P = \dfrac{E}{t}$ $\therefore E = P \times t$ $E = 1500\ W \times (20 \times 60\ s)$ $E = 1\,800\,000\ J$ (or 1.8×10^6 J)

Placement of fuses and circuit breakers

Fuses and circuit breakers are **placed in series with devices** to protect them from excessive currents that can **overheat** and **destroy the components**, and which could cause an **electric fire**.

Fuses may be placed in a **distribution box**, within a **plug** (see Figure 11.20), or built into a **device**.

Three-core flex used in domestic wiring

A **three-core flexible cable** is used to connect appliances to the electrical mains. It contains individually **insulated** and **colour coded** wires: the **live, neutral and earth** wires (see Figures 11.15 and 11.16).

Several of our appliances are sourced from the UK, Europe, the USA and Canada (see Table 11.6).

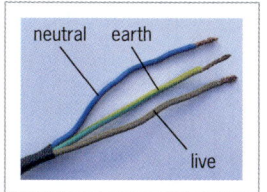
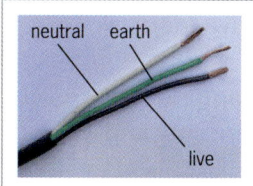

Figure 11.15 Colour code used in the UK and Europe

Figure 11.16 Colour code used in the USA and Canada

Table 11.6 Domestic wiring codes

	LIVE	NEUTRAL	EARTH (GROUND)
UK and Europe	Brown	Blue	Green and yellow striped
USA and Canada	Black	White	Green or bare copper (USA)/green or green and yellow striped (Canada)

Live and neutral wires

Under **normal operation**, the current flows in the appliance via the **live and neutral wires** with no current flowing in the earth wire (see Figure 11.17).

Earth wire and short circuits

Figure 11.18 shows that during a **short circuit**, **current increases** as it takes the **path of less resistance** through the **case** of the appliance via the **live and earth** wires. The higher current 'blows' the fuse:

- **protecting the appliance** as it breaks the circuit,
- **protecting the user** from electric shock as the case is no longer electrified.

An appliance having a **non-conducting** case **does not usually require an earth wire**.

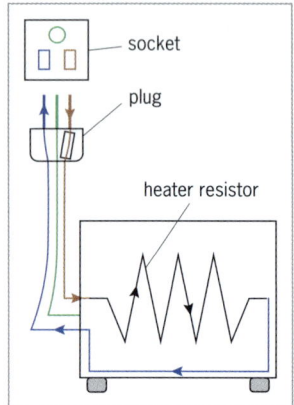
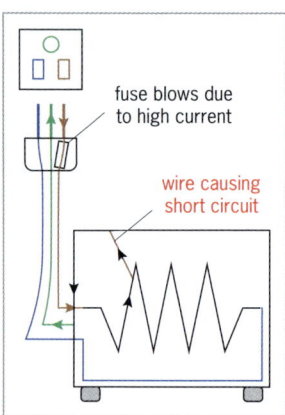
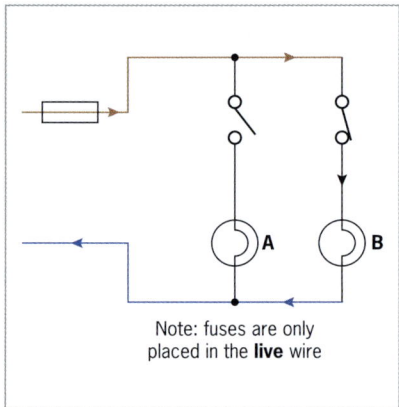

Figure 11.17 Current flow under normal operation

Figure 11.18 Current flow during short circuit

Figure 11.19 Lighting circuits usually have no earth wire

Wiring a three-pin plug with a fuse

1. **Remove** approximately **4 cm** of the plastic **flex casing**.
2. **Strip 1 cm** of the **insulation** from **each of the three cables** (live, neutral and earth).
3. For each cable, **tightly twist its strands**, **twirl** the strands using the screws in **a clockwise** direction around the appropriate terminal and **tighten** so that **no strands** are left **exposed**.

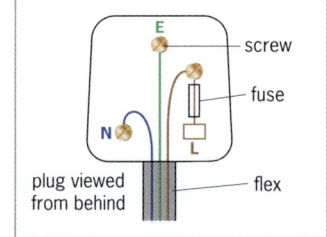

Figure 11.20 Wiring a three-pin plug with a fuse

Thick wires are required for circuits with high currents

High powered circuits draw **large currents** and so require **thick wires** to prevent the **overheating** of **electrical components**, the **melting of insulation** and the **production of electrical fires**. Examples include the following.

- **Heavy duty appliances** – Heaters, electric stoves, washing machines, clothes dryers and power tools.
- **Circuits having several power outlets** – The total current is the sum of the currents from each outlet.
- **Electrical power grid** – These cables carry large currents over long distances.

Electrocution and the danger of electrical shock

Electrocution is execution by electricity.

Our skin is a poor conductor when dry, but when wet (for example with sweat), increased conductivity can cause intense electrical shock resulting in **burns** and even **electrocution**. A voltage that is high enough to penetrate our skin and reach our body fluids (blood etc.) has a similar effect, because the fluids are good conductors.

If a part of the body touches a live wire as another part is grounded, current will flow between them.

Energy conservation measures

Energy conservation is the effort to reduce wasteful energy consumption by using fewer energy services and/or more efficient energy services.

Conserving energy is **beneficial** for the following reasons.
- Fuel expenses are reduced.
- Pollution is reduced.

Faulty appliances can **waste energy** for the following reasons.
- Poor lubrication can lead to heat loss due to friction.
- Corrosion can lead to current leakage.
- Faulty thermostats can lead to excessive heating or cooling.

Reducing energy consumption in the home

- Use **efficient**, **certified appliances** including refrigerators, LED and LCD TV screens and LED lighting.
- Install **PV systems** to produce electricity and solar water heaters to heat water (see pages 127 to 128).
- **Switch off appliances** and lights when not in use and avoid unnecessarily opening the fridge.
- Use **pressure cookers** and covered saucepans; adjust heat sources to a minimum when boiling.
- **Wash full loads** in the washing machine; dry clothes using lines or racks (not electric dryers).
- Reduce conduction into and out of homes by **double glazing windows**, adequately **insulating roofs and ceilings**, and building with **hollow concrete blocks**. Reduce incoming radiation by placing **hoods** over windows, using **curtains**, **painting walls white** and planting **nearby trees**.

Artificial sources of light

Daylight is the light we receive from the Sun. It contains a **full spectrum of frequencies** which includes all the **colours of the rainbow**.

Artificial light is the light we obtain from our filament lamps, fluorescent tubes and LEDs. These sources make colours **less vibrant** as they emit a **limited range** of **frequencies** but **lack certain frequencies**.

Fluorescent tube and compact fluorescent lamp (CFL) – Figure 11.21 shows how electrons accelerated by a voltage can produce light in a fluorescent tube. A CFL uses the same principle and is designed to fit into the sockets of filament lamps.

Light-emitting diode (LED) – A semiconductor that emits light when current flows through it.

Incandescent (filament) lamp – A **high resistance filament coil** made of **tungsten** that emits light when heated by a current to about **2500 °C** in a **sealed glass enclosure** containing an **inert gas** or **vacuum**. Several countries have **banned** these lamps since they **waste** a great deal of energy as heat.

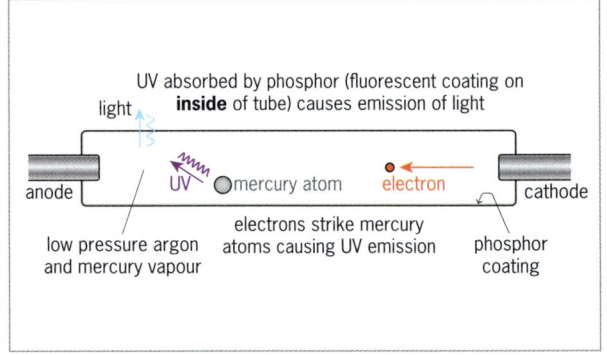

Figure 11.21 *Function of fluorescent tube*

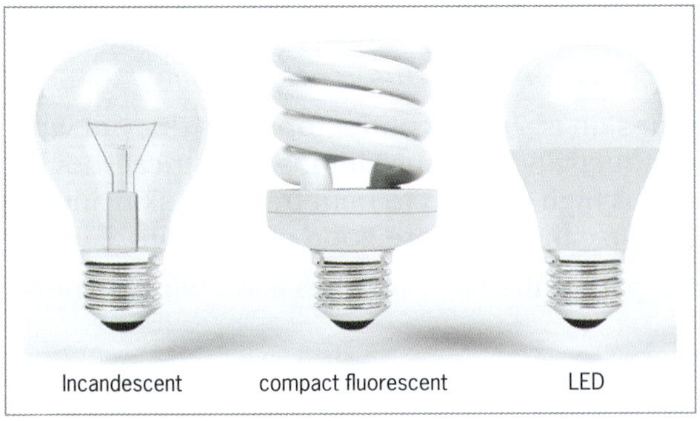

Figure 11.22 *Filament lamp, CFL and LED*

Table 11.7 Comparison of lamps used for lighting

Characteristic	Fluorescent	LED	Filament
Lifespan	Longer	Longest	Relatively short
Efficiency	High	Very high	Poor
Wasted heat produced	30% heat	Very little heat	90% heat
Contains toxic mercury	Yes	No	No
Similarity to daylight	Yes	'Daylight-type' for some	No. Produces a warm yellow light
Shadow	Diffused	Sharp	Diffused
Brightness control	Difficult	Some are dimmable	Controlled by varying voltage

Revision questions

1
  a Distinguish between conductors and insulators.
  b Name a good conductor which:
   i is solid but not a metal   ii is a solution.
  c Name TWO semiconductors and TWO insulators.

2 Draw a circuit diagram of a 1.5 V cell in series with TWO resistors, a switch, an ammeter to measure the current and a voltmeter to measure the voltage across ONE of the resistors.

3 Complete the following sentence.
  Ammeters have .... resistance and voltmeters have .... resistance.

4 Define:
  a potential difference (pd)   b current   c power.

5 A pd of 12 V exists across a resistor of resistance 4 Ω. Calculate:
  a the current
  b the power consumed by the resistor
  c the energy used in 5 minutes.

6 a State Ohm's law.

  b Use Table 11.8 to plot a graph of voltage vs current and calculate the resistance from its gradient.

  Table 11.8 *Voltage and corresponding current through a resistor*

Voltage/V	0.0	3.0	6.0	9.6	13.5	18.0
Current/A	0.0	1.0	2.0	3.2	4.5	6.0

7 List THREE advantages of parallel connection of devices in a domestic wiring system.

8 a Calculate the equivalent of 1 kW h in J.

  b Why do we use kW h instead of J when measuring electrical energy used in the home?

9 What is the reading on the analogue meter shown in Figure 11.23?

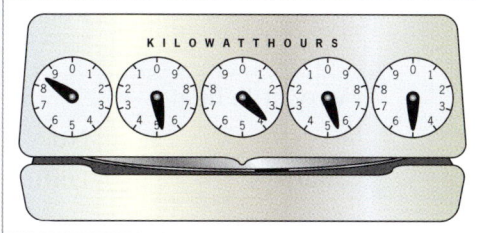

Figure 11.23 *Analogue electric meter*

10 a Distinguish between a fuse and a circuit breaker.

  b What is meant by the 'current rating' of a fuse?

  c Fuses of current ratings 5 A, 14 A, 20 A and 30 A are available.

  For a device rated at 120 V, 1500 W:

  i Calculate the current it should normally take.

  ii State, giving a reason, which fuse is suitable.

11 a State the electrical colour code used in the UK for the live, neutral and earth wires.

  b Which TWO wires carry the current under normal operation of a device?

  c If a short circuit occurs and a fuse 'blows', which TWO wires would then have carried the current?

  d One end of the earth wire is connected to the ground. What is the other end connected to?

12 Describe how you would wire a three-pin plug.

13 What is essential of the wires of a circuit carrying large currents?

14 a What is meant by the term 'electrocution'?

  b Compare the electrical conductivity of dry skin with:

  i wet skin   ii blood.

  c We can be protected from electrical shock if we are wearing rubber shoes. Explain.

15 a Define the term energy conservation.

   b State the TWO main reasons that energy conservation is important.

   c State TWO ways that energy can be wasted by faulty electrical appliances.

   d State THREE ways that energy consumption in the home can be reduced.

16 a Explain why coloured objects appear different in daylight than in artificial light.

   b What is the approximate temperature of the filament in an incandescent lamp as it emits light?

   c What colour is the light emitted by a filament lamp?

17 Complete Table 11.9.

Table 11.9 *Characteristics of various lamps*

Characteristic	Fluorescent	LED	Filament
Lifespan	Longer		Relatively short
Efficiency	High		
Wasted heat produced		Very little heat	
Contains toxic mercury			No
Shadow		Sharp	Diffused
Brightness control			

Electrical and fire hazards

Hazards associated with electricity and methods of preventing them

Illegal connection to the power grid

Illegal installers are usually uncertified and so the systems they design may overload, causing fires and/or electrocution. **Laws and fines** are put in place and **certified inspection** is mandatory.

Flying kites near power lines and/or in rainy conditions

Avoid flying kites in the rain or kites with metal parts. Release your kite if it drifts close to power lines. **Call the appropriate authorities** to remove it if it becomes entangled in power lines since it can become a good electrical conductor and can lead to electrocution.

Picking fruit or trimming trees near power lines and/or in rainy conditions

Trees in contact with power lines may be electrified and so we should never pick fruit from them. Never use metal poles or metal ladders to pick fruit in wet conditions even if not in contact with power lines, as lightning may strike at the tree and readily conduct through the metal. **Call authorised trained experts** to remove people stuck in power lines or in contact with trees touching power lines.

Working with electrical equipment in damp conditions

NEVER work on electrical equipment in the **rain**. Always wear **insulated boots** and **gloves**.

Faulty electrical equipment and overloading electrical circuits

Always **maintain electrical equipment** in good condition since **loose electrical contacts** or **damaged power cords** can produce **short circuits**, resulting in electrical fires and electrocution (see Figure 11.24).

Never **connect too many appliances** (particularly high-powered ones) to the same circuit or outlet bar since the **total current** may be sufficient to **melt the insulation** and cause electrical fires and/or electrocution. **Extension cords** should be **thick** to **take large currents** if necessary.

Figure 11.24 *Overloading a circuit*

CRT (cathode ray tube) TVs and screens

CRT TVs (bulky, older-designed TVs) can store **high voltages** that can cause electrocution. They use a **vacuum tube** enclosed by a **glass screen** containing **toxic lead.** Should the screen break, the glass fragments will implode and then rebound at high speeds, providing a hazard to anyone nearby. These devices should therefore only be **repaired by qualified technicians** and **recycled by a regulated body.**

EMF radiation from computers, cell phones, tablets and microwave ovens

EMF (electromagnetic field) radiation can cause minor **heating**, **headaches** and **fatigue. Avoid using CRT TVs** since they emit significant EMF radiation. Use **speakerphone or Bluetooth** instead of holding a cell phone to your head, and keep laptops away from your abdomen to avoid heating affecting your reproductive health. However, EMF radiation is **low-frequency** and **low-energy radiation** and is **non-ionising to our body cells**.

Microwave ovens emit EMF radiation. They produce **heating effects** but there is no evidence that they cause cancer. The oven's metal casing and the mesh grille in its glass door absorb the waves, preventing their escape. Only **qualified technicians** should repair these circuits as they retain high voltages.

Other hazards and methods of preventing them

X-ray machines and radioactive materials

Strongly energetic, **high frequency**, **ionising radiation** from these sources **damages body cells** and **causes cancer. Lead-lined absorbers** should be worn to protect parts of the body. Radioactive materials should be stored in **lead containers** and **labelled as dangerous.**

Faulty gas supplies

Combustible gases are **toxic** and can **ignite** when in contact with hot surfaces. Gas lines and the associated equipment should be properly **maintained**. The supply should be **turned off immediately** should there be a fire.

Overheating cooking oils

Monitor heated cooking oil to prevent it from igniting and causing **violently splattering flames.**

Treating the victim of an accident resulting in electrical shock

- **Switch off the electrical supply** or, if he or she is still attached to it, **separate the victim** from it using an insulated material.
- **Assess** the **victim's condition** and **seek medical help** with a phone on **speaker mode beside the patient**.
- If the victim's **heart has stopped beating** (cardiac arrest), and/or if he or she has **stopped breathing** (respiratory arrest), **cardiopulmonary resuscitation** (CPR) should be performed (see page 117).

Causes of electrical burns

- **Touching a high voltage** electrified object **whilst being 'grounded'** at some other point.
- Being in the **path of an 'arc blast'** produced between broken wires of a powerful electrical circuit.

Treating a victim with 1st degree burns

- Wash hands with **antibacterial soap** and then **remove clothing** not stuck to the burn.
- Soak the area with cool **water** (ice will cause further damage) and clean with mild soap.
- Soothe with aloe vera, cover with a **sterile non-adhesive** gauze and take a **pain reliever**.

1st **degree** burns only affect the **epidermis**. More **severe burns** should be **seen by a doctor**.

2nd **degree** burns damage the **dermis** and can **cause blisters**.

3rd **degree** burns harm **all layers** of the skin.

4th **degree** burns affect **all layers** of the skin along with **bones**, **muscles** and **tendons**.

Fires and various methods of extinguishing them

The fire triangle

The elements **FUEL**, **OXYGEN** and **HEAT** constitute the fire triangle (see Figure 11.25). They are required **together** to **start** and **sustain** a fire. Table 11.10 shows how these elements may be removed.

Figure 11.25 *The fire triangle*

Table 11.10 *Removing elements of the fire triangle*

Element	Method of removal
Oxygen	Cover with an inert material, for example, sand or carbon dioxide (CO_2), to separate oxygen from the fuel.
Heat	Add a material with a high capacity for absorbing heat, for example, cold water.
Fuel	Clear the area of the material being burnt or disconnect the supply if the fuel is a gas.

Some types of fires

Electrical fires – These are formed due to **short circuits** or **overloaded** circuits (see Figure 11.24).

Chemical fires – These occur in **laboratories** or **factories** and may release **toxic gases**. Figure 11.26 shows that firefighters use **atmospheric respirators** and **tanks of clean air** to extinguish these fires.

Bush fires – These are generally widespread in **dry periods** where **thick bush** can rapidly burn. **Natural winds** together with **convection winds** produced by the hot, rising gases increase the rate of combustion. Scorched regions suffer **plant and animal loss** and **stress ecosystems. The topsoil** erodes due to **wind in the dry season** and **water in the wet season**. These fires are extinguished by **qualified firefighters** who **clear the bush upwind** of the fire. Figure 11.27 shows that large fires are extinguished by aircraft dropping **water bombs** and **fire-retardant chemicals**.

Figure 11.26 *Extinguishing a chemical fire using an atmospheric respirator*

Figure 11.27 *Extinguishing a bush fire using water bombs*

Types of fire extinguishers

A **fire extinguisher** is a **small** and **portable, hand-held** vessel containing a wet or dry **chemical agent under pressure** which can be discharged to extinguish small fires.

Figure 11.28 shows types of fire extinguishers.

Classes of fires

Table 11.11 shows that the extinguishing agent used to stop a fire depends on the class of the fire.

Figure 11.28 *Types of fire extinguishers*

Table 11.11 *Extinguishing agents used with various classes of fire*

Class of fire	Example of fuel	Extinguishing agent	Additional information
A Common combustible solids	Bush, rubber, cloth, paper, trash, plastic	Water, foam, ABC powder, sand	CO_2 is unsuitable since gases absorb very little heat. Clearing unburnt material removes fuel.
B (i) Flammable liquids	Gasoline, oil, grease, paint, solvents	CO_2, foam, ABC powder, sand	Water cannot be used since it will splatter, spreading the fire.
B (ii) Flammable gases	Propane, butane, methane	ABC powder	ABC powder smothers the flames but the gas supply MUST be switched off.
C Electrical equipment	Faulty electrical equipment	CO_2, ABC powder	Water conducts electricity and so can cause electrocution.
D Flammable metals	Lithium, sodium, potassium, magnesium	Dry powder specific to the type of metal	Water and CO_2 are unsuitable since they may react with these reactive metals.
K Organic oils and fats	Cooking oils, animal fats	Wet chemical agent, fire blanket	Fire blankets are useful but may not be effective.

Table 11.12 *Summary of extinguishing agents used for different classes of fire*

Class of fire		Water	Foam	Carbon dioxide	ABC powders	Specific powders	Wet chemical	Fire blanket
A	Common combustible solids	✓	✓		✓		✓	✓
B	Flammable liquids		✓	✓	✓			✓
	Flammable gases				✓			
C	Electrical equipment			✓	✓			
D	Flammable metals					✓		
K	Organic oil/fat						✓	✓

Methods of extinguishing a fire

To extinguish a fire, we must remove at least one of the three elements of the 'fire triangle'.

Removing heat – Use a **cold** substance with **high heat absorption** properties, such as water, foam or a wet chemical agent. All extinguishing agents remove heat to some extent.

Removing oxygen – Table 11.12 shows that several agents are used to displace oxygen.

- **ABC powders** are dry powders used on classes **A**, **B** and **C** fires.
- **Special dry powders** are used on class **D** metal fires to withstand extremely high temperatures.
- **Fire blankets** are **flexible** and **heat resistant** and **are made of unreactive material** (usually woven fibreglass). They can be used for small class **A** and **B** fires and for class **K** fires.
- **Wet chemical agents** form a **soap-like barrier** over hot cooking oils and fats of class **K** fires.

Removing fuel – Switch off electrical supplies or gas supplies. Clear the area of combustible materials.

Conventional protective gear/wear in work and in sport

Table 11.13 shows that protective gear is used to **increase productivity** and to **protect** us from **physical injury**, **infection**, **dangerous chemical exposure**, **harmful radiation** and **hot materials**.

Table 11.13 *Applications of protective gear/wear in work and sport.*

Gear/wear	Protection	Examples of users
Gloves	• Thick, padded, heavy duty, leather gloves protect the hands from being **cut** and **bruised**.	• Construction workers
	• Gloves made of sterile latex, nitrile rubber, PVC or neoprene protect from **germs** and **viruses**.	• Doctors, nurses, chemists
	• Anti-corrosive gloves protect the hands from **corrosive chemicals**.	• Farmers, gardeners
	• Heat insulating gloves protect the hands from **hot objects**.	• Cooks
Boots	• Heavy duty leather boots, usually with a steel cover over the front, protect the feet from being **cut** and **bruised**.	• Construction workers, farmers, gardeners
	• Waterproof and anticorrosive rubber boots protect against some **pathogens** and **dangerously reactive chemicals**.	• Fire fighters, gardeners
	• Light duty sports boots protect the feet from being bruised.	• Cricketers, footballers, hikers

Gear/wear	Protection	Examples of users
Goggles	• Protect the eyes from **airborne particles**, **splashing chemicals** and **water**.	• Swimmers, chemists, gardeners
	• Dark goggles or visors protect the eyes from **sparks** and **ultraviolet radiation**.	• Welders
Earmuffs	• Protect the ears from **loud noises** and/or **low temperatures**.	• Construction workers, skiers, airport workers
Helmets	• Protect the head from **cuts**, **bruises** and **skull** and **brain damage** by absorbing shock and preventing penetration.	• Construction workers, cyclists, climbers
	• Some have visors and cages to protect the face from **airborne objects**.	• Cricketers, hockey players
Coats/aprons	• Protect against **caustic chemicals**, **hot materials** and **ionising radiation** (for example, **X-rays**) in **laboratories**, **hospitals**, **industrial sites** and **kitchens**.	• Doctors, nurses, chemists, factory workers, cooks, surgeons
Respirators	• **Filter particles** such as **dust** and **smoke** from the air.	• Construction workers, fire fighters
	• Some have **canisters** attached to **filter harmful gases** and **toxic sprays**.	• Painters, fumigators
	• Some **provide clean air** for environments with limited air.	• Astronauts, fire fighters
Chest guards	• Protect against obtaining **broken bones** and **damaged organs**.	• Racing drivers
Back braces	• Support and **protect the spine** when lifting **heavy objects**.	• Construction workers, dock workers, weightlifters
Groin boxes	• Protect the groin from **fast-moving cricket balls**.	• Cricketers
Sports pads	• Protect the **elbows**, **knees** and **shins** from the impact of **collisions**.	• Cricketers, skateboarders

Revision questions

18 List THREE ways we can help a victim who has received a severe electrical shock.

19 List TWO ways that someone can receive electrical burns.

20 What action should be taken in the following cases?

   a A kite drifts near to a power line.

   b A kite becomes tangled in a power line.

   c Stormy conditions approach with heavy rain as we are flying our kite.

21 **a** Why can faulty circuits in microwave ovens be fatal if not repaired by qualified technicians?

   b How are electromagnetic field (EMF) waves prevented from exiting a microwave oven?

   c Name TWO other common devices that emit EMF waves.

   d Contrast the frequencies and ionising capabilities of X-rays with those of EMF waves.

   e Which of X-rays and EMF waves is considered more harmful?

22 **a** Explain the significance of the elements of the fire triangle.

b Complete Table 11.14 by giving ONE example of EACH of the materials (fuels) being burnt and place ticks (✓) in the appropriate boxes to indicate the type/types of suitable extinguishers.

Table 11.14 *Extinguishers for various types of fire*

Class of fire		Fuel	Water	Foam	Carbon dioxide	ABC powders	Specific powders	Wet chemical	Fire blanket
A	Common combustible solids	wood	✓	✓		✓		✓	✓
B	Flammable liquids								
	Flammable gases								
C	Electrical equipment								
D	Flammable metals								
K	Organic oil/fat								

23 Which element of the fire triangle is mainly removed in EACH of the following cases?

a Clearing the area of combustible material.

b Covering with an inert material.

c Adding a material of high specific heat capacity.

24 **a** State an action that may be taken to protect against X-rays and radioactive materials.

b List TWO dangers presented by faulty gas supplies.

c Why is water unsuitable for extinguishing a fire of:

   i flammable liquids?   **ii** electrical equipment?   **iii** combustible metals?

25 Complete Table 11.15.

Table 11.15 *Protection from and users of various types of gear/wear*

Gear/wear	Protection from	Example of user
Gloves made of sterile latex		
Heavy duty, steel-tipped leather boots	Damage to feet, particularly toes	
Dark goggles or dark visors		Welder
Respirators with attached canisters		Fumigator

12 Temperature control and ventilation

An understanding of heat transfer processes together with the thermal properties of materials is essential for the design of common devices such as thermostats, stoves, refrigerators and water heaters. Techniques of providing adequate ventilation are also important as the hot and humid Caribbean climate contributes to health issues and to the deterioration of structures.

Processes of heat transfer

Conduction is the process of heat transfer through a medium by the **relaying of energy** between **particles colliding** with each other.

Convection is the process of heat transfer within a medium by **the movement** of particles between regions of **different density**.

Thermal radiation is the process of heat transfer by means of **electromagnetic waves**.

Conduction

Conduction explained by kinetic theory

Non-metals

When one end of a non-metallic bar is heated, the particles there absorb energy and **vibrate faster**. The increased kinetic energy **relays** by **collisions** between the particles along the bar. Since **temperature** is proportional to the **kinetic energy** of the particles, the bar becomes hotter (see Figure 12.1a).

Metals

Metals contain vibrating **cations** that transfer energy similarly to the particles of non-metals. However, metals also possess **free-moving electrons**. When heated, these electrons **translate faster**, gain **kinetic energy** and collide into cations with **greater force**, **relaying** the energy along the bar (see Figure 12.1b). This **extra mode** of conduction makes **metals better conductors** than non-metals.

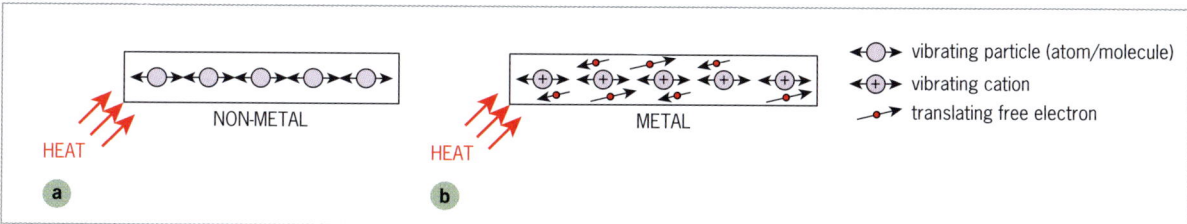

Figure 12.1 *Conduction and kinetic theory*

Comparing thermal conduction through solids

Figure 12.2a shows three rods of equal length, each with one end in a water bath and a piece of wax placed at the other end. On heating the water bath, the wax melts first from the copper and last from the wood, showing that copper is the best conductor of the three materials and wood the worst.

Figure 12.2b shows paper, tightly wrapped around the region where rods of copper and wood are joined. On heating the region, the paper only chars over the wood. Unlike copper, wood is a poor thermal conductor and so heat conducted to it **collects at the interface**, **charring the region**.

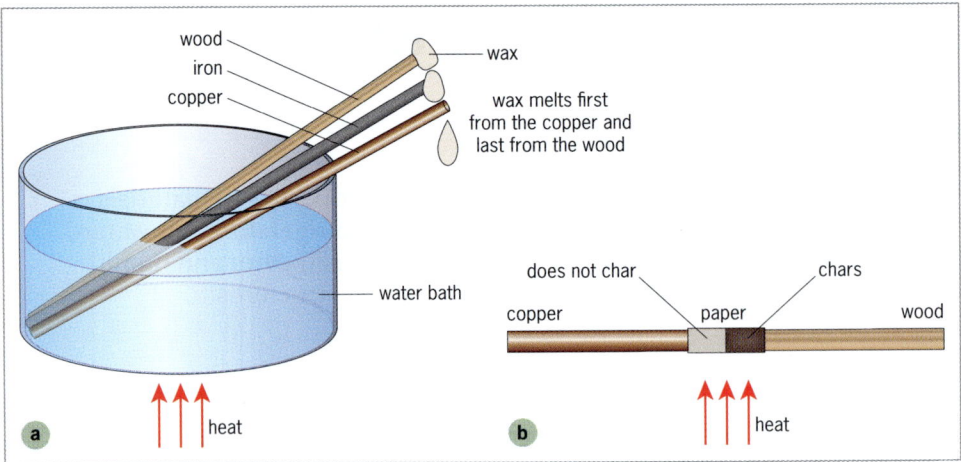

Figure 12.2 *Comparing thermal conduction through solids*

Water is a poor thermal conductor

Figure 12.3 shows a piece of ice wrapped in a heavy copper mesh submerged in water. When the **water surface** is **heated**, **heat transfers to the ice only by conduction** as heat **cannot flow downward by convection**. The water surface boils rapidly while the ice remains solid, showing that water is a **poor conductor**.

Good and poor thermal conductors

Metals are **good thermal conductors** and are used in many devices.
- Pots, pans and the bases and sides of kettles are made of copper, aluminium and stainless steel.
- Large boilers are generally made of steel.
- Radiators are usually made of aluminium to conduct heat efficiently to their outer surfaces.

Non-metals are **poor thermal conductors.**
- Handles of cooking utensils, pots and pans, and garden tools are made of plastic or wood.
- Heat-resistant silicone gloves (lined with cotton for comfort) are used to handle hot dishes.
- Water heater tanks, refrigerators and ovens lagged with polyurethane reduce outward conduction.
- Igloos reduce outward conduction of heat since ice is a poor thermal conductor

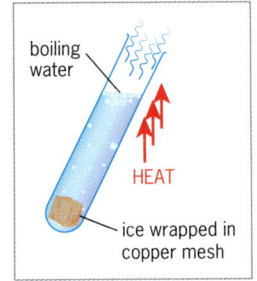

Figure 12.3 *Water is a poor conductor*

Air is a **poor thermal conductor.**
- Woollen jackets have air pockets to reduce heat conduction outward from the body.
- Expanded polystyrene contains air and so is used to insulate floors, ceilings and walls.
- Roofs made with leaves, such as those of palm trees, trap air and so can insulate shelters.
- Feathers, air and fur insulate animals from the cold by trapping air.
- Concrete blocks have air pockets to reduce conduction into and out of buildings.

Convection

Convection explained by kinetic theory

When a **liquid** or **gas** is heated, its molecules **acquire more kinetic energy**, **spread further** and become **less densely packed**. Figure 12.4 demonstrates the formation of a **convection current** that results due to **differences in density**.

Demonstrating convection in liquids and gases

Liquids

Figure 12.5a shows the path of a convection current in a liquid, revealed by the **purple solution** from the dissolved **crystal** as it is **heated from below**.

Small aluminium flakes used instead of soluble crystals will also show the path of the current.

Gases

Figure 12.5b illustrates how a fan is made to spin by a convection current in air.

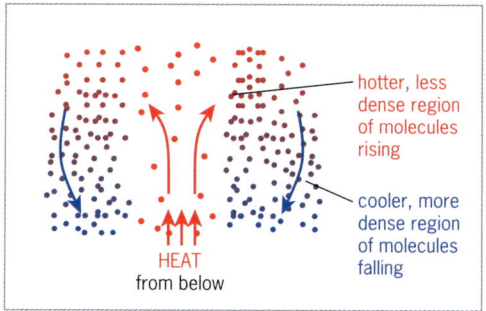

Figure 12.4 *Convection and kinetic theory*

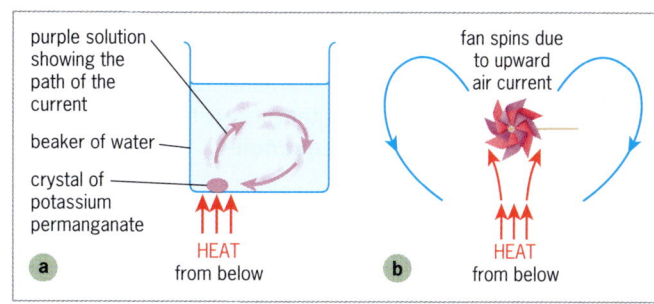

Figure 12.5 *Demonstrating convection*

Phenomena involving convection

Land and sea breezes moderate coastal climate

Day

During the day, the Sun's **radiation** warms the land more than it warms the sea. Air in contact with land is heated by **conduction** from the land, becomes **less dense** and **rises** by **convection**. **Cooler, denser** air then **blows onshore** to take its place, preventing the coast from becoming too hot (see Figure 12.6a). **Inland regions become very hot** because air there cannot rise as there is no nearby cooler air to take its place.

Night

At night, the land **radiates** more than the sea and so cools faster. Air over the sea is heated by **conduction** from the sea, becomes **less dense** and **rises** by **convection**. **Cooler, denser** air blows **offshore** to take its place, preventing the coast from becoming too cold (see Figure 12.6b). Air over **inland regions is cold** and still since there is no nearby rising air for it to take the place of.

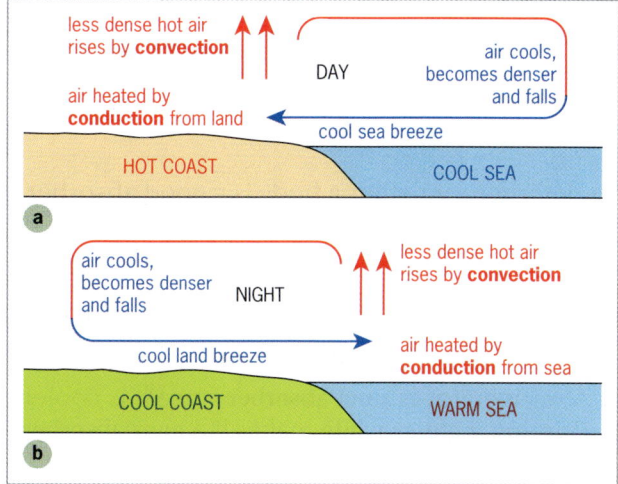

Figure 12.6 *Land and sea breezes*

Other applications of convection currents

- Air conditioners are placed near the ceiling so that the cooler, denser air can fall.
- Heaters are placed on the bases of vessels so that the heated, less dense fluids (including air) can rise.

12 Temperature control and ventilation

- Water conducts heat to ice which floats in it. The cooler, denser water falls and forces warmer water upward to be chilled.
- Water evaporating from lakes and oceans rises by convection, cools, and condenses as rain.
- Figure 12.7 shows how miners can draw fresh air into tunnels by generating convection currents.

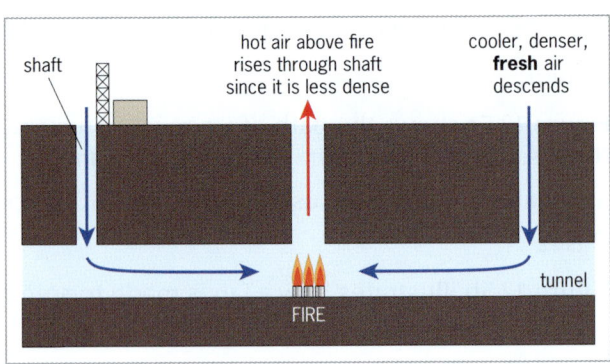

Figure 12.7 *Fresh air from convection currents*

Thermal radiation

Thermal radiation is **electromagnetic radiation** (such as infrared, visible light and ultraviolet) emitted by a body **due to its temperature** (see Chapter 8). **Hotter bodies emit higher frequencies**.

Emitters and absorbers of thermal radiation

- Bodies that are **hotter than their surroundings** are **net emitters** (emit more than they absorb).
- Bodies that are **cooler than their surroundings** are **net absorbers** (absorb more than they emit).
- A **good absorber** is a **poor reflector** and vice versa.

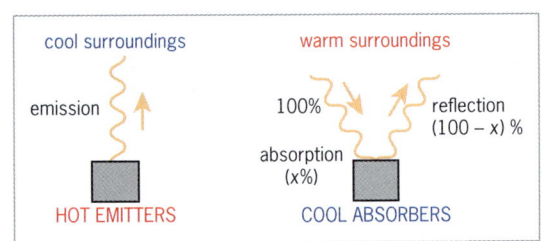

Figure 12.8 *Emitters and absorbers of thermal radiation*

Effect of colour and texture on emission and absorption of thermal radiation

Table 12.1 *The colour and texture of a surface affects its absorption and emission of thermal radiation*

Matt/dull/rough/black	Shiny/smooth/polished/silver (or white)
Good absorbers (poor reflectors)	Poor absorbers (good reflectors)
Good emitters	Poor emitters

To determine whether a body is a good absorber or good emitter of thermal radiation, use the following steps.
- Determine whether the body is a **net absorber** or **net emitter** (see Table 12.1).
- Then examine its **surface texture** and **colour** to determine how good an absorber or emitter it is.

Examples include the following.
- A refrigerator is a **net absorber** since it is cooler than its surroundings.
  Painting it **glossy white** makes it a **poor absorber** (good reflector).
- An oven is a **net emitter** since it is hotter than its surroundings.
  Painting it **glossy white** makes it a **poor emitter.**

Applications utilising the processes of heat transfer

The solar water heater

Heating cycle

During the day, hot water heated in the panel rises by **natural convection** to the tank, and cool water in the tank falls to the panel to be heated. Large tanks may be placed at **ground level** but will require an **electric pump** to force hot water downward to it from the panel.

Usage cycle

Turning on the hot tap causes the water mains to push water from the tank to the user. Figure 12.9 explains the function of the solar water.

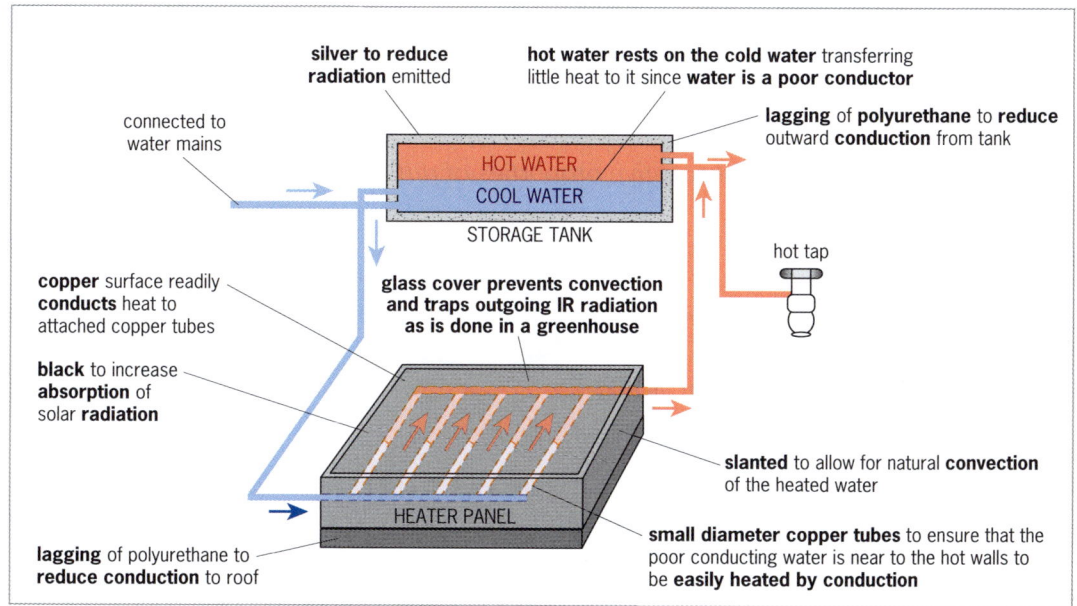

Figure 12.9 *Solar water heater*

Other applications of heat transfer processes

- Figure 12.10 shows how a **vacuum flask** reduces heat transfer to or from its contents.
- **White houses** keep us **cool in summer** since they are then **poor absorbers** of thermal radiation from the **hotter surroundings**.

  They keep us **warm in winter** since they are then **poor emitters** of thermal radiation to the **cooler surroundings**.
- **Cricketers** wear **white** clothes to be **poor absorbers** of solar radiation from the **hotter surroundings**.
- **Fire fighters** should wear **silver** suits to be **poor absorbers** of intense thermal radiation from the hotter surroundings.
- **Astronauts** wear **silver** suits so that when in the path of solar radiation, the **intense rays** are **poorly absorbed** preventing them from overheating. When not in the path of solar radiation the suits act as **poor emitters**, keeping the astronauts warm.

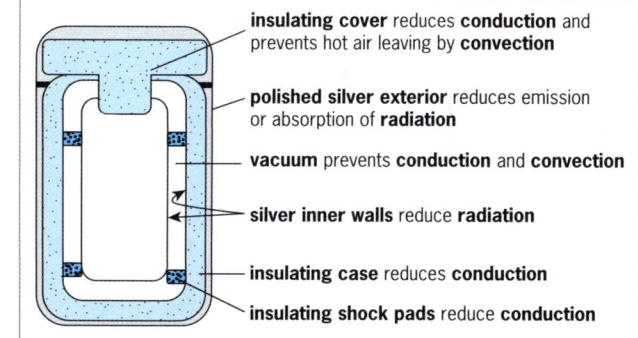

Figure 12.10 *Vacuum flask*

Comparing the emissive and absorptive properties of different surfaces

Figure 12.11a illustrates that although the matt black surface and the shiny silver surface are both at the temperature of the hot water, the matt black surface is emitting more radiation.

Figure 12.11b shows that although both **inner** surfaces receive equal radiation, the matt black surface absorbs more (reflects less), causing it to become hotter and to melt the wax on its **outer** surface.

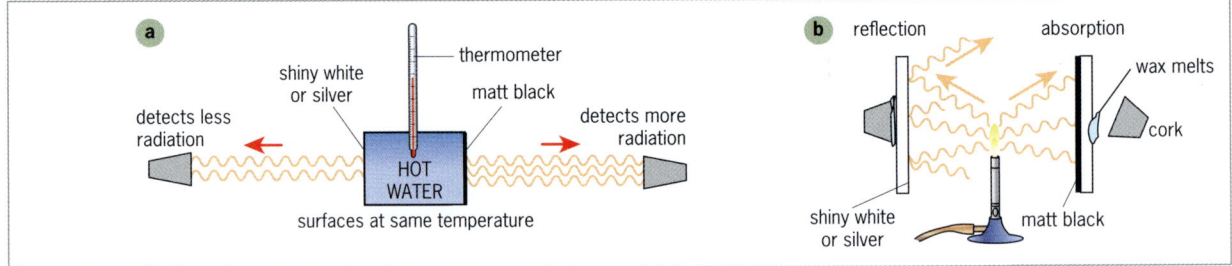

Figure 12.11 *Comparing the emissive and absorptive properties of different surfaces*

The bimetallic strip

A **bimetallic strip** *consists of two strips of different metals joined along their length.*

The metal that expands more when heated will also contract more when cooled. Figure 12.12 shows the effect of heating and cooling a straight bimetallic strip. **Copper**, **aluminium**, **zinc**, **tin** and **brass** have high **expansivity** (expand significantly when heated) while **steel** and **invar** have low expansivity.

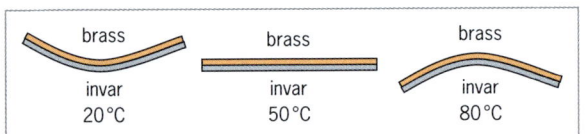

Figure 12.12 *Effect of temperature on bimetallic strip*

Thermostats in household appliances

A **thermostat** *is a device that automatically regulates temperature, or that activates or deactivates a device at a certain critical temperature.*

Thermostat used in an electric oven

Figure 12.13 shows an electric thermostat. As the heater warms, the bimetallic strip curves to the right, separates the contacts and breaks the circuit, preventing the temperature from rising above a set value. By **advancing the screw**, we can change this set temperature since more heat would then be required to break the circuit. With the heater disconnected, the strip cools and straightens until it reconnects the circuit and restarts the heating. **By switching** the positions of the **brass and invar**, the thermostat can be used to prevent a fridge from becoming too cold.

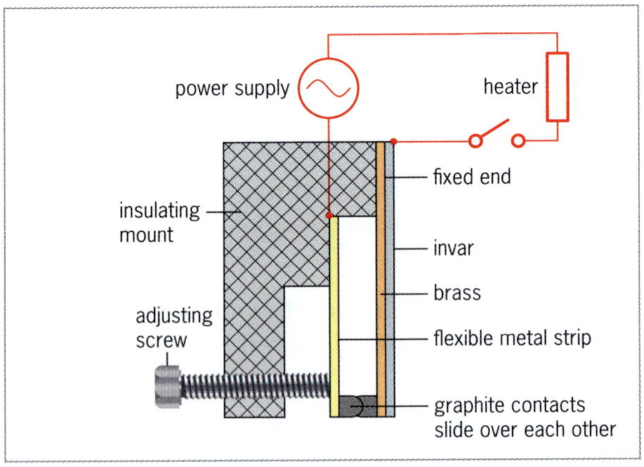

Figure 12.13 *Thermostat used in an electric oven*

Thermostat of an electric iron

Figure 12.14 shows that this works like the thermostat of an electric oven. Advancing the screw forces the end of the strip downward. Additional heating is then required before it curves sufficiently to break the circuit.

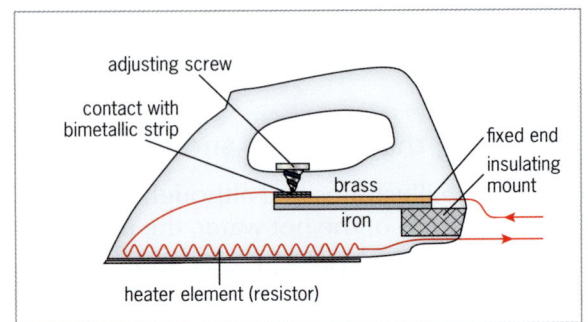

Figure 12.14 *Thermostat used in an electric iron*

Thermostat of a gas oven

Figure 12.15 shows a thermostat used in a gas oven. As the temperature rises, the brass tube expands more than the invar rod. This pulls the valve head to the left, narrowing the channel. The gas flow reduces, and the temperature stabilises.

By advancing the adjustable knob, the valve head moves to the left and reduces the flow of gas further. This feature allows the thermostat to be set for different stabilising temperatures.

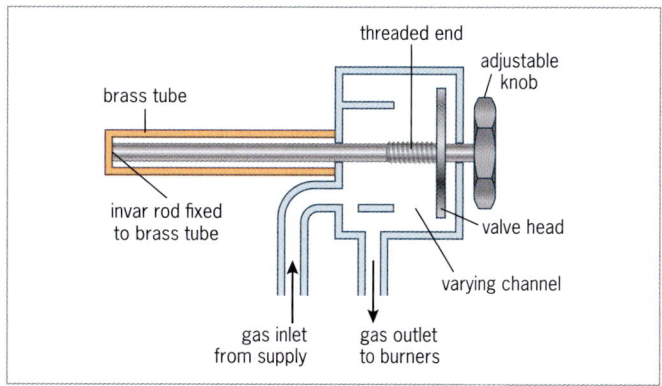

Figure 12.15 *Thermostat used in a gas oven*

Temperature and thermometers

Temperature is the degree of hotness of a body measured on a chosen scale.

- The **temperature** of a body **increases** with the **speed** and **kinetic energy** of its **particles**.
- **Thermometers** are instruments that measure temperature. The **SI unit** of temperature is the **kelvin (K)**, but temperature is also expressed as **degrees Celsius (°C)**.
- The **freezing point** of water is **0 °C** and its **boiling point** is **100 °C**.

Liquid-in-glass thermometers

Laboratory mercury thermometer and alcohol thermometer

Figures 12.16 and 12.17 show that these thermometers are typically used over **different temperature ranges**. The range of the **mercury thermometer** is suitable for most work in the **school laboratory**.

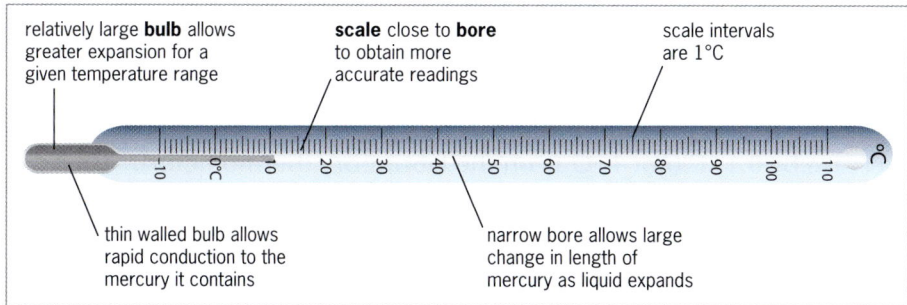

Figure 12.16 *Liquid-in-glass laboratory mercury thermometer*

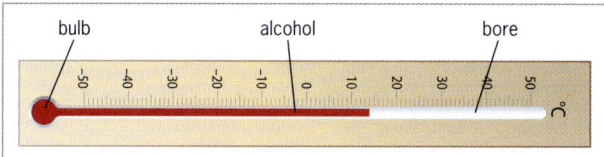

Figure 12.17 *Liquid-in-glass alcohol thermometer*

Advantages of using mercury instead of alcohol in a thermometer

- Mercury, being a **metal**, has **higher conductivity** and so **responds faster** to temperature change.
- Relative to alcohol, mercury **requires less heat** to raise its temperature. The temperature of the **body being measured** is therefore **hardly affected** by the measurement.

- Mercury has a **higher boiling point** and so can **measure higher temperatures** before it vaporises.
- Mercury is **bright silver** and can be **seen easily** whereas alcohol is colourless and must be tinted.

Advantages of using alcohol instead of mercury in a thermometer
- Alcohol is **cheaper.**
- Alcohol is **non-toxic** in moderate amounts, but mercury can be highly toxic.
- Since the **expansivity** of alcohol is **six times that of mercury**, an alcohol thermometer with the same sized bulb can be longer with the degree **intervals spread further apart** on the scale.
- Alcohol has a **lower freezing point** and so can **measure lower temperatures** before it solidifies.

Example 1

Calculate the temperature indicated by the thermometer in Figure 12.18.

Solution

25 cm represents a change of 100 °C

∴ 1 cm represents a change of $\frac{100\ °C}{25}$, that is, 4 °C

∴ 15 cm represents a change of (4 × 15) °C = 60 °C

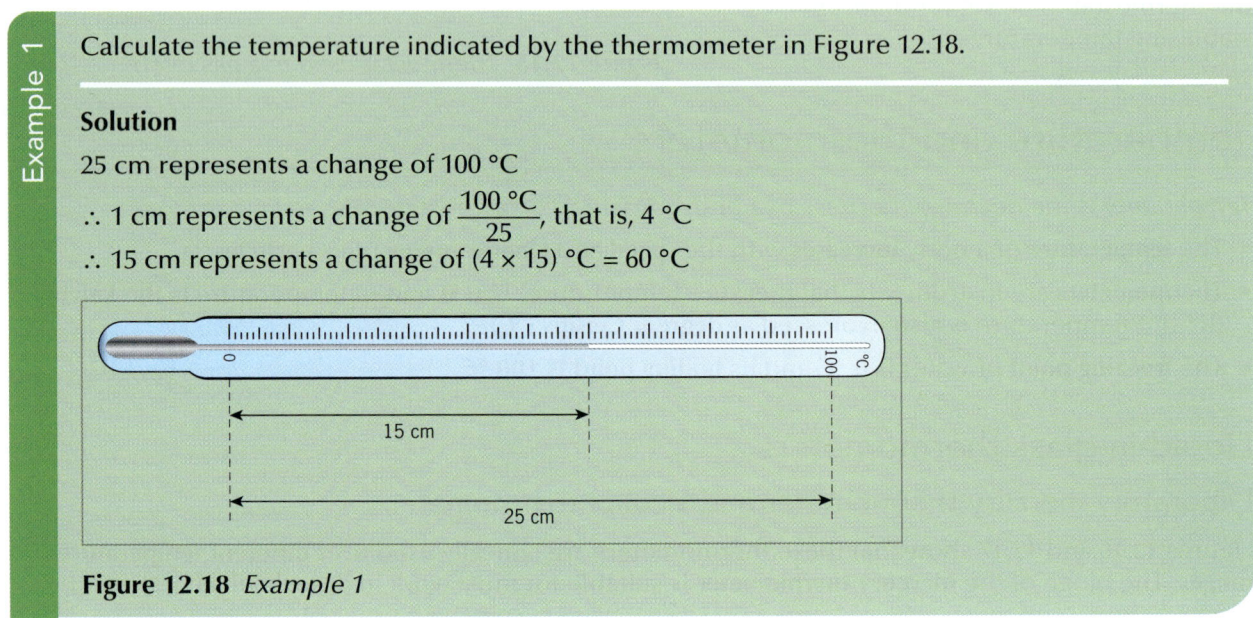

Figure 12.18 *Example 1*

Clinical liquid-in-glass mercury thermometer

- The temperature range of living humans is from **35 °C to 43 °C**.
- The markings on its scale increase by **0.1 °C** (not 1 °C as on the laboratory thermometer).
- On removal from the patient, the mercury quickly **cools** and **recedes** to the bulb, **breaking** at the **constriction** and leaving the mercury above it to be read.
- Alcohol is **unsuitable** in these thermometers since **it would not break at the constriction**.
- Due to **mercury's toxicity** these thermometers are being **replaced** by **digital thermometers**.

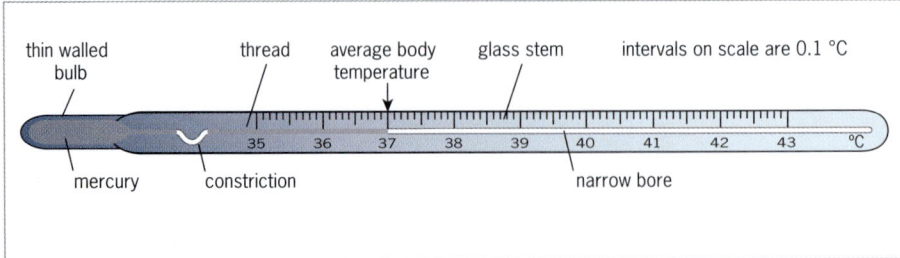

Figure 12.19 *Clinical thermometer*

Digital thermometers

These thermometers have **temperature sensors** that produce **voltages** to output the corresponding temperature readings. There are several types of digital thermometers dependent on different sensor capabilities. Digital thermistor thermometers and digital infrared thermometers are two such types.

Advantages of digital thermometers are that they are capable of **rapid response**, are generally **accurate**, and their output is **easily read on a digital display**. Their main disadvantage is that they **depend on batteries** that must be periodically replaced and which increase waste.

Digital thermistor thermometer – The tip, which contains a temperature-sensitive resistor, is placed on the patient. Figure 12.20 shows such a thermometer.

Digital infrared thermometer – Detects temperature on receiving infrared radiation. It is **hygienic** to use because it makes **no contact with the patient**.

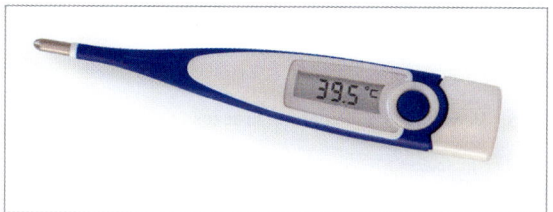

Figure 12.20 *Digital thermistor thermometer*

Temperature regulation in humans

Humans must maintain a **constant** internal body temperature of about **37 °C** for **enzymes** to function properly, and **sweating** is important in reducing a person's body temperature if it **rises** above normal (see pages 50 to 51). As the **water** in sweat **evaporates**, it takes heat energy from the body, thereby **cooling** the body. This is because converting a liquid to a vapour requires **energy** to overcome the attractive forces between the liquid molecules. This energy is known as the **latent heat of vaporisation**.

Latent heat of vaporisation is the amount of heat energy that is needed to convert a unit mass of a liquid into a vapour without changing its temperature.

Water has a **high** latent heat of vaporisation because the attractive forces between its molecules are **strong**. When water in sweat evaporates, it removes large amounts of **heat energy** from the skin.

Effect of changes in body temperature on metabolic rate

Chemical reactions occur constantly in all living cells to sustain life. These reactions are referred to as the body's **metabolism** and the rate at which they occur is called the **metabolic rate**. Metabolic rate is a measure of the amount of **energy per unit time** that a person needs to keep the body functioning. **Changes** in a person's **body temperature** can affect this rate because the rates of chemical reactions and of enzyme activity are both affected by temperature (see page 110).

- If the body temperature **rises** above normal, e.g. during exercise, illness or when the environmental temperature is high, a person's metabolic rate **increases** because as temperature increases, the rate of chemical reactions and enzyme activity increase.
- If the body temperature **drops** below normal, e.g. when the environmental temperature is low, a person's metabolic rate **decreases** because as temperature decreases, the rate of chemical reactions and enzyme activity decrease.

Ventilation

Ventilation is the process by which clean air is provided to a space as stale air is removed from it.

To remain healthy and comfortable, the air around us should have the following properties.
- Contain adequate **oxygen**.
- Have adequate **humidity**.
- Be at an adequate **temperature**.
- Be free of **dust**, **pollen** and **toxic chemicals**.
- Be free of **mould** and **bacteria**.

*Natural ventilation is ventilation produced by **natural breezes** or **natural convection currents**.*

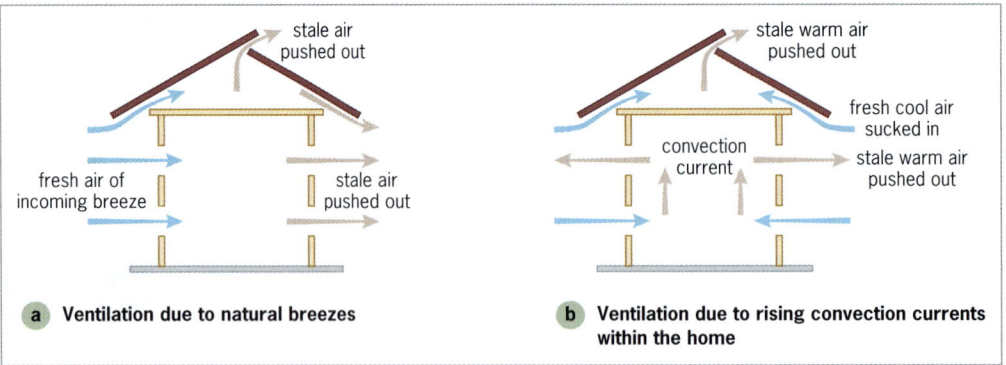

Figure 12.21 *Design features promoting natural ventilation*

Mechanical ventilation is ventilation produced by appliances such as fans and air purifiers.
- **Bathroom exhaust vents** use fans to remove foul air.
- **Kitchen exhaust vents** above cookers use fans to remove hot air and contaminants.
- **Fans** cool us by evaporating sweat from our skin and by removing warm, moist air from around us.
- **Air conditioners** replace warm, moist air with cool, less humid air.
- **Dehumidifiers** are used in very humid rooms such as cellars to prevent the build-up of mould.
- **Air purifiers** use filters to remove airborne pollen, mould spores, bacteria, viruses and smoke.
- **Humidifiers** are used in rooms where the air is too dry.

Problems of inadequate ventilation of enclosed spaces
- **Respiring replaces oxygen** with carbon dioxide and water vapour, making breathing difficult.
- **Exhaled water vapour causes increased humidity**, reducing the cooling effect of evaporation.
- **Microorganisms** thrive in warm, humid environments, creating an unhealthy, musty odour.
- **Toxins and dust** interact negatively with our bodies.

Temperature and humidity

*Relative humidity is the amount of water vapour in the air expressed as a percentage of the total amount it can hold **at that temperature**.*

Caribbean climates are hot maritime climates with relative humidities **between 60% and 90%**, which is far above the **recommended level** of **between 30% and 50%**.

Negative effects of high humidity
- Leads to **decay of buildings**, especially wooden structures.
- Promotes development of **fungi** and **mould**, which can lead to allergic reactions.
- Promotes the development of **bacteria** and **viruses.**
- Produces **distasteful odours.**
- Encourages the growth of **dust mites.**
- Can cause **dehydration** and **heatstroke** if excessive perspiration cannot evaporate.

Negative effects of low humidity

- Can **dehydrate mucous membranes** of the nose and throat.
- Can cause materials to **shrink**, become **brittle** and **crack**.
- Can cause materials to become **electrically charged**, resulting in the **attraction of dust** and the destruction of semiconducting devices.

Revision questions

1. Define:
   a conduction   b convection   c thermal radiation.

2. a Use kinetic theory to explain conduction in non-metals.
   b Explain why metals conduct better than non-metals.
   c Describe how you can demonstrate that water is a poor thermal conductor.

3. a Explain why coastal temperatures are moderated by sea breezes during the day.
   b Explain why air conditioner wall units should be mounted more than 2 m above the floor.

4. Create a table to compare the absorptive and emissive properties of dull black and shiny silver surfaces.

5. a What colour clothes will keep you warmer if you stand:
      i near to a bush fire?   ii in a freezer at an ice-cream factory?
   b Give a reason for each answer in a.

6. a Explain the following features of a solar water heater in terms of the **type** of **heat transfer** applicable.
      i The storage tank is placed above the heater panel.
      ii The storage tank is shiny silver, but the copper tubes in the heater panel are black.
      iii Polyurethane is placed under the heater panel and on the inner walls of the storage tank.
      iv Hot water rests on top of cold water in the tank without transferring much heat to it.
   b Give TWO reasons for having a glass sheet over the heater panel.

7. Answer the following questions about a thermos flask in terms of the type of heat transfer applicable.
   a What is the importance of the vacuum?
   b Why are the inner-facing walls of the vacuum shiny silver?
   c How is conduction reduced in this system?

8   **a** What is a thermostat?

b When heated, copper, aluminium, zinc and tin expand significantly while steel and invar expand very little. Select two pairs of metals from this list which are suitable for making a bimetallic strip.

c In Figure 12.22, in which direction will end A move when it is:

    i heated?   **ii** cooled?

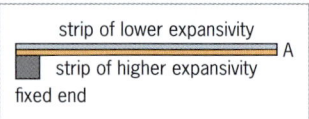

Figure 12.22 *Revision question 8*

9 Explain each of the following.

a Mercury is more suitable than alcohol for use in a liquid-in-glass school laboratory thermometer.

b A liquid-in-glass CLINICAL thermometer must contain mercury and not alcohol.

c The range of a liquid-in-glass thermometer is usually between 35 °C and 43 °C.

d The intervals on the scale of a clinical thermometer must be very small; typically, 0.1 °C.

10   **a** Name TWO types of digital thermometer.

b What can be said of the sensitivity, accuracy and ease of reading these thermometers?

c Why are digital thermometers now replacing clinical liquid-in-glass thermometers?

11 Provide a definition for 'latent heat of vaporisation' and explain the significance for the human body of water's high latent heat of vaporisation.

12 Outline the effects that any changes in body temperature have on the body's metabolic rate.

13   **a** Define ventilation.

b Distinguish between natural ventilation and mechanical ventilation.

14   **a** List FIVE properties of a well-ventilated room.

b List FOUR ways that enclosed spaces can result in inadequate ventilation.

15   **a** Define relative humidity.

b Why do Caribbean climates have high humidities?

c State:

    i FOUR negative effects of high humidity

    ii TWO negative effects of low humidity.

Exam-style questions – Chapters 8 to 12

1 **a)** **i)** Describe the energy transformation occurring as a horizontal spring is stretched by a person pulling on it. **(1 mark)**

   ii) Complete the paragraph below to describe the energy changes in a hydro-electric power station.

   Water trapped in an elevated reservoir has .... energy. When released, it flows downhill gaining .... energy. Friction due to the motion of the water causes a transformation to .... energy. The water, however, continues to move rapidly and presses on the fins of a turbine, causing it to turn and transferring .... energy to it. This motion relays to a generator which transforms the energy mainly to .... energy with a little wasted as .... energy. **(3 marks)**

b) A force of 20 N pushes an object of weight 40 N through 4 m in the direction of the force. Calculate:

   i) the work done   **ii)** the energy used. **(3 marks)**

c) Explain why the weight of the object is not considered in the calculation. **(2 marks)**

d) There is a need to increase the efficiency and reduce the pollution caused by the internal combustion engine.

   i) Describe THREE design features that car manufacturers can implement to achieve this. **(3 marks)**

   ii) List THREE ways drivers can reduce the energy their vehicles consume. **(3 marks)**

   Total 15 marks

2 **a)** Figure 1 summarises the process of photosynthesis.

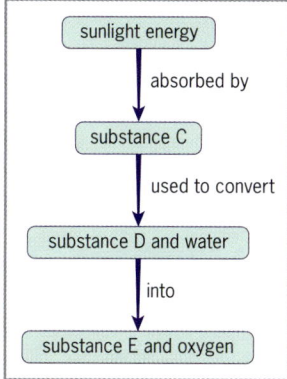

Figure 1 *The process of photosynthesis*

   i) Identify substances C, D and E. **(3 marks)**

   ii) Write a balanced chemical equation to summarise the process of photosynthesis. **(2 marks)**

   iii) During photosynthesis, light energy is converted into another form of energy. What form is this? **(1 mark)**

b) Table 1 shows the food sources of several organisms found in the ocean.

Table 1 *Food sources of some organisms found in the ocean*

Organism	Food source
zooplankton	phytoplankton
shrimp	phytoplankton
jellyfish	zooplankton and shrimp
crab	shrimp
sea turtle	crab and jellyfish

    i) Using only the information contained in Table 1, construct a food web for the organisms. **(2 marks)**

    ii) Identify from the food web ONE herbivore and ONE secondary consumer. **(2 marks)**

    iii) Decomposers are not usually shown in food webs, however they are essential in any ecosystem. What are decomposers and what is their role within an ecosystem? **(2 marks)**

c) Provide a detailed explanation of why the number of organisms decreases at successive levels in a food chain. **(3 marks)**

Total 15 marks

3 a) i) What are enzymes? **(2 marks)**

    ii) The enzyme amylase breaks down starch during chemical digestion. Identify TWO regions of the digestive system where amylase is active. **(2 marks)**

    iii) Manon carried out an experiment to determine the effect of pH on the activity of amylase. To do this, she added 2 cm^3 of starch suspension to a solution containing 2 cm^3 of amylase solution and 5 drops of iodine solution, and recorded the time for the blue-black colour to disappear. She repeated this 4 times, each time adjusting the pH of the amylase/iodine solution. Her results are shown in Figure 2.

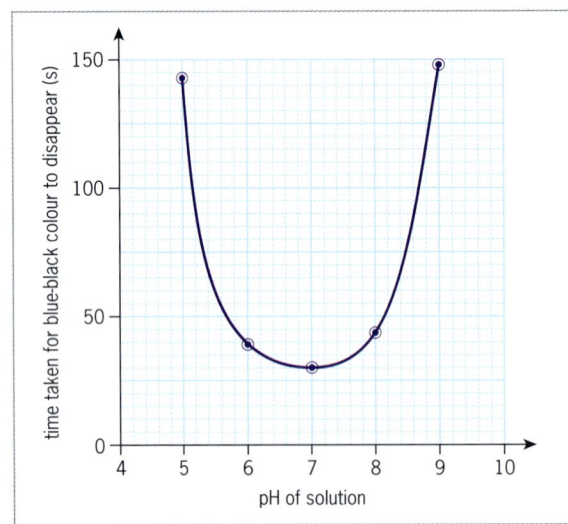

Figure 2 *Graph showing the time taken for the blue-black colour to disappear at different pH values*

Account for the shape of the graph between:

- pH 5 and pH 7
- pH 7 and pH 9. **(4 marks)**

    iv) What is the optimum pH for amylase? **(1 mark)**

b) i) Provide a suitable definition for the term 'mechanical digestion'. **(2 marks)**
   ii) Teeth are important in mechanical digestion. Explain how the structure of a molar tooth is adapted to its function. **(2 marks)**
   iii) As people age, they often begin to lose some of their teeth. Explain how this would affect their ability to digest their food. **(2 marks)**

Total 15 marks

4 a) Distinguish between respiration and breathing. **(2 marks)**
   b) Figure 3 shows part of the human respiratory system.

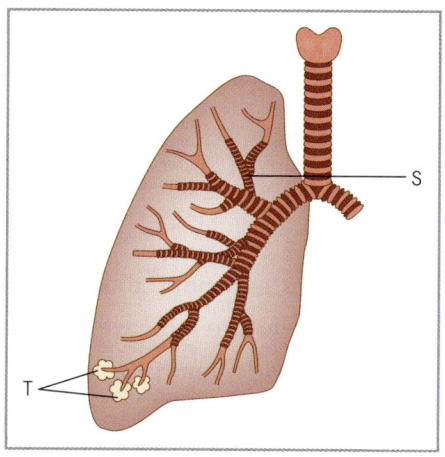

Figure 3 *Part of the human respiratory system*

   i) Name the structures labelled S and T. **(2 marks)**
   ii) Identify TWO features of the structures labelled T that make them efficient in carrying out gaseous exchange. **(2 marks)**

   c) A study was carried out into the link between smoking levels and lung cancer. The researchers recorded the number of deaths from lung cancer per 100 000 individuals in the population and the smoking levels of each person who died. The results for men aged 60 to 69 years are given in Table 2 below.

Table 2 *Death rates from lung cancer among men aged 60 to 69 years related to smoking levels*

Level of smoking		Death rate per 100 000 population
Never smoked		12
Smoked 20 cigarettes per day for:	30 years	234
	40 years	487
Smoked 40 cigarettes per day for:	30 years	576
	40 years	608

   i) Which individuals are most at risk of dying from lung cancer? **(1 mark)**
   ii) Give TWO conclusions that the researchers could have drawn from the results. **(2 marks)**
   iii) Identify ONE measure authorities can put in place to reduce the costs associated with smoke-related illnesses. **(1 mark)**

d) Laila used to be a very good sprinter; however, she started to smoke heavily and now finds that whenever she tries to run she quickly becomes breathless and at times she collapses.
   i) Explain what causes Laila to collapse. (3 marks)
   ii) Other than lung cancer, suggest TWO ways in which Laila's heavy smoking contribute to her breathlessness. (2 marks)

Total 15 marks

5 Tashae lives in a community which is severely affected by acid rain.
   a) What is acid rain? (2 marks)
   b) List the TWO MAIN gases that produce acid rain. (2 marks)
   c) i) What framework of activity usually produces the acidic gases mentioned in b)? (1 mark)
      ii) Describe how these gases are transformed to acids. (4 marks)
   d) Describe FOUR ways that acid rain may have harmed Tashae's environment. (4 marks)
   e) How can acidic gases be removed at industrial sites? (2 marks)

Total 15 marks

6 a) What is the term used to reference the increase in temperatures around the planet? (1 mark)
   b) i) Explain how the 'greenhouse effect' results in this temperature increase. (4 marks)
      ii) Name TWO greenhouse gases OTHER THAN carbon dioxide, methane and chlorofluorocarbons. (2 marks)
   c) State ONE way that EACH of the gases carbon dioxide, methane and chlorofluorocarbons can be generated. (3 marks)
   d) State FIVE reasons why the Caribbean nations need to be concerned about the effects of climate change due to rising atmospheric temperatures. (5 marks)

Total 15 marks

7 a) i) Complete Table 3, which Zac has drawn up to calculate the total energy used for the week in kW h. The oven is used every day for 2 hours and the lamps are on every night for 8 hours.

Table 3 *Energy used in 1 week by different appliances*

Appliance	Power (kW)	Time (h)	Energy (kW h)
One 1500 W electric oven			
Seven 60 W filament lamps			
Total energy (kW h)	–	–	

(7 marks)

   ii) Zac is purchasing a toaster oven and a LED 65-inch TV. State, giving a reason, which will consume the greater power. (1 mark)

b) On 31st March, a customer's meter reading was 47 538 kW h and on 31st April it was 47 689 kW h. Calculate the electricity bill for the period given the following added information.
   - The first 100 kW h is charged at $0.50 per kW h and additional usage is charged at $0.40 per kW h.
   - There is also a fuel adjustment charge of $0.03 per kW h and a monthly rental fee of $15.00. (7 marks)

Total 15 marks

8 a) Abi lives in a region which presents a threat of devastating bush fires.
   i) Explain why bush fires usually progress rapidly and are generally widespread. **(3 marks)**
   ii) Describe the impact of these fires on the ecosystems they affect. **(2 marks)**
   iii) Outline methods of extinguishing bush fires. **(3 marks)**

  b) i) Identify TWO common locations of chemical fires and specify (with justification) TWO necessary pieces of protective gear required to safely extinguish them. **(4 marks)**
   ii) Provide TWO reasons why electrical fires occur and explain why water cannot extinguish them. **(3 marks)**

Total 15 marks

9 a) i) Distinguish between heat and temperature. **(1 mark)**
   ii) State the SI unit of temperature. **(1 mark)**
   iii) State another unit of temperature commonly used and closely related to the SI unit. **(1 mark)**

  b) **Figure 4** *A mercury liquid-in-glass thermometer*

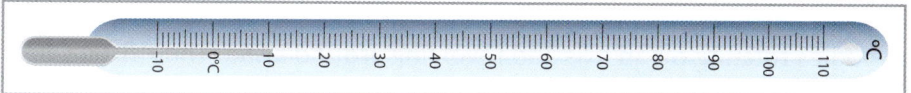

   Figure 4 *shows a mercury liquid-in-glass thermometer.*

   i) Why is the casing around the bulb made of very thin glass? **(1 mark)**
   ii) Why is the bore very narrow? **(1 mark)**
   iii) Why is the thermometer suitable for work in the school laboratory? **(1 mark)**
   iv) Why is the bulb large? **(1 mark)**

  c) State THREE advantages and THREE disadvantages of choosing mercury instead of alcohol as the liquid in a thermometer. **(6 marks)**

  d) The distance between the 0 °C mark and 100 °C mark on a mercury thermometer is 25 cm. Calculate the temperature reading when mercury rises 18 cm from the 0 °C mark. **(2 marks)**

Total 15 marks

Module 3 – Our planet

13 The universe and our solar system

Celestial bodies

The universe is extremely vast, and humans have always been curious to learn of the various celestial bodies of which it is comprised. Table 13.1 describes some of its components.

Table 13.1 *Some important celestial bodies of the universe*

Component	Description
Earth	A surface of 70% water and 21% oxygen sustains life at its suitable temperature.
Star	A luminous sphere composed of plasma and held together by gravitational forces.
Sun	An average star and the closest to Earth.
Planets	**Large** bodies rotating on an axis and revolving around a star. Their **strong gravitational fields** cause them to be **spherical** and to **clear their orbital paths** of most other bodies.
Planetoids	Known as **dwarf planets**: Ceres, Pluto, Haumea, Makemake and Eris. They also rotate on an axis and orbit a star. They are **small** with **weak gravitational fields** and so are not spherical and cannot clear their paths of other bodies.
Asteroids	Small rocky masses that orbit the Sun. Their very weak gravitational fields prevent them from being spherical and from having an atmosphere.
Asteroid belt	An orbital path around the Sun between **Mars and Jupiter** containing millions of asteroids and ONE dwarf planet, **Ceres**.
Meteoroids	**Remnants of comets** and broken pieces of other celestial bodies that orbit the Sun.
Meteors	**Bright trails** (shooting stars) due to friction on meteoroids speeding through the atmosphere.
Meteorites	**Remnants** of meteoroids which have **fallen to Earth**.
Comets	Celestial bodies composed of frozen water, carbon dioxide (CO_2), methane and ammonia mixed with rock and dust orbit the Sun in eccentric (elongated) paths. When near the Sun, the ice vaporises and forms a tail of gas and dust directed away from the Sun that can form **meteor showers** on nearby planets.
Kuiper belt	A path around the Sun beyond Neptune where comets and planetoids orbit.
Galaxy	A group of billions of stars.
Milky Way	The galaxy which includes our Sun as one of more than **200 billion stars**. It is a typical **spiral galaxy** of dust and hot gases with a **diameter** of **100 000 light years** (light takes over 100 000 years to cross it) and has a huge **black hole** at its centre.
Universe	The entire cosmic system, composed of all the matter and energy of billions of galaxies.

How do bodies stay in orbit?

*A **satellite** is a body that orbits another body of larger mass.*

Satellites may be natural or artificial (made by humans). The Earth is a natural satellite of the Sun, and a weather satellite is an artificial satellite of the Earth.

Gravitational forces keep satellites in orbit

The gravitational force depends on the **mass**, **speed** and orbital **radius** and **acts towards the centre of the circle** (see Figure 13.1). A satellite of **greater mass and/or speed** requires a **greater force**; one of **greater orbital radius**, requires a **smaller force**.

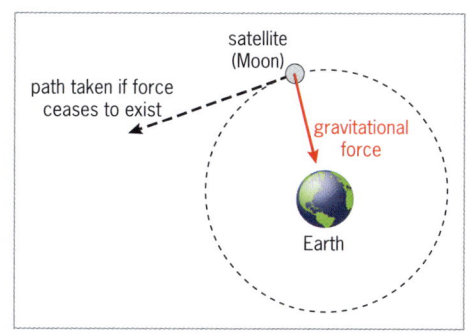

Figure 13.1 *Gravitational forces keep satellites in orbit*

Some artificial satellites of the Earth

***Geostationary satellites** orbit with a period of 24 hours in the same direction as the Earth revolves and are always directly above the same point on Earth (see Figure 13.2) (see Figure 13.3).*

Uses

- **Communications** by radio, television and telephone since their signals can be located easily.
- **Monitoring hurricanes** as they are always above the same region of the Earth.

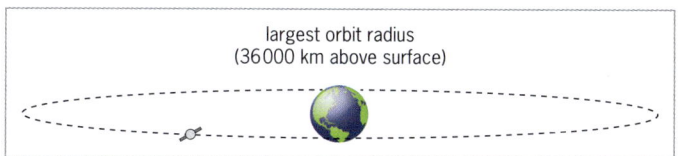

Figure 13.2 *Geostationary satellite*

***Polar satellites** have orbits almost parallel to Earth's longitudinal lines and so can obtain data in an east-west direction as the planet rotates, and in a north-south direction as they proceed along their orbits (see Figure 13.3).*

They orbit very **close to Earth** and so can produce images of **good resolution**. Short orbital periods of about 90 minutes enable them to make several observations of the Earth's surface in just 24 hours.

Uses

- **Mapping and surveying**.
- **Monitoring weather patterns**.
- **Military purposes** such as locating soldiers and enemy sites.

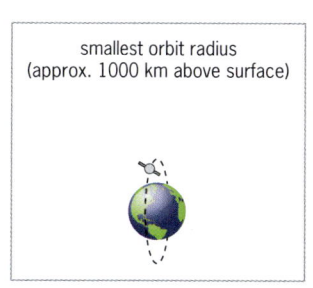

Figure 13.3 *Polar satellite*

***Satellite navigational systems** are those which can determine the **precise location** of a user at a **given time** by processing signals issued by a network of satellites (see Figure 13.4).*

The Global Positioning System (GPS) of satellites of the USA is one such system. It has **30 satellites** orbiting in **six paths** with **orbital periods of 12 hours** (see Chapter 15, page 204).

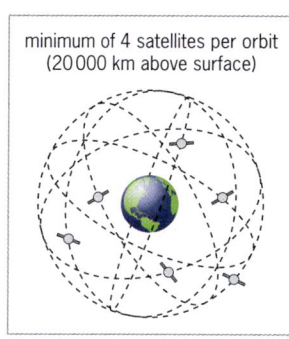

Figure 13.4 *GPS satellites*

13 The universe and our solar system

Our solar system

Planets and their moons, planetoids, asteroids, meteoroids, meteors and comets all orbit the Sun.

Terrestrial planets – Mercury, Venus, Earth and Mars are **smaller** and **warmer** with a **rocky surface**.

Gas giants – Jupiter, Saturn, Uranus and Neptune are **larger** and **colder** and are composed of gaseous **hydrogen** and **helium** and **liquid ammonia**. Each has a **ring system** of dust and/or ice around it.

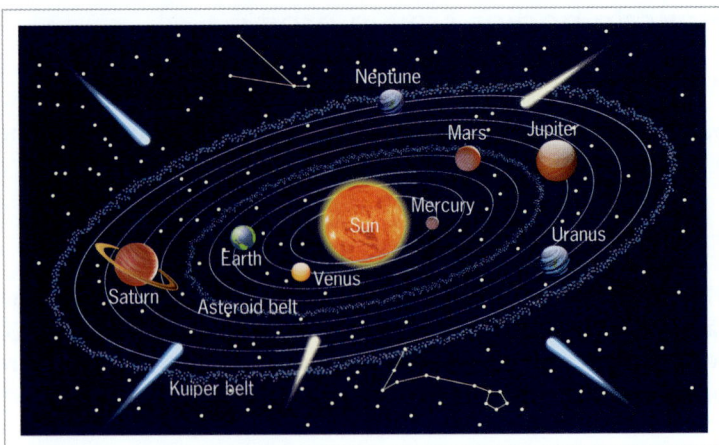

Figure 13.5 *Our solar system*

Table 13.2 *Characteristic features of the planets of our solar system*

Planet	Characteristic features
Mercury	**Smallest** planet; has **no atmosphere**; **rotates very slowly** so the hemisphere facing the Sun becomes very hot and the opposite hemisphere, very cold.
Venus	**Hottest** planet as its **dense atmosphere of carbon dioxide** traps heat by the **greenhouse effect**; **brightest** planet since it is covered by a **cloud of sulfuric acid** which reflects strongly; **spins in retrograde** (opposite direction to other planets); has thousands of volcanoes.
Earth	**Largest terrestrial** planet; **supports life** due to presence of water and suitable temperature.
Mars	Known as the 'red planet' due to **iron** in its surface of **rock, dust and ice**; has the **largest dust storms** and **largest volcano**; has a **scorched surface** due to the sterilising effect of ultraviolet radiation through its **thin atmosphere** consisting mainly of **carbon dioxide**.
Jupiter	**Largest** planet; has the **shortest day**; has a 'Great Red Spot' due to a permanently raging storm; has 95 moons, one (**Ganymede**) is larger than Mercury; has a **ring system of dust**.
Saturn	Has a **profound ring system of ice and dust**; has the **most moons (146)**; one moon (**Titan**) is larger than Mercury!
Uranus	**Appears on its side** as it has a very tilted rotational axis; has a **ring system of ice and dust**.
Neptune	**Ring system of ice and dust**.

Table 13.3 *Other features of the planets of our solar system*

Planet	Distance from Sun (km)	Diameter (km)	No. of known moons	Mean temp. °C
Mercury	60 million	5000	0	167
Venus	110 million	12 000	0	464
Earth	150 million	13 000	1	15
Mars	230 million	7000	2	−65
Jupiter	780 million	143 000	95	−110
Saturn	1430 million	121 000	146	−140
Uranus	2870 million	51 000	28	−195
Neptune	4500 million	50 000	16	−200

The effect of other bodies on the Earth

Day and night

As the Earth spins on its axis, the hemisphere that faces the Sun receives daylight and the hemisphere on the side away from the Sun experiences night. This is due to the **rectilinear propagation** (straight-line motion) of light.

The seasons

The axis of rotation of the Earth is tilted at 23.5°. The northern hemisphere therefore receives more hours of daylight than the southern hemisphere for six months (summer) and then more hours of darkness than the southern hemisphere for the next six months (winter), as shown in Figure 13.6.

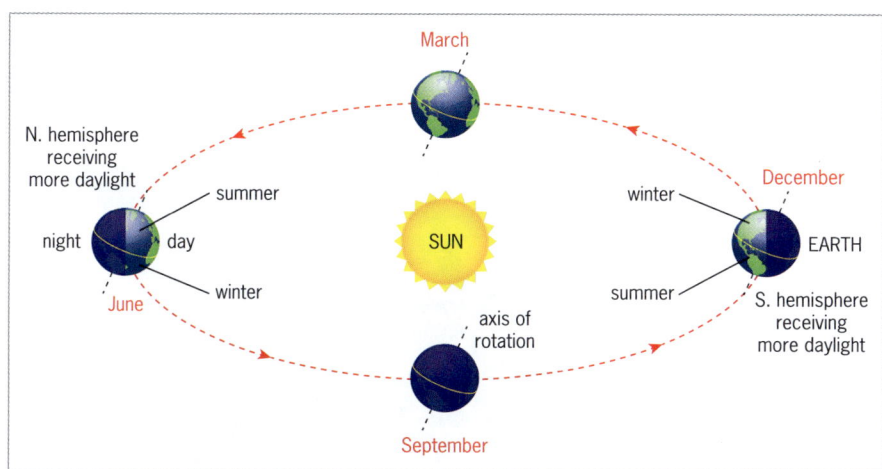

Figure 13.6 Day, night and the seasons of the year

Viewing the Moon

The Moon is **non-luminous** and so can only be seen from Earth by **reflection** of light from the Sun. Light from the Sun reaching the Moon at an angle of incidence 'i' will have the same angle of reflection 'r' as it proceeds to Earth. Figure 13.7 shows that we can see the Moon during the day or during the night.

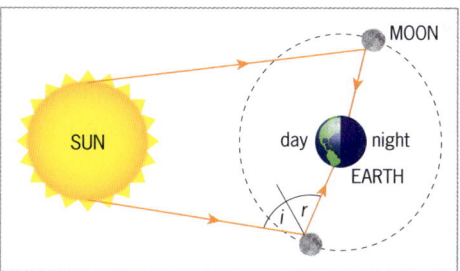

Figure 13.7 Viewing the Moon by reflection

Eclipse of the Moon (lunar eclipse)

Due to the rectilinear propagation of light, shadows can be cast by objects that intercept rays. Figure 13.8 shows that occasionally, the Moon enters the Earth's shadow (umbra) instead off passing beyond its apex, and therefore cannot reflect light to Earth. This can only occur at '**full moon**' (see Figure 13.10 page 174), i.e. when the Sun and Moon are on opposite sides of the Earth.

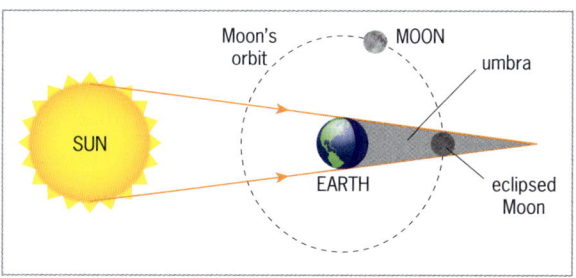

Figure 13.8 Eclipse of the Moon

13 The universe and our solar system

Eclipse of the Sun (solar eclipse)

Occasionally, the Moon's orbit can pass through the rays directed from the Sun to the Earth and can cast a shadow on the Earth, as in Figure 13.9. Depending on a person's position on Earth, a total eclipse, partial eclipse or no eclipse of the Sun may be observed.

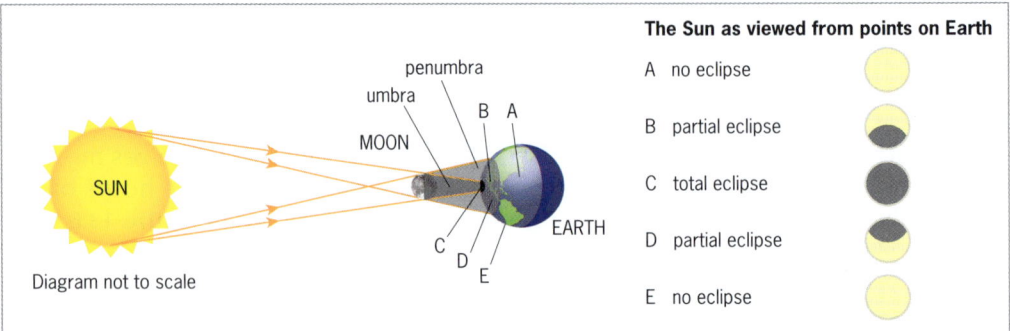

Figure 13.9 *Eclipse of the Sun*

The Moon's orbit

The Moon takes approximately 30 days to orbit the Earth. The **inner ring** of Figure 13.10 shows that the **side facing the Sun is always lit** and the opposite side is always in darkness. The **outer ring** shows the Moon as **seen from Earth**.

- Gibbous and crescent refer to the Moon's shape.
- Waxing and waning refer to the Moon's reflection of more and less light, respectively.

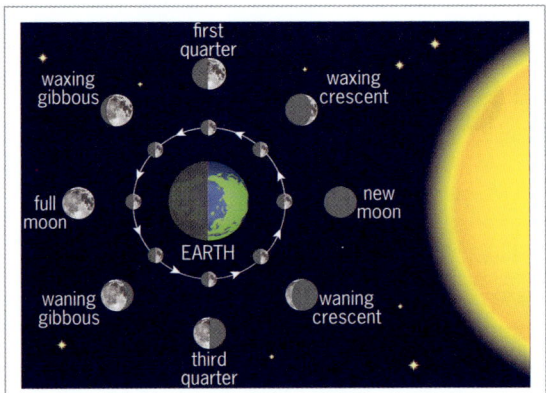

Figure 13.10 *Observing one cycle of the Moon*

Human exploration of the universe

Tables 13.4 gives reasons for space exploration and Table 13.5 lists characteristics of outer space that make its exploration challenging. These challenges mainly relate to personal exploration although there are difficulties with unmanned exploration such as messaging delays between space vehicles and Earth stations.

Table 13.4 *Reasons for space exploration*

Reason	Details
Curiosity	The need to understand the evolution of the universe and our place in it.
Research	Some scientific research is best done in space to eliminate variables such as gravity.
Asteroid threat	Approximately every 10 000 years an asteroid collides with Earth. Exploration of and experimentation in space may help to avoid future catastrophic collisions.
Mining	Useful substances can be mined from planets, moons and asteroids to provide fuel, construction materials, fertilisers and life support for explorers.
Technology	Expanding technologies and creating new industries.
Unity	Fostering peaceful cooperation with other nations provides benefits of unity.
Migration	Population increase and climate change may cause us to seek a less harmful habitat.

Table 13.5 *Characteristics of outer space that make its exploration challenging*

Challenge	Action required
Lack of atmosphere	Constantly, **oxygen** must be added and **carbon dioxide** removed from the air in the spacecraft. **Temperature** and **pressure** levels must be carefully maintained. **Radio communication** is used since sound cannot pass through a vacuum.
Weightless-ness	Astronauts need to adjust to **locomotion without weight**. **Urine** is passed to sealed containers. **Food** in semi-liquid form is sucked to the mouth and forced to the stomach using muscles. To prevent **muscle atrophy** and **reduced bone mass**, astronauts should **exercise** 2 ½ hours per day, 6 days per week.
Food and water	**Dehydrating foods** before storing them reduces mass and **avoids spoilage**. They can be rehydrated with recycled water from **sweat, breathing** and **urine** before consumption.
Cosmic radiation	**Special clothing** is used to protect against **high radiation levels** in space. **Special construction design** is required to protect space vehicles from radiation.
Psychological stress	Being confined and isolated far from home strongly affects mental health and so **motivational therapy** is required.
Technology	Complex and costly technologies are required since failures can be disastrous.
Messaging delays	Communication delays due to distances are managed by **predictive and autonomous technologies** that act without waiting on instructions from Earth.

The International Space Station (ISS)

Table 13.6 *The International Space Station*

Description	The ISS (Figure 13.11) is the **largest artificial satellite**. It was built by space agencies from USA, Russia, Canada, Japan and Europe, each having a control centre on Earth. It has living quarters, docking ports and solar panels. A **crew of six** carry out experiments and maintains the station.
Orbit	A **low orbit** of height **400 km** and period **90 minutes**.
Function	An **intermediate port** for space exploration and for **scientific research** in a space environment.

Figure 13.11 *International Space Station*

The Hubble Space Telescope and the James Webb Space Telescope

Figures 13.12 and 13.13 show the Hubble Space Telescope (HST) and the James Webb Space Telescope (JWST). They **produce images** that help us to understand the universe including **current events** and **events billions of light years in our distant past**. They inform us of the **birth and death of stars** and planetary systems, the **merging of galaxies**, the physics of **black holes** and the **collisions of celestial bodies**.

The **JWST** is the **largest** and **most powerful** telescope ever constructed. It can view **older** and **more distant** galaxies than the HST and helps in the study of the **formation of the universe**. It investigates the atmospheres of **exoplanets** (planets outside our solar system) to determine their **suitability for sustaining life**.

Figure 13.12 *Hubble Space Telescope*

Figure 13.13 *James Webb Space Telescope*

Table 13.7 *Some differences between the HST and the JWST*

Aspect	Hubble Space Telescope	James Webb Space Telescope
Launched	25th April 1990	25th December 2021 – relatively new.
Radiation detected	Ultraviolet radiation, visible light and **some infrared radiation**.	**Wide range of infrared radiation**, which **avoids interference** from ultraviolet radiation and visible light and allows viewing more distant events through dust and cosmic clouds that block visible light.
Mirrors	**Smaller** receiving mirrors.	**Larger** receiving mirrors.
Location	**Low Earth orbit**, 540 km above surface.	1 500 000 km from Earth, **does not orbit Earth**.
Servicing	Maintained by **astronauts**.	Its distance from Earth is beyond human space flight capability and so it **cannot be serviced by astronauts**.

Exploring Mars

Venus is closer to us than Mars but is **too hot** and is burdened with **sulfuric acid rain**. Consequently, scientists are eager instead to explore whether life ever existed on Mars. Future **colonisation** may be necessary should Earth become uninhabitable due to **climate change** or **over-population**. We may also be able to mine for **important minerals** and develop **new technologies** and **scientific theories**.

Table 13.8 outlines factors of the Martian atmosphere that make exploration of its surface difficult.

Table 13.8 *Atmosphere on Mars*

Feature	Details
Composition	**CO_2 (95%)**, N_2 (≈3%), Ar (2%), O_2 (< 0.2%), small quantities of H_2O, CO and noble gases.
Density	Only 2% of the atmospheric density on Earth (a **very thin atmosphere**).
Pressure	Only 1% of the atmospheric pressure on Earth (**very low pressure**).
Temperature	**Colder than Earth** (−60 °C) since it is **further** from the Sun than Earth and since 'greenhouse warming' is insignificant due to its very thin atmosphere.
Storms	**Dust storms** commonly occur on Mars.

Spacecraft used to obtain information of Mars

Flybys get close enough to a celestial body to **gather data** and **relay it to Earth**.

Orbiters continuously **orbit a planet** collecting data from the **surroundings** (including from **landers** and **rovers** – see later) and then **relay it to Earth**. Orbiters of a planet can also **act as flybys** of its moons.

Landers land **on the planet** and gather information. They then **transmit it to orbiters** which **relay it to Earth**. Some landers **deliver rovers** to the planet.

Rovers are **remotely controlled**, **robotic**, **motorised** vehicles (including small helicopters). Rovers can investigate the **geology**; **climate and radiation** levels; the possibility of **microbial life**. They may also **collect samples** for further analysis when returned to Earth.

Important Mars rovers

Sojourner (National Aeronautics and Space Administration, NASA – **landed in 1997**)

This was less than 1 m long and stayed within 12 m of the lander. It lasted 4 months, analysed rocks, produced hundreds of images, and investigated the weather and the possibility that life can exist on the planet.

Spirit and *Opportunity* (NASA – **landed in 2004**)

These were a pair of rovers (known as 'the twins') placed on opposite sides of Mars to investigate its geology. They **confirmed that water existed** on the planet. Spirit ceased functioning in 2010, Opportunity in 2018.

Curiosity (NASA – **landed in 2012**)

This is the size of an SUV and still investigates the **geology**, **climate** and **potential for life** on Mars. In 2018 it **discovered organic molecules** that indicated the possibility that life may have existed on Mars.

Perseverance (NASA – **landed in 2021**)

This investigates if the **Jezero Crater** was once a lake that could support life. It is the most technologically advanced rover, having a **7-foot arm** holding a **drill**, and **23 cameras** uniquely designed for **navigation, viewing the landscape** and **analysing rocks and regolith** (loose soil covering).

Ingenuity (NASA – **landed with *Perseverance***)

A helicopter **used as an engineering test** for future helicopters to be sent to Mars to help in retrieving soil and rock samples.

Zhurong (Chinese National Space Administration, CNSA – **landed in 2021**)

Investigates the geology of the surface and evidence of past or current life.

Figure 13.14 *Perseverance*

Exploring Jupiter

Landers are not used on Jupiter since, being composed of gaseous hydrogen and helium, its **surface is not solid**.

Juno is an **orbiter** of Jupiter since 2016 gathering information on the **atmosphere**, **interior structure**, **magnetic field, many moons, dust rings** and **auroras**. It also does **flybys** on several of its moons.

Europa Clipper will be an **orbiter** of Jupiter from 2030 and will conduct **flybys** of Europa (one of Jupiter's moons).

JUICE will be an **orbiter** of Jupiter from 2031 and will conduct **flybys** of three of Jupiter's icy moons (**Europa, Ganymede and Callisto**) to determine if they have oceans beneath their ice which can **possibly host life**. JUICE will then become an **orbiter** of **Jupiter's largest moon, Ganymede**.

Revision questions

1. **a** What percentage of the Earth's surface area is water?
   b What percentage of the Earth's atmosphere (by volume) is oxygen?

2. Distinguish between planets and planetoids.

3. Give THREE characteristics of an asteroid.

4. Distinguish among meteoroids, meteors and meteorites.

5. What are comets?

6. Between the orbits of which planets is the asteroid belt located?

7. **a** What is at the centre of the Milky Way galaxy?
   b What is the diameter of the Milky Way galaxy in light-years?

8. **a** Define a satellite.
   b Draw a diagram showing the type and direction of the force that keeps a satellite in orbit.

9. Define and give TWO uses of each of the following:
   a geostationary satellites   **b** polar satellites.

10. Distinguish between 'terrestrial planets' and 'gas giants'.

11. For each of the following, identify the relevant planet of our solar system.
    a Smallest.   **b** Largest.   **c** Hottest surface.
    d Orbited by 146 moons.   **e** Known as the 'red planet'.
    f Known for its 'Great Red Spot'.   **g** Brightest when seen from Earth.
    h Appears to be spinning on its side.   **i** Covered by a dense atmosphere of carbon dioxide and sulfuric acid.
    j Known for its profound ring system of ice and dust.

12. Describe the alignment of the Sun, Moon and Earth during a lunar eclipse.

13. **a** Draw a single diagram showing the position of the Moon relative to the Earth and Sun during:
    i a full moon   **ii** a new moon   **iii** a first quarter moon.
    b What percentage of the Moon's surface reflects light to Earth during a full moon, new moon and first quarter moon, respectively?

14. **a** Give THREE reasons humans desire to explore space.
    b Describe THREE challenges of exploring space and how humans are overcoming them.

15. The following questions are about the International Space Station (ISS).
    a From which countries are the space agencies that combined to build the ISS?
    b What are the orbital height and period of the ISS?
    c What are its main functions?
    d How many crew members does it generally have?

16 The following questions are about the Hubble Space Telescope and the James Web Space Telescope.

  a List THREE types of celestial occurrences that the images from these telescopes inform us of.

  b Which telescope:

  i is the newer?

  ii detects a wide range of infrared radiation?

  iii can be maintained by astronauts?

  iv investigates atmospheres of exoplanets (planets outside our solar system)?

17 **a** Give TWO reasons why scientists are eager to investigate other planets.

  b Why do scientists seek to land on Mars instead of our nearest neighbour, Venus?

18 Table 13.9 lists features of the Martian atmosphere. Complete the table by selecting answers from within the brackets.

Table 13.9 Question 18

Feature	Correct description of feature
Most abundant gas (CO_2, O_2, H_2, CO, N_2)	
Density (very high, same as on Earth, very low)	
Pressure (very high, same as on Earth, very low)	
Temperature (–60 °C, 27 °C, 60 °C, 300 °C)	

19 **a** Briefly describe the functions of THREE types of spacecraft (other than rovers) used to investigate other celestial bodies.

  b Name:

  i the most technologically advanced Mars rover

  ii two other Mars rovers.

  c Name the crater being investigated by the rover mentioned in **b i** and give the main reason for the investigation.

20 **a** Why are space vehicles not sent to land on Jupiter?

  b Name the orbiter currently investigating Jupiter.

  c What will be the missions of: **i** Europa Clipper? **ii** JUICE?

14 The terrestrial environment

Our terrestrial environment is a dynamic system of land, air, water and living organisms, shaped by natural processes, human activity and Earth's position relative to the Sun and Moon, all of which sustain biodiversity on Earth.

The Earth's atmosphere and weather

Air masses

Air masses are extensive bodies of air with approximately uniform characteristics of temperature and humidity at any given latitude.

Air masses affect the weather and transport pollutants over thousands of kilometres.

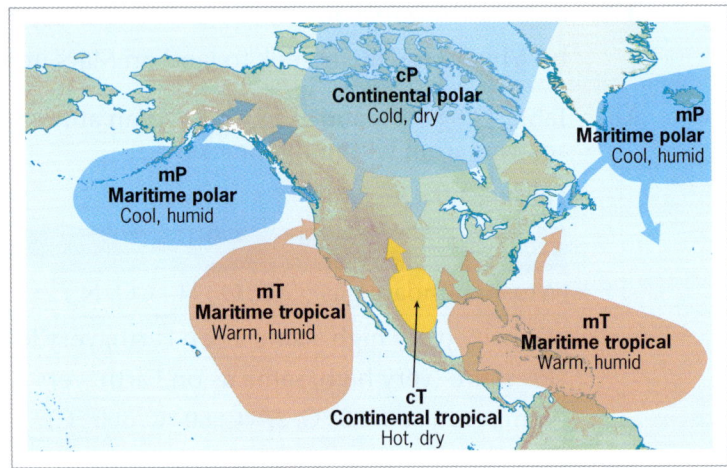

Figure 14.1 *Air masses affecting North America and the Caribbean*

Table 14.1 *The main types of air masses*

Type	Region found on Earth	Properties
Maritime tropical	Tropical and sub-tropical seas and oceans such as the **Caribbean Sea**	Hot/moist
Continental tropical	Large arid land areas such as **deserts** in southwest of USA (i.e. in **California**, **Nevada**, **Arizona** or **New Mexico**)	Hot/dry
Maritime polar	Polar oceans such as the **Atlantic, east of Newfoundland**	Cool/moist
Continental polar	Large land areas near to poles, such as **Canada**	Cool/dry

Table 14.2 *Pollutants transported by air masses*

Pollutant	Associated problem
Harmful industrial gases	**Acidic gases** (carbon monoxide, nitric oxide, nitrogen dioxide and sulfur dioxide) produce **acid rain** and cause **respiratory problems** (see Table 10.3 page 126).
Landfill fumes	Stress and **respiratory problems** due to hydrogen sulfide and methane; **soil** and **water contamination**; **foul odours**; threat of **global warming** and spontaneous **explosions** due to methane; transport of **harmful bacteria**.
Particulate matter	Produced by **industrial processes**, **dust storms** and **volcanoes**. Leads to **respiratory problems, reduction in photosynthesis** and **transportation of harmful bacteria**.
Radioactive fallout	Materials from nuclear explosions can be transported to contaminate ecosystems of animals and plants, causing cancers and radiation sickness.

Saharan dust storms

Low pressure over the hot Sahara produces rising air, carrying sand up to 5 km high. Air masses transport and deposit the sand across the Atlantic to the Caribbean and the Americas.

Benefits
- Provides important nutrients, such as iron to **marine bacteria** and **phosphorus** to **Amazonian soils**.
- **Weakens tropical cyclones**.

Problems
- **Spreads bacteria** causing **deadly diseases** (e.g. cholera).
- **Causes respiratory** and **allergic reactions**.

Saharan dust has a complex effect on global warming, producing varying weather patterns.
- **Cools** the planet by **reflecting incoming radiation** into space.
- **Warms** the planet by **absorbing outgoing radiation** and then **emitting it** back to **Earth**.
- Acts as **condensation nuclei**, increasing possible **precipitation**.

Volcanic ash and gases

These shoot **tens of kilometres** into the air and can travel **thousands of kilometres** from the volcano.

Benefits
- Ash **brings nutrients** for the plants to grow, and **minerals** to **balance pH** and **improve soil structure**.
- Finely crushed **basalt improves crop yields**.

Problems
- Blocks sunlight and so **reduces photosynthesis**.
- Produces **respiratory illness**.

Local fronts and their effect on weather

*A **front** is a boundary or transition zone where air masses of different temperatures meet.*

Warm, less dense air rises above cooler air. If humid, it cools, condenses and produces rain.

Table 14.3 Fronts and their weather map symbols

Type of front	Cold front	Warm front	Stationary front	Occluded front
Map symbol	▲▲▲▲	●●●●	▲●▲●	▲●▲● or ▲●▲●

Cold fronts and warm fronts

*A **cold front** is the boundary where a cold polar air mass advances below a warm tropical air mass.*

*A **warm front** is the boundary where a warm tropical air mass advances over a cold polar air mass.*

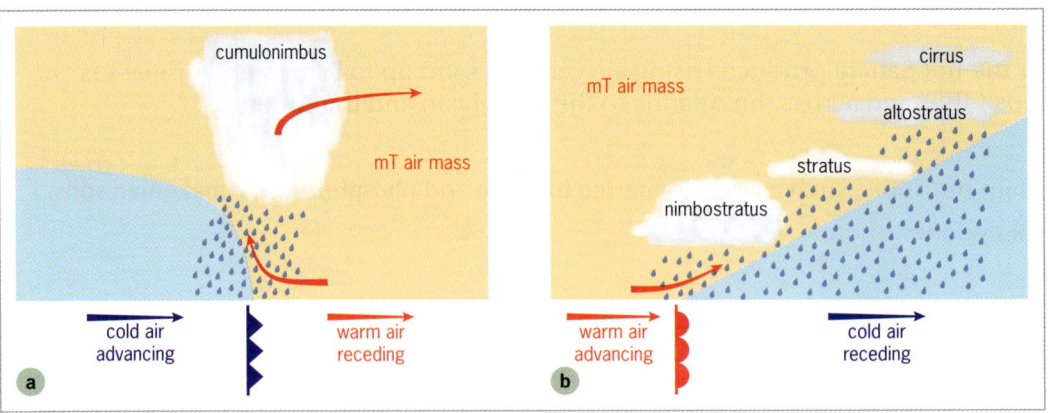

Figure 14.2 Weather at a cold front and at a warm front

Table 14.4 Comparing the weather produced by cold fronts and by warm fronts

	Cold front	Warm front
Action	Cold air **ploughs under** warm air.	Warm air **glides over** cold air.
Gradient	Warm air forced upward – **steep** slope.	Warm air forced upward – **gentle** slope.
Temperature	Region at front becomes **cooler**.	Region at front becomes **warmer**.
Clouds	**Vertical cumulonimbus** clouds rise **rapidly**.	**Horizontal stratus** clouds form **slowly**.
Showers	**Brief** and **intense** (usually thunderstorms).	**Continuous**, **light to moderate**.
Region affected	**Small** due to steep gradient.	Relatively **large** due to gentle gradient.
Frontal speed	**Fast** compared to warm fronts.	**Slow** compared to cold fronts.

* Showers occur if the rising warm air contains moisture, as it will then cool and condense.

Stationary front

*A **stationary front** occurs at the boundary between two air masses of different temperature when neither air mass can displace the other.*

Weather at the stationary front is generally fair to partly cloudy but can also be rainy for long periods. The front will **remain stationary** if the **winds are blowing parallel to it** (see Figure 14.3).

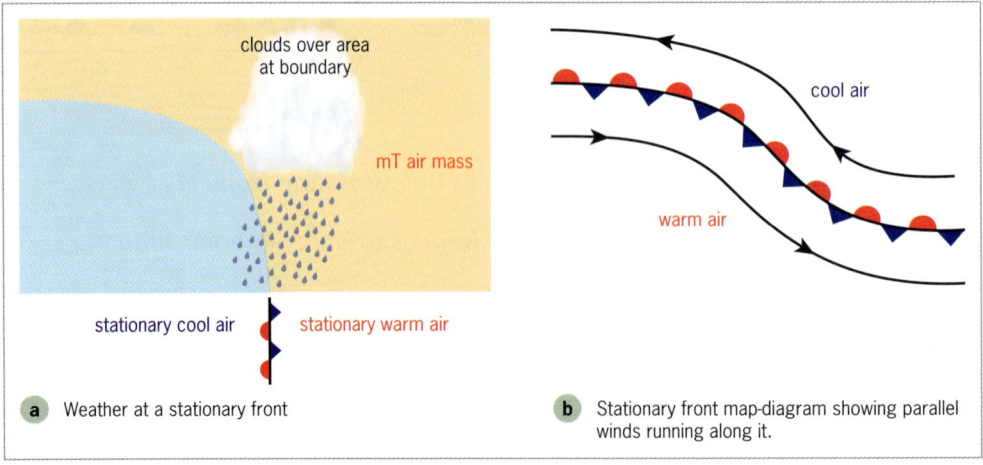

a Weather at a stationary front

b Stationary front map-diagram showing parallel winds running along it.

Figure 14.3 Weather at a stationary front

Occluded front

*An **occluded front** occurs when a cold front catches up with a warm front, raising the warm air between them upward and away from the ground.*

Cold fronts advance faster than warm fronts. The cold air squashes the warm air onto the cold air ahead. Warm air rises at the interfaces, forming cumulonimbus and stratus clouds as shown in Figure 14.4a. The last stage of the storm occurs when the cold front catches the warm front and lifts the hot air completely off the ground, producing major precipitation as shown in Figure 14.4b.

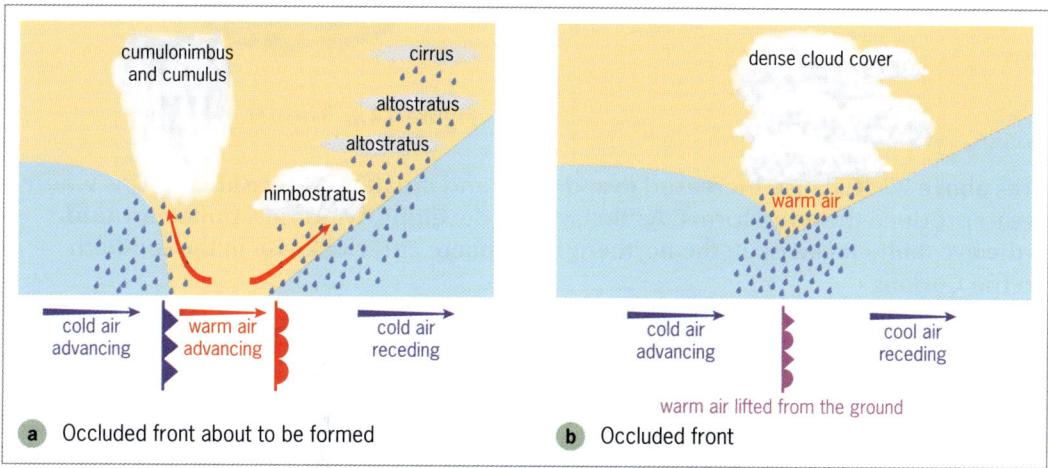

Figure 14.4 *Weather at an occluded front*

Due to the **Earth's rotation**, **winds** in the northern hemisphere experience a **'Coriolis force'** which causes them to blow **anticlockwise** around a **low-pressure centre**.

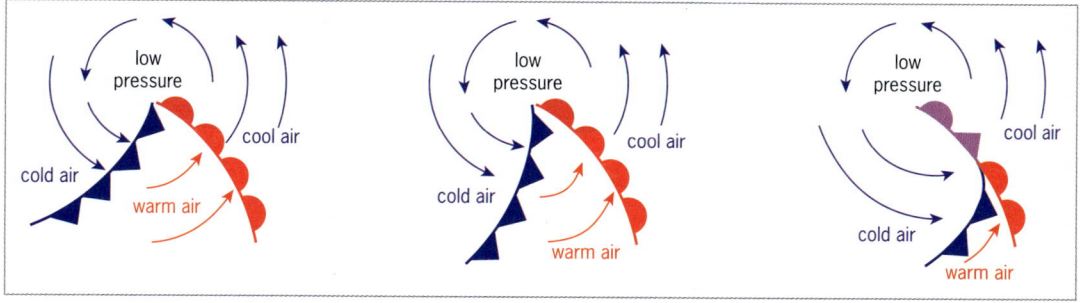

Figure 14.5 *Map diagram showing the formation of an occluded front*

Weather patterns in the Caribbean

The seasons

- **Wet season:** May/June to November/December; temperature, humidity and rainfall are highest.
- **Dry season:** December/January to April/May; temperature, humidity and rainfall are lowest.
- **Atlantic hurricane season:** Officially runs between June 1st to November 30th.

Temperature and rainfall

Temperature

Day: 30 °C to 34 °C.
Night: 20 °C to 26 °C.

Rainfall

Warm, humid northeast and southeast trade winds converging at the ITCZ (Intertropical Convergence Zone) are forced upward, cool and produce clouds, heavy rainfall and thunderstorms.

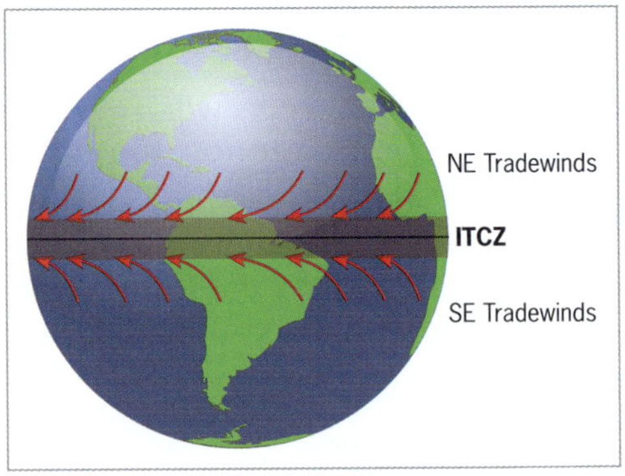

Cyclones and tropical cyclones

*A **cyclone** is a large-scale wind system that rotates around a low-pressure centre.*

Figure 14.6 *The ITCZ*

Tropical cyclones

Ocean temperatures **above 27 °C** cause increased **evaporation** and create **low-pressure** regions where **warm moist air rises** to produce thunderstorms. As the system develops, winds are pulled around the **calm centre of the eye (anti-clockwise in the northern hemisphere** and **clockwise in the southern hemisphere)** due to the **Coriolis force** created by the **Earth's rotation**. The **moist air cools and condenses**, releasing **latent heat** of vaporisation and **energising the updraft**. Water vapour, now **sucked in at an increased rate**, produces **torrential rains** and **high winds**, **strongest in the eyewall**. Within the **eye**, there is a **calm** with little precipitation. As the eye of a hurricane passes, the wind **reverses direction**.

A **storm surge** may occur as the hurricane's **eye pulls ocean water onto the coast**. Despite the **reduced wind speed** from land **friction** and **less oceanic evaporation**, rainy weather persists for some time afterward.

Figure 14.7 *Satellite view of a hurricane*

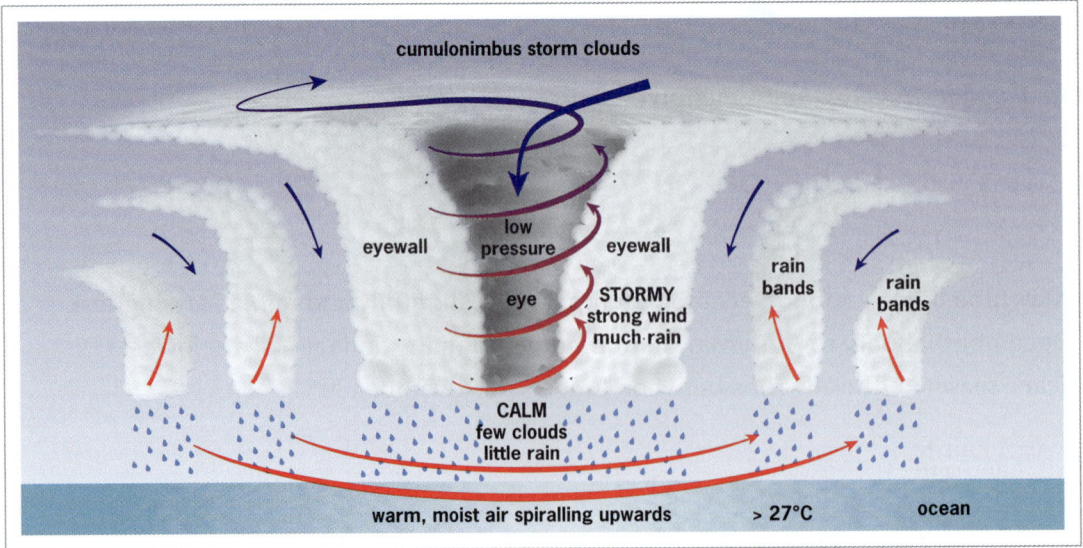

Figure 14.8 *Cross-section through a hurricane*

Table 14.5 *Classification of tropical cyclones*

Classification	Tropical depression	Tropical storm	Hurricane
Wind speed (km/h)	up to 62	63 to 117	greater than 117
Wind speed (mph)	up to 38	39 to 73	greater than 73

Hurricane preparedness

- Medical kits, food and clean water, batteries, flashlights and radios should all be stocked.
- Generators or battery packs should be available in case electricity is cut off.
- Doors, windows and loose objects should be shuttered or strapped.
- A full tank of gasoline, a full bottle of cooking gas and a gas stove should be on hand.
- Important documents and personal possessions should be stored in a waterproof container.
- One should stay tuned to the news and know the location of the nearest shelter in case evacuation is necessary.
- Shelter should be provided for animals.

After the hurricane

- Drinking water should be boiled or sterilised using tablets.
- News updates should be followed to learn of problems such as flooded areas or fallen trees.

Tides and their effects

Tides are the changes in sea levels on Earth caused by the gravitational attractions of the Sun and the Moon on the Earth together with the rotation of the Earth.

- **Tides can erode** beaches, cliffs and dunes by **repeated bombardment**, the **removal of sediments** and the **undercutting of structures** which eventually collapse.
- Coastal erosion intensifies during **storms** and when **humans disrupt coastal ecosystems**.

High tides and low tides

High tides occur on the sides of the Earth closest to the Moon and furthest from the Moon as shown in Figure 14.9. Since the Earth takes 1 day to rotate once on its axis, there are **two high tides** and **two low tides** occurring at most points on its surface **daily**.

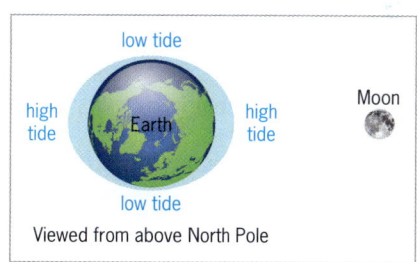

Figure 14.9 *High and low tides*

Spring tides and neap tides

- **Spring tides** occur at **new moon** and **full moon**, when the **gravitational attraction** of the Earth by the Sun and the gravitational attraction of the Earth by the Moon have a **maximum combined effect**. See Figure 14.10.

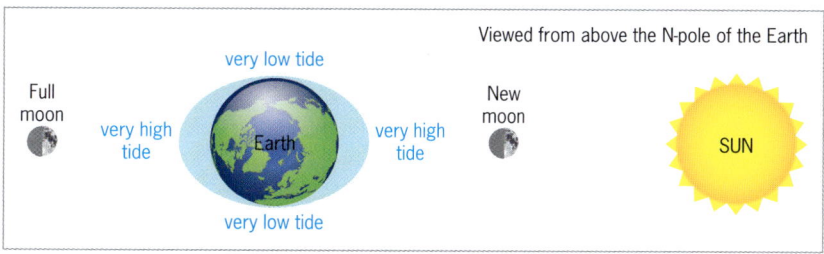

Figure 14.10 *Spring tides*

- **Neap tides** occur at **first quarter** and **third quarter moon**, when these attractions have a **minimum combined effect**. See Figure 14.11.

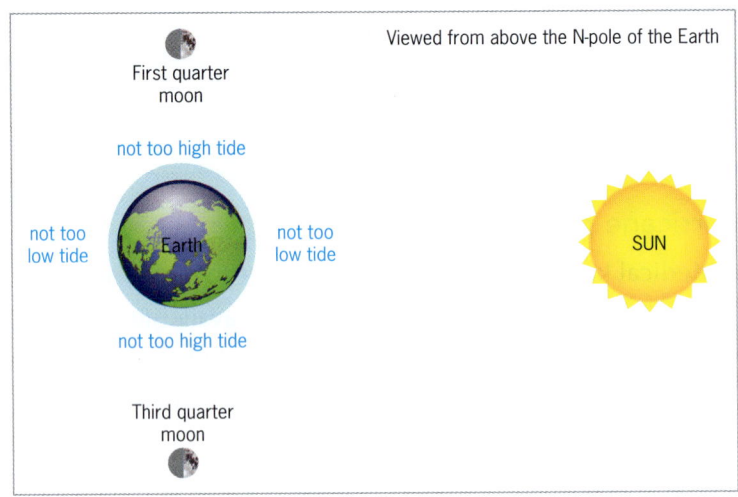

Figure 14.11 *Neap tides*

Tidal waves and tsunamis

*A **tidal wave** is a regularly occurring **shallow-water** wave caused by the **gravitational interactions** of the Sun, Earth and Moon.*

Tidal waves can be intensified by frictional drag from winds. They are not tsunamis.

*A **tsunami** is a large **deep-water** wave produced by a **sudden disturbance** of a **large volume of water** by an event such as a coastal landslide, an underwater volcano or earthquake, or a huge meteorite crashing into the ocean.*

A **strong**, **sudden** disturbance displaces a **large volume** of water **far below the surface**. A wave of amplitude about **1 m** in deep water travels in all directions at speeds up to **800 km/h**. On approaching land, **friction slows the wave significantly**, but its amplitude can increase to about **30 m**. **Coastal waters** are **sucked outward and upward** and towering waves slam the coastline for **several hours**. The destruction reaches as far as **15 km inland** and as the **water retreats**, it **drags debris** causing more damage.

Rescue operations are usually put in place quickly. Communities should **survey damage** and **rebuild** the infrastructure. **Seawalls** and **channels** should be built to protect against tsunamis and careful **monitoring** of **seismic activity** should be used with adequate **evacuation routes** and **alarms**.

Table 14.6 *Comparing tidal waves and tsunamis*

	Tidal wave	Tsunami
Cause	Changing tides and gravitational forces.	Earthquakes, volcanoes, asteroid collisions.
Nature	Predictable and regularly occurring.	Generally unpredictable and sudden.
Impact	Not as destructive since the energy is carried only by water near the surface.	Extremely destructive since the energy is carried by the entire volume of water down to the seabed.

Volcanoes and earthquakes

Volcanic eruptions

*A **volcano** is an opening in a planet's crust from which molten lava, rock fragments, ashes, dust and gases are ejected from far below the surface.*

__Fumaroles__ are small vents which emit only steam and hot gases such as CO_2, SO_2, H_2S, HF, and HCl.

The Earth's core is filled with hot **magma** that pushes outward, causing **tectonic plates** to drift. When these plates **diverge** or **converge**, magma rises to the surface through cracks. Viscous magma can **solidify** and block **vents**, increasing the pressure until **explosive** eruptions occur. These release **pyroclastic** materials (rocks, cinder and ash), **gases** and **molten lava**, forming towering clouds of smoke.

__Calderas__ are bowl shaped depressions formed as __empty__ magma chambers collapse.

Table 14.7 Factors that result in different types of volcanic eruptions

Cause	Effect
Viscosity of magma	**Viscous** lava contains much **silica**. Unlike **non-viscous** or **basaltic** lava, it is thick and slow moving, forming **steep slopes**. It solidifies in vents causing **explosive** eruptions.
Tectonic plate setting	**Convergence** – when an oceanic plate collides with a continental plate, it subducts, creating deep trenches, powerful seismic activity and mountain ranges (Figure 14.15). **Divergence** – if tectonic plates separate, magma forces into the gap to fill the space.
Water	**High water content** results in strong steam pressure and violent explosions.
Gases	**High concentration of gases** results in strong pressure and violent explosions.

Main types of volcanoes

Cinder cone, shield and composite volcanoes

Table 14.8 Comparing the main types of volcanoes

	Cinder cone	Shield	Stratovolcano (Composite)
Size	Small / one vent / wide crater	Massive / multiple vents	Massive / parasitic cones
Viscosity of lava	Only slightly basaltic / flows slowly / blocks vent / *1	Basaltic / non-viscous / flows easily / does not block vent	Viscous / contains much silica / blocks vent
Gentle or explosive	Explosive / much steam and gases. Usually erupts only once.	Not explosive / flows easily and has very little gas	Explosive due to much steam and gases / *2
Steepness	Steep with loose material	Gentle as a warrior's shield	Steep
Example	Paricutin, Mexico (inactive)	Kilauea, Hawaii	Soufrière Hills, Montserrat

*1 High pressure causes the lava to **spatter**, forming a loose material as it cools in the air. As the gas pressure drops, the lava seeps under the loose material through the sides of the cone (see Figure 14.12).

*2 Composite volcanoes have an **explosive stage** as it clears its vent of pyroclastic material and then a **non-explosive stage** as lava flows easily from its cleared vent. This produces **layers of ash, rock and cinder** followed by **layers of solidified lava** as shown in Figure 14.14.

Kick'em Jenny – a submarine volcano

Kick'em Jenny is a **submarine volcano** situated **8 miles north of Grenada** (see Figure 14.15). It lies along the **Lesser Antilles Volcanic Arc** (see Figures 14.15 and 14.16) where a series of volcanic islands exist and has a **maritime exclusion zone** of radius **1.5 km** surrounding it. Its lava is mainly **basaltic** and it produces **explosive as well as gentle** eruptions. In 1939, it caused a small Tsunami as it sent debris almost **300 m** into the air.

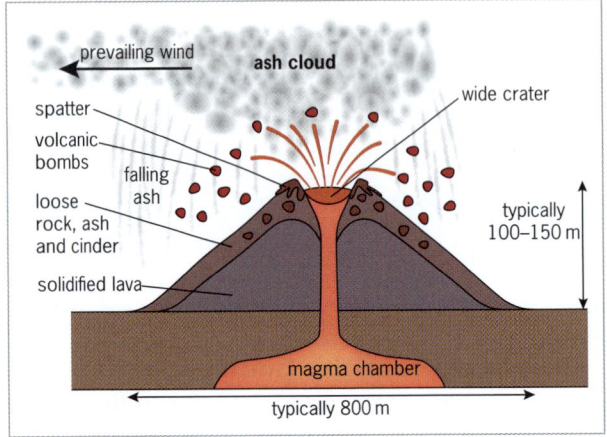

Figure 14.12 *Cinder volcano*   Figure 14.13 *Shield volcano*

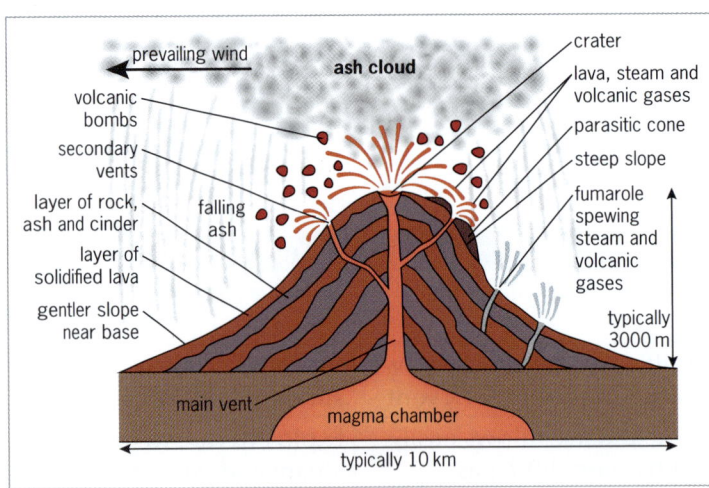

Figure 14.14 *Composite or stratovolcano*

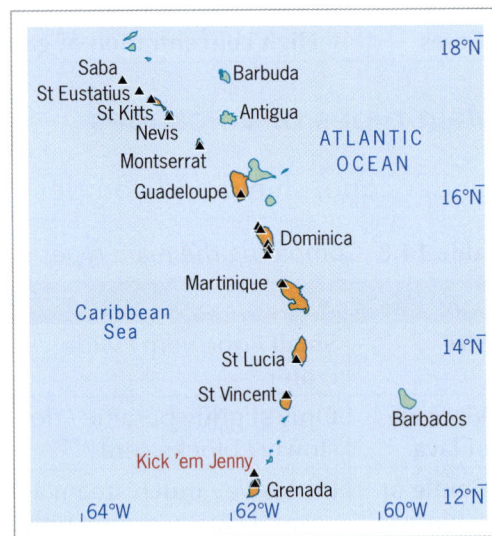

Figure 14.15 *Caribbean Volcanic Arc*

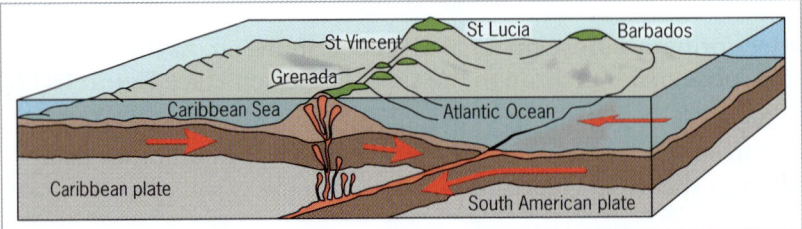

Figure 14.16 *Formation of the Lesser Antilles Volcanic Arc*

Ecological consequences of volcanoes

Table 14.9 *Ecological consequences of volcanoes*

Negative	Positive
Lava instantly destroys anything in its path.	Volcanic materials can contain important minerals.
Ash causes respiratory illness and falls on leaves, reducing photosynthesis.	Tourism can be boosted by volcanic landscapes such as calderas.
Landslides cause massive destruction.	Weathered lava eventually becomes fertile.
Destroyed landscapes reduce tourism.	Hot springs provide geothermal energy.
Animals killed, pollute the environment.	Hot springs can be therapeutic.

Earthquakes

Earthquakes are produced when the forces between tectonic plates result in a sudden burst of energy in the form of **shock waves** known as seismic waves.

The **hypocentre** is the point where an earthquake originates.

The **epicentre** is the point at the Earth's surface directly above the hypocentre.

A **seismograph** is an instrument that measures the vibrations of the ground.

The **Richter scale** is a scale of 1 to 10 used to compare earthquakes.

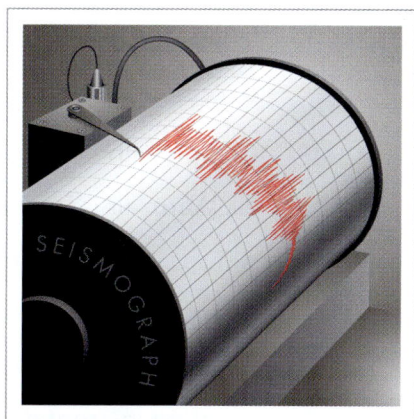

Figure 14.17 *Seismograph*

Table 14.10 *Interpreting the Richter scale*

Value	Typical effect	Value	Typical effect
< 5	Vibrates and rattles indoor objects	8–9	Damages earthquake-resistant buildings
5–6	Damages poorly constructed buildings	> 9	Causes near-total destruction, even to landscapes
6–8	Damages well-built structures		

Table 14.11 *Relationship between volcanoes and earthquakes*

Volcanoes	Earthquakes
Occur along the edges of tectonic plates.	Occur along the edges of tectonic plates.
Magma forced upward as plates interact.	Vibrations produced as plates interact.
Pressure from the magma causes **earthquakes**.	Cracks created act as magma vents for **volcanoes**.

Revision questions

1. **a** What are air masses? **b** List the FOUR main types of air masses.

2. List FOUR pollutants transported by air masses together with an associated problem of EACH.

3. **a** How does Saharan dust get into the air masses that bring it to the Caribbean and Americas?
   b List TWO advantages and TWO disadvantages of Saharan dust reaching the Caribbean.

4. List TWO advantages and TWO disadvantages of volcanic ash and/or lava.

5. Define the following: **a** cold front **b** warm front **c** stationary front **d** occluded front.

6. Complete the following: At a cold front, cold air .... warm air. Warm air is forced upward with a .... gradient at a cold front. Tall .... clouds occur at a cold front. Warm fronts move .... than cold fronts. Showers occur at a front if the warm air contains .... At the boundary of a warm front, various types of .... clouds are formed. A stationary front does not move if the winds blow .... to the front.

7. **a** In what rotational direction do winds blow around a low-pressure region in the northern hemisphere?
   b Name the force that produces this deflection of the winds and state what causes it.

8. Over what range of months do the following seasons occur?
   a Wet season. **b** Dry season. **c** Hurricane season.

9. Briefly describe how the ITCZ affects the weather in the region near to the equator.

10. **a** What is a tropical cyclone?
    b List the three classifications of tropical cyclones in the order they develop to become stronger.

11. **a** What are tides?
    b How many high tides and how many low tides generally occur in each day at a particular location?
    c Sketch a diagram showing positions of the Sun, Moon and Earth at a spring tide.
    d How are spring tides different than neap tides?

12. **a** Define: **i** a tidal wave **ii** a tsunami.
    b List THREE ways we can reduce the hazard of the impact of a tsunami.
    c Complete the following. In deep water, the amplitude of a tsunami is about .... and it travels .... As it approaches land, its .... decreases due to .... and its amplitude ....

13 Copy and complete Table 14.12.

Table 14.12 *Question 13*

	Tidal wave	Tsunami
Cause		
Predictability		
Impact		

14 Define: **a** volcano **b** fumarole **c** caldera.

15 a State whether high concentrations of each of the following increase or decrease the viscosity of lava:
   i silica **ii** basalt.
b State the effect of increasing the viscosity of lava on each of the following:
   i steepness of volcanic slope **ii** speed of lava flow
   iii possibility of blocking the vent **iv** explosive nature of eruptions.
c Of the shield, cinder cone and stratovolcano, state which:
   i is the smallest. **ii** is the tallest. **iii** is the widest.
   iv has only one vent. **v** emits lava with the highest silica content.
   vi can be called a 'spatter cone'.
   vii has profound alternate layers of solidified lava and pyroclastic material.

16 a What type of volcano is Kick'em Jenny?
b Which island is it nearest to, and how far is it from that island?
c How are persons protected from the dangers of this volcano?
d What is the name of the volcanic arc on which it exists?

17 State THREE negative and THREE positive ecological consequences of volcanoes.

18 a How are earthquakes produced?
b Distinguish between the hypocentre and the epicentre.
c What is: **i** a seismograph? **ii** the Richter scale?

19 What do the following values on the Richter scale indicate?
a Less than 5. **b** Between 6 and 8.

15 Water and the aquatic environment

Water is essential for life on Earth. The bodies of living organisms contain between 60% and 70% water and water covers about 71% of the Earth's surface. Water provides a **habitat** for aquatic organisms and humans **use** water in very many different ways. Water can also become **polluted** by human activities.

Properties of water

Water is the only natural substance that exists in **all three** physical states: solid, liquid and gas, at the temperatures found on Earth. Based on its **chemical composition**, water can be classified into **pure water**, **fresh water**, found in lakes, ponds, rivers and streams, and **seawater**, found in oceans and seas.

Pure water has several unique **chemical** and **physical properties**, which are affected by the presence of **dissolved salts**.

Table 15.1 *The chemical properties of pure water, fresh water and seawater compared*

Property	Pure water	Fresh water	Seawater
Chemical composition	Contains water (H_2O) molecules only.	Contains a very low concentration of dissolved salts.	Contains a higher concentration of dissolved salts than fresh water.
Salinity (concentration of dissolved salts)	Zero.	Less than 0.5 parts per thousand (ppt) or 0.05%.	Approximately 35 parts per thousand (ppt) or 3.5%.
pH	7.0 (neutral).	Typically 6.5 to 7.5 (very slightly acidic to neutral to very slightly alkaline).	Typically 7.5 to 8.5 (slightly alkaline).

Note: The composition of dissolved salts and pH of fresh water depends on various factors, e.g. the composition of the surrounding rocks and soil, use of the surrounding land and human activity.

The presence of **dissolved salts** in water **decreases** the melting point of pure water and **increases** its boiling point, density and electrical conductivity.

Table 15.2 *The physical properties of pure water, fresh water and seawater compared*

Property	Pure water	Fresh water	Seawater
Melting point at standard pressure	0 °C	Approximately 0 °C	Typically −1.9 °C
Boiling point at standard pressure	100 °C	Approximately 100 °C	Typically 100.7 °C
Density at standard temperature and pressure	1 g cm^{-3}	Approximately 1 g cm^{-3}	Typically 1.025 g cm^{-3}
Electrical conductivity	Very low	Slightly higher than pure water	High
Taste	Tasteless	Almost tasteless	Salty

Effects of fresh water and seawater on aquatic organisms

Aquatic organisms are faced with challenges created by the differences in concentration between the fluids inside their bodies and the surrounding water.

- **Fresh water** is **more dilute** than the body fluids of organisms living in the water; therefore, they are at risk of water **entering** their bodies and body cells by **osmosis**.

- **Seawater** is **more concentrated** than the body fluids of many organisms living in the water, except invertebrates; therefore, they are at risk of water **leaving** their bodies and body cells by **osmosis**.

To prevent the above, aquatic organisms, especially **fish** who are in danger of water entering or leaving their bodies through their **gills**, have developed certain adaptations to prevent this.

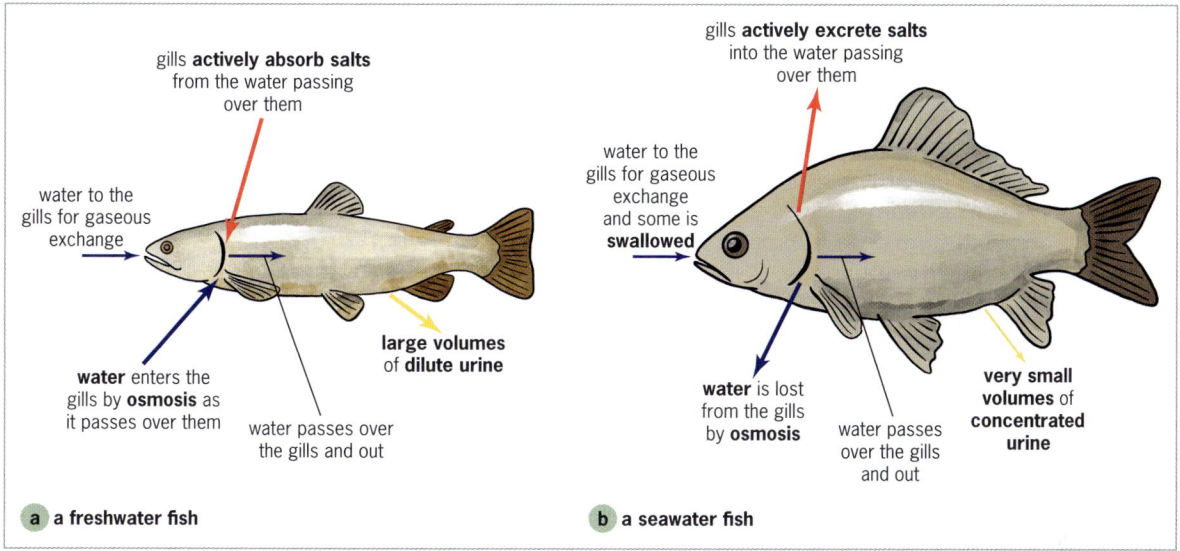

Figure 15.1 *Effects of fresh water and seawater on fish*

Hard and soft water

Water for domestic, industrial and agricultural purposes can be classified as **hard** or **soft**.
- **Hard water** contains **high** concentrations of dissolved **calcium** and **magnesium salts**, mainly calcium and magnesium hydrogencarbonate and calcium and magnesium sulfate. It does not lather easily with **soap**. When soap is added to hard water, insoluble, greasy **scum** forms (see page 239). Soap only lathers when all the calcium and magnesium ions have been converted to scum.
- **Soft water** contains very **low** concentrations of dissolved calcium and magnesium salts and lathers easily with soap; it does not form scum. Rainwater is soft.

Advantages and disadvantages of hard and soft water

Table 15.3 *Advantages and disadvantages of hard and soft water*

Type of water	Advantages	Disadvantages
Hard water	- When drunk, it is good for building strong bones and teeth due to the presence of calcium ions, and good overall health due to the presence of magnesium ions. - It does not dissolve lead from pipes, so does not contribute to lead poisoning in homes that still have lead water pipes. - It tastes better than soft water.	- It forms **scum** with soap, which discolours clothes and forms an unpleasant grey, greasy layer around sinks, baths and showers. - It **wastes** soap because soap only lathers and cleans when all the calcium and magnesium ions have to be removed as scum. - It causes **limescale (calcium carbonate)** to be deposited in boilers, hot water pipes, kettles and other household appliances. This can block pipes, waste electricity and reduce the lifespan of household appliances.

Type of water	Advantages	Disadvantages
Soft water	- It has good cleaning power because it does not form scum. - It does not waste soap because it lathers immediately. - Limescale does not build up in boilers, hot water pipes, kettles and other household appliances. This keeps plumbing systems efficient, saves energy and maintains the lifespan of household appliances. - It is gentle on the skin and hair.	- It does not help build strong bones and teeth and is not as good for overall health as hard water because it lacks calcium and magnesium ions. - It dissolves lead from lead water pipes, so it can cause lead poisoning. - If produced by softening hard water using an ion exchange process, it has a higher content of sodium ions than normal, which may lead to hypertension when drunk.

Figure 15.2 *Limescale (calcium carbonate) deposited on the heating element of a kettle*

Softening hard water

There are two **types** of water hardness: **temporary hardness** and **permanent hardness**.

- **Temporary hardness** is caused by dissolved **calcium** and **magnesium hydrogencarbonate**. It is found in limestone-rich areas and the hardness **can** be removed by **boiling** the water.
- **Permanent hardness** is caused by dissolved **calcium** and **magnesium sulfate**. Permanent hardness **cannot** be removed by **boiling**.

Hard water can be converted into **soft water**, or softened, by removing the dissolved calcium and magnesium ions.

- **Boiling** – This removes **temporary hardness** only. It causes dissolved calcium and magnesium hydrogencarbonate to decompose and form **insoluble** calcium and magnesium carbonate. These insoluble carbonates settle within the water or are deposited on surfaces as **limescale** or **kettle fur**, and this removes dissolved calcium and magnesium ions from the water.
- **Adding washing soda (sodium carbonate)** – This causes dissolved calcium and magnesium ions to form **insoluble** calcium and magnesium carbonate, which settle, thereby removing the dissolved ions.
- **Distillation** – Boiling the water and condensing the steam forms **pure distilled water** and leaves any dissolved salts behind.

Uses of water

Uses of water in life processes

Water's **unique properties**, especially its ability to **dissolve** a large number of substances, make it important to living organisms.

- Water **dissolves** chemicals so that **metabolic (chemical) reactions** can occur in cells. Enzymes, which catalyse (speed up) these reactions, also need to be dissolved to function.
- Water **dissolves** substances so that they can be **transported** around the bodies of living organisms, e.g. water in blood plasma dissolves and transports substances around the human body, and a solution of sucrose and amino acids is transported through the phloem of plants (see page 44).
- Water **dissolves** waste substances so that they can be **excreted** from the bodies of living organisms.
- Water present in digestive juices **dissolves** the digestive enzymes for **digestion**, and **dissolves** the products of digestion so they can be **absorbed** into the body fluids from the intestines.
- Water acts as a **reactant** in certain chemical reactions occurring in living organisms, e.g. **digestion** in animals and **photosynthesis** in green plants.

Uses of water in the home

- Water is used directly for **drinking** and it can be used to make various beverages, e.g. tea.
- Water is used for **cooking**. It is used to wash and prepare fresh produce, to boil, steam or poach foods and as an ingredient in many food items, e.g. soups, stews, sauces and gravies.
- Water is used for **washing** and **cleaning** purposes. It is used for bathing or showering, brushing teeth, washing hands, dishes and clothes, cleaning floors, surfaces and windows, flushing toilets and for washing patios, driveways and cars.

Water wastage and **water conservation** must be constantly monitored and addressed for **environmental sustainability**. Overusing and **wasting water** can lead to a decrease in worldwide water availability and water security, an increase in food scarcity and starvation, and the loss of habitats for freshwater organisms. To **conserve** water in the home:

- **Meter** all domestic water supplies.
- Regularly check for **water leaks** and immediately **repair** any leaks found.
- Install **water-saving devices** and **appliances**, e.g. **low-flow** shower heads and taps, **low-flush** or **dual-flush** toilets, **aerators** on taps and **water-efficient** washing machines and dishwashers.
- **Do not** leave taps running when brushing teeth, shaving, washing dishes and defrosting food.
- Use **grey water**, i.e. water that has been used for washing dishes, clothes or fresh produce, to flush toilets, wash cars and water gardens.
- Collect **rainwater** and use it to wash cars and water gardens.

Uses of water in agriculture

- Water is used to **irrigate crops**. This maximises yields, especially in areas with insufficient rainfall.
- Water is essential for **hydroponics**, i.e. growing crop plants in a nutrient-rich solution (see Table 2.2, page 16).
- Water is essential for **aquaculture** and **mariculture**. **Aquaculture** involves farming **aquatic organisms** in aquatic environments, i.e. fresh water, brackish water (usually found in estuaries) and seawater environments. **Mariculture** involves farming **marine organisms** in seawater environments. Organisms commonly farmed include tilapia, trout, salmon, sea bass, oysters, mussels, clams, shrimp, prawns, lobsters and seaweed. They can be farmed in ponds, lagoons, man-made tanks, cages and pens.

Figure 15.3 *Cages used to farm fish*

- Water is essential for rearing **livestock**, mainly for drinking and maintaining cleanliness.
- Water is used to dissolve **pesticides** so they can be sprayed on crops.

Other uses of water

- **Recreational activities** – Water is used for activities such as swimming, boating, surfing, water skiing, paddle boarding, scuba diving, cruising, water parks, fishing and ice skating (when frozen).
- **Firefighting** – Water is used to fight fires. It cools the burning material, and the steam produced when the water is heated displaces air containing oxygen from around the fire.
- **Generation of electricity** – **Thermoelectric** power plants use a source of fuel, e.g. oil, to boil water. The steam produced turns turbines, which drive a generator that converts the mechanical energy from the turbines into electrical energy. **Hydroelectric** power plants use the kinetic energy of flowing water to turn the turbines. **Tidal power**, created by the rise and fall of the tides, and **wave power**, created by ocean waves, can also be used to generate electricity.

Fishing methods used in the Caribbean

Fishing is a vital part of the Caribbean economy because it provides **employment** and **food**. Most commercial fishing is carried out at sea using various methods.

Table 15.4 *Fishing methods used in the Caribbean*

Method	Description
Handpicking	Shellfish, including crabs, oysters and sea eggs (sea urchins), are **handpicked** from intertidal zones or the seabed by free divers and scuba divers.
Spearfishing and harpoon fishing	**Spears** and **harpoons** are held by hand, thrown or shot from special guns. Spears are used mainly by free divers or scuba divers to catch smaller fish, e.g. snappers. Harpoons are used from boats to catch larger fish, e.g. tuna.
Netting	Fish are caught within the mesh of **nets**. The mesh size determines the size of the fish caught. Methods of net-fishing include: • **Cast netting** – A **circular net** with weights around its edge is thrown in shallow water close to shore. As the net sinks, it covers any fish below and traps them as the net edges are drawn together. • **Purse seining** – A **wall of netting**, with weights at the bottom and supported by floats at the top, is launched from a boat in a **circle** around a shoal of fish. A drawstring is pulled to close the bottom of the net and the fish are hauled aboard the boat. • **Trawling** – One or two boats tow a large **trawl net** through the water. Mid-water trawling catches fish in the open ocean. Bottom trawling catches fish close to or on the seabed and can damage the seabed.
Lining	Fish are caught using **lines**. Methods of lining include: • **Hand line** or **rod and line** – A baited hook on a line is thrown into the water. As a fish bites the bait, the hook catches in its mouth and the fish is pulled or reeled in. • **Long-lining** – A long **main line** is held floating horizontally near the surface. Shorter, vertical, **branch lines** with baited hooks are attached to it at intervals to catch fish swimming below the main line. Long-lining can kill seabirds and turtles.
Fish pots and traps	**Baited cages**, usually made of chicken wire attached to a wooden frame, are placed on the seabed with a surface buoy attached. Fish and shellfish enter through a cone-shaped funnel or trap door, and cannot get back out. Lost traps can turn into 'death traps' for fish.
Fish farming	Freshwater and seawater fish are raised commercially in **ponds**, **lagoons**, **tanks** or **enclosures**, e.g. cages and pens, submerged in natural bodies of water (see page 195).

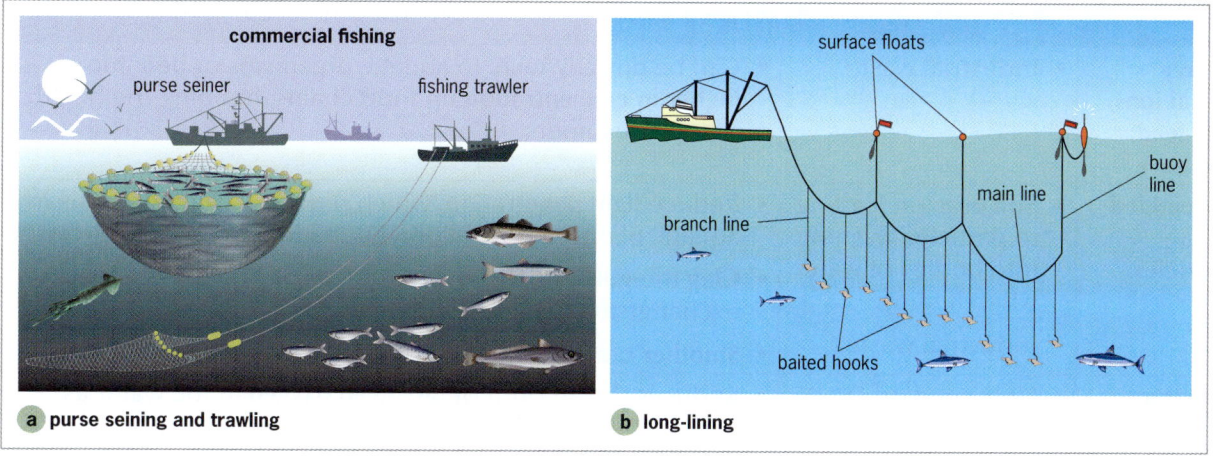

Figure 15.4 *Purse seining, trawling and long-lining*

Water pollution

Pollution is the contamination of the natural environment by the release of unpleasant and harmful substances or forms of energy into the environment.

*A **pollutant** is any unpleasant and harmful substance or form of energy that causes contamination of the natural environment when released.*

Water can be **polluted** by pollutants being **released directly** into **bodies** of water, e.g. lakes, streams, rivers, estuaries and oceans, from industrial, agricultural and domestic sources, or by being **washed off** the land into the water. These pollutants can harm aquatic organisms.

Table 15.5 *Sources of water pollutants and their effects on aquatic life*

Pollutant	Source	Harmful effects
Nitrates and phosphates (plant nutrients)	• Fertilisers used in agriculture. • Synthetic detergents. • Improperly treated sewage.	• Cause **eutrophication**, in which these added plant nutrients enrich the aquatic environment and cause the rapid growth of green plants and algae. This growth causes the water in lakes, ponds and rivers to turn green, which reduces light penetration and photosynthesis. When the plants and algae die, they are decomposed by **aerobic bacteria**, which multiply and use up **dissolved oxygen** in the water. This causes other aquatic organisms, e.g. fish, to die.
Pesticides, including insecticides, fungicides and herbicides	• Used in agriculture to control pests, pathogens and weeds. • Used to control vectors of disease, e.g. mosquitoes.	• Can be directly **toxic** to aquatic organisms, e.g. fish. • Can accumulate in the tissues of aquatic organisms and become **higher in concentration** up food chains where they can harm the health of the top consumers, e.g. large fish such as tuna, sharks and marlin, fish-eating mammals such as dolphins, whales and manatees, and fish-eating birds such as pelicans and eagles.
Oil	• Oil spills originating from oil tankers, offshore oil rigs and pipeline leaks. • Industrial activities, e.g. oil refining. • Runoff from roads and car parks.	• Chemical constituents of oil can be directly **toxic** to aquatic organisms, e.g. fish. • Forms **slicks** on the surface of water, which prevent **oxygen** from dissolving for aquatic organisms and block out **light** for aquatic plants. • **Coats** sea birds and mammals, causing birds to be unable to fly and both to be unable to keep warm. • **Smothers** and **kills** plants and animals living on the shore.

Pollutant	Source	Harmful effects
Heavy metal ions, e.g. lead, mercury	• Industrial waste.	• Can be directly **toxic** to aquatic organisms or become **higher in concentration** up food chains, harming the health of top consumers.
Suspended solid particles	• Soil erosion leading to particles washing into bodies of water when it rains. • Industrial waste.	• Reduce **light penetration**, which reduces photosynthesis in aquatic plants and coral polyps, leading to reduced growth. • **Clog** the gills of fish, which reduces or prevents gaseous exchange and can result in death. • **Smother** bottom-dwelling organisms, e.g. corals.
Organic matter	• Untreated or improperly treated sewage and industrial waste. • Manure and other farmyard waste.	• Reduces the amount of **dissolved oxygen** in the water as aerobic bacteria multiply and decompose the organic matter, which uses oxygen. Aquatic organisms die due to a shortage of oxygen, and their decomposition depletes dissolved oxygen levels further.

Figure 15.5 *Eutrophication*

Figure 15.6 *A shore crab covered in oil*

The effects of water pollution on aquatic ecosystems

Water pollution can have a detrimental impact on **coral reefs**, **mangrove swamps** and other **wetland ecosystems**, e.g. freshwater swamps, marshes and bogs. Loss of organisms from these ecosystems disrupts food chains and webs, leads to a loss of biodiversity and can destroy the ecosystems.

- **Coral reefs** are affected by reduced light penetration and smothering caused by **solid particles**, the toxic effects of **chemical pollutants**, e.g. pesticides, heavy metals and oil, and **reduced oxygen** levels caused by eutrophication, oil and organic matter, all of which kill coral polyps and reef organisms.
- **Mangrove swamps** are affected by **oil** smothering the roots of the mangrove trees, the toxic effects of **chemical pollutants** and **reduced oxygen** levels caused by eutrophication, oil and organic matter, all of which lead to the death of the trees and mangrove organisms, e.g. oysters, crabs and fish.
- Other **wetland ecosystems** are affected by **sediment** building up, which reduces the depth of the water and alters the habitat for organisms living in the wetlands, e.g. frogs, toads, snails, crayfish and fish. The toxic effects of **chemical pollutants** and **reduced oxygen** levels caused by eutrophication, oil and organic matter also lead to the death of wetland plants and animals.

Purifying water for domestic use

Sources of water for domestic use

Water used in homes comes from **surface water sources**, e.g. **springs**, **rivers**, **lakes** and **reservoirs**, and **groundwater sources**, e.g. **aquifers**. To make the water **potable**, i.e. safe to drink and use in food preparation, it must be **treated** to remove harmful contaminants, e.g. bacteria, viruses, dissolved chemicals and suspended solid particles. This usually occurs in **large-scale water treatment plants**. After treatment, the water should be clear and not discoloured, and have an acceptable taste and odour.

Seawater and **brackish water** (slightly salty water) can also serve as **sources** of potable water, especially in coastal areas. To make the water potable, it must be **desalinated**, i.e. have the salts removed.

After treatment or desalination, the water is piped to homes where it can be further **purified** to ensure that it is absolutely safe for human consumption (see Table 15.6).

Desalination of seawater for domestic use

Two main methods of **desalinating** water are **distillation** and **reverse osmosis**. These use large amounts of **energy**, therefore seawater and brackish water are used as water sources in locations where other sources are insufficient.

- **Distillation** – The water is boiled and the steam is condensed to form **pure distilled water** (see page 245). The dissolved salts and other impurities are left behind.
- **Reverse osmosis** – The water is forced through a **semi-permeable membrane** under pressure and the dissolved salts and other impurities remain behind on the pressurised side. It uses less energy than distillation.

Water purification methods in the home

Water can be **purified** in the home if the **quality** of tap water is questionable or it might contain **contaminants**, e.g. microorganisms or chemicals such as heavy metal ions and pesticides.

Table 15.6 *Methods used to purify water in the home*

Method	Explanation
Boiling	Water is brought to a **rolling boil** for 1 to 3 minutes to kill harmful microorganisms.
Filtration	Water is passed through a **filter** to remove unwanted particles. Domestic filters such as filter jugs and inline water filters usually contain a **filter cartridge**: • **Fibre mesh filters** contain a mesh of synthetic or glass fibres of varying pore size. • **Ceramic filters** contain a porous ceramic material of variable pore size. The smallest pores can filter out bacteria and protozoans. • **Activated carbon filters** usually contain **activated charcoal** and can remove many dissolved chemical contaminants, dissolved organic compounds, odours and unpleasant tastes.
Chlorination	**Chlorine tablets** can be added to water or 2 drops of **chlorine bleach** can be added to each litre of water. These release **chlorine** which kills harmful microorganisms.
Distillation	Water is boiled to produce steam, which is condensed to form **pure distilled water**, and any dissolved salts, other impurities and microorganisms are left behind.
Additives	Certain substances, e.g. **powdered alum (potassium aluminium sulfate)**, can be added to water to cause fine suspended particles to clump together and settle out.

Figure 15.7 *New cartridges for a domestic water filter on the left and used cartridges on the right*

Revision questions

1. Distinguish between fresh water and seawater in terms of:
   a. their chemical properties
   b. their physical properties.

2. Outline the challenges faced by seawater fish and explain how their bodies are adapted to overcome these challenges.

3. Distinguish between hard water and soft water, and suggest TWO advantages and TWO disadvantages of having hard water in the home.

4. Explain THREE ways of softening hard water.

5. Explain the significance of water's ability to dissolve a large number of substances to living organisms.

6. List THREE ways water is used:
   a. in the home
   b. in agriculture.

7. Describe how EACH of the following methods is used to catch fish in the Caribbean:
   a. harpoon fishing
   b. long-lining
   c. trawling.

8. Define the following terms:
   a. pollution
   b. pollutant.

9. Identify TWO sources of EACH of following water pollutants and outline the effects of EACH on aquatic organisms:
   a. pesticides
   b. oil.

10. What is eutrophication and what effects does it have on aquatic life?

11. a. Identify FIVE different types of water pollutants that can lead to the death of fish.
    b. Name THREE different ecosystems in the Caribbean that are affected by water pollution.

12. a. Discuss the different sources of water used in homes and explain the meaning of the term 'potable'.
    b. Explain THREE ways in which water can be treated in the home to make it safe to drink and TWO ways in which seawater can be desalinated for domestic use.

Conditions for flotation

Density, upthrust and flotation

Density is the mass per unit volume of a body.
Mass is the quantity of matter making up a body.

Finding the density of an irregularly shaped stone

The **mass** of a stone can be measured on a balance and its **volume** can be measured as the overflow produced when immersed in a displacement can filled to its spout, as shown in Figure 15.8.

$$\text{density} = \frac{\text{mass}}{\text{volume}} \qquad \text{i.e. } \rho = \frac{m}{V} \qquad \text{Density can be expressed in } \frac{g}{cm^3} \text{ or } \frac{kg}{m^3}$$

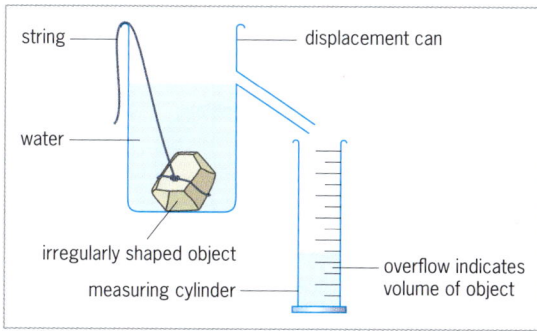

Figure 15.8 *Finding the density of a stone*

Weight is the gravitational force of a planet or other large body on an object.

weight (*W*) = mass (*m*) × gravitational field strength (*g*) ... *W* = *mg*

weight (*W*) = mass (*m*) × acceleration due to gravity (*g*)

On Earth, the gravitational field strength is **10 N/kg** and the acceleration due to gravity is **10 m/s²**. These can be used interchangeably: Since *W* = *mg*, an object of **mass 1 kg** is of **weight 10 N**.

The principle of Archimedes and the principle of flotation

*The **principle of Archimedes** states that when a body is wholly or partially immersed in a fluid it experiences an **upthrust** equal to the **weight of the fluid displaced** (pushed away).*
upthrust = weight of fluid displaced

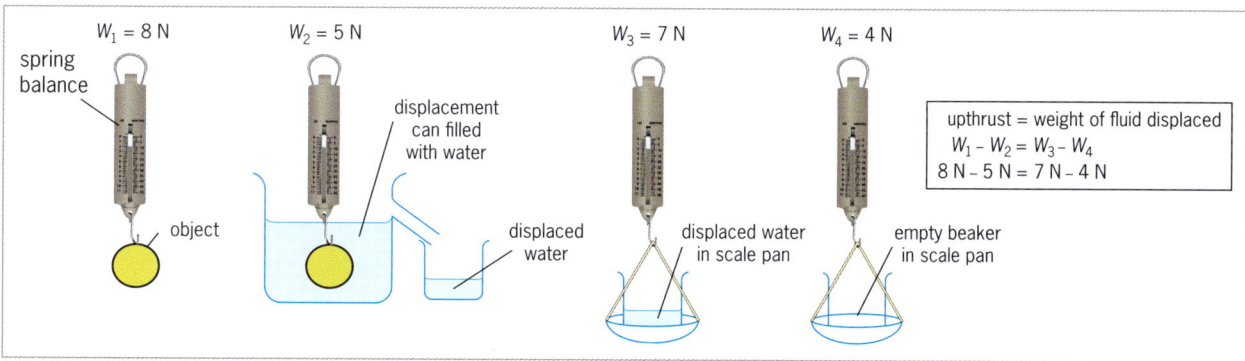

Figure 15.9 *Demonstrating the principle of Archimedes*

*The **principle of flotation** states that the weight of a floating body is equal to the upthrust on it.*
upthrust = weight of fluid displaced = weight of floating body

Sinking, floating and rising

Sinking – An object **sinks** if its **density is greater than that of the fluid it displaces**. The weight of the object is then greater than the upthrust (weight of fluid it displaces).

Floating – A floating ship displaces an amount of water equal to its weight in order to provide the necessary upthrust. Since **fresh water** is **less dense** than **seawater**, more fresh water is displaced to provide the balancing upthrust for a ship travelling from the ocean into the mouth of a river where the water is less dense. The same occurs when a ship travels from **cold water** to **less dense warm water**.

Rising – A hot air balloon rises since the weight of the cooler, denser air that it displaces provides an upthrust greater than the weight of the balloon and its contents. For the same reason, the balloon will also rise if the hot air within it is replaced with a **low-density gas** such as **helium**.

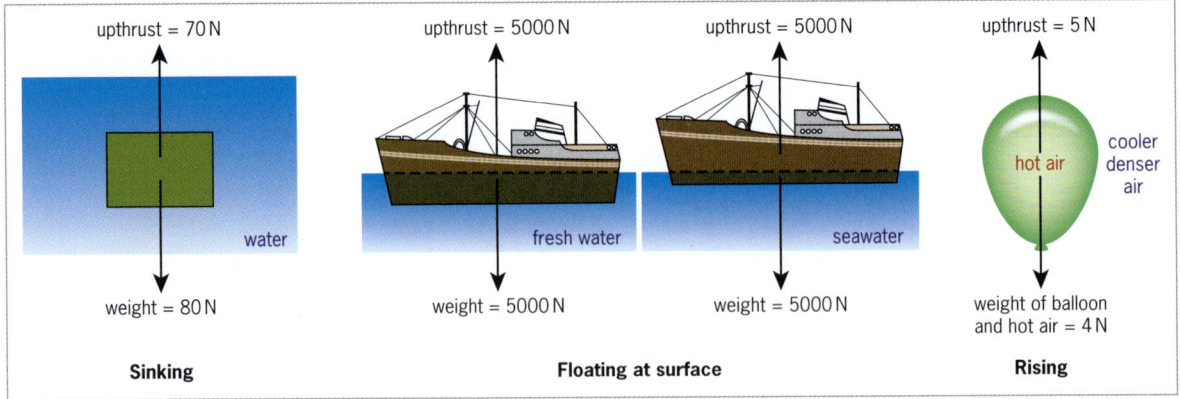

Figure 15.10 *Sinking, floating and rising in fluids*

Upthrust and submarines

Submarines can alter their weight by taking water into, or forcing water from, their **ballast tanks** as shown in Figure 15.11. The *darker blue* region of the diagram shows the **water in the tank** and the *hatch lines* indicate the **volume of water displaced**.

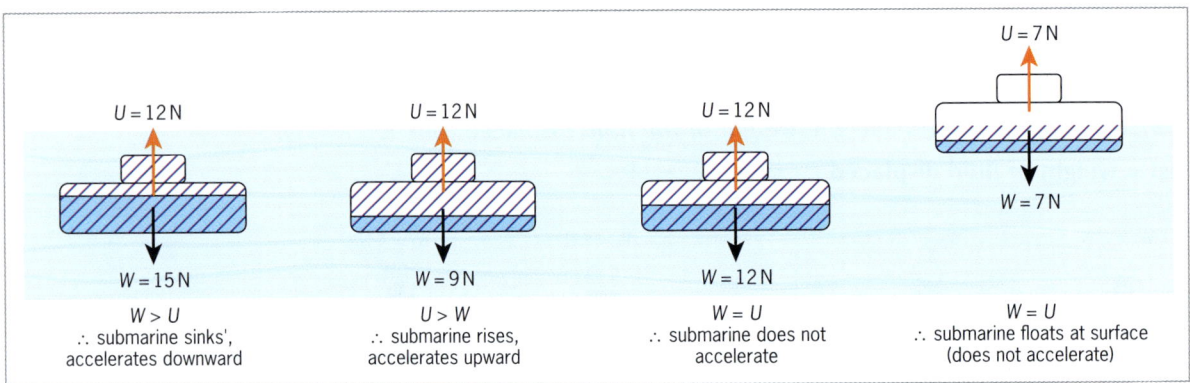

Figure 15.11 *Upthrust and submarines*

Sinking and rising in relation to density

Sinks by accelerating downward: density of object > density of fluid in which it is placed.

Rises by accelerating upward: density of object < density of fluid in which it is placed.

Example 1

A cube of volume 400 cm³ and density 6 g/cm³ is submerged in water.

a Calculate: i its mass   ii its weight.

b Determine the volume of water displaced.

c Given that the density of water is 1 g/cm³, calculate:

    i the mass of water displaced   ii the weight of water displaced   iii the upthrust.

Solution:

a   i   For the cube, density = $\dfrac{\text{mass}}{\text{volume}}$ $\therefore$ mass = density × volume

$$\text{mass} = 6 \, \frac{\text{g}}{\text{cm}^3} \times 400 \text{ cm}^3 = 2400 \text{ g}$$

ii First express mass in kg: $\text{mass} = 2400 \text{ g} = \frac{2400}{1000} \text{ kg} = 2.4 \text{ kg}$

∴ weight = mass × gravitational field strength (or acceleration due to gravity)

$\text{weight} = 2.4 \text{ kg} \times 10 \frac{\text{N}}{\text{kg}} = 24 \text{ N}$

b Volume of water displaced = volume of cube = 400 cm³

c i For the water displaced, $\text{density} = \frac{\text{mass}}{\text{volume}}$ ∴ mass = density × volume

$\text{mass} = 1 \frac{\text{g}}{\text{cm}^3} \times 400 \text{ cm}^3 = 400 \text{ g}$

ii First express mass in kg: $\text{mass} = 400 \text{ g} = \frac{400}{1000} \text{ kg} = 0.4 \text{ kg}$

∴ weight = mass × gravitational field strength (or acceleration due to gravity)

$\text{weight} = 0.4 \text{ kg} \times 10 \frac{\text{N}}{\text{kg}} = 4 \text{ N}$

iii Upthrust = weight of water displaced = 4 N

Plimsoll line

A *plimsoll line* is a line marked on a ship that indicates the maximum depth to which it should sink in water in which it is loaded.

Figure 15.12 shows plimsoll lines with a key corresponding to the type of water at the loading port.

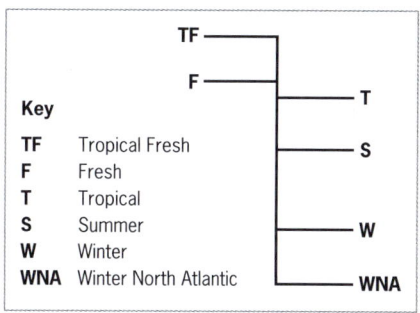

Key
TF Tropical Fresh
F   Fresh
T   Tropical
S   Summer
W   Winter
WNA Winter North Atlantic

Figure 15.12 *Plimsoll lines*

Navigational devices used at sea

Compasses

A *magnetic compass* consists of a magnetised needle mounted on a pivot such that it can spin freely and point along the Earth's magnetic field to the Earth's *'magnetic north'*, see Figures 15.13 and 15.14.

A *gyrocompass* consists of a fast-spinning disc which together with the rotation of the Earth determines the direction of geographical *'true north'*, see Figure 15.15.

The gyrocompass is preferred to the magnetic compass for navigation since it is more accurate – the steel hulls of ships distort the Earth's magnetic field.

Figure 15.13 *Magnetic compass*

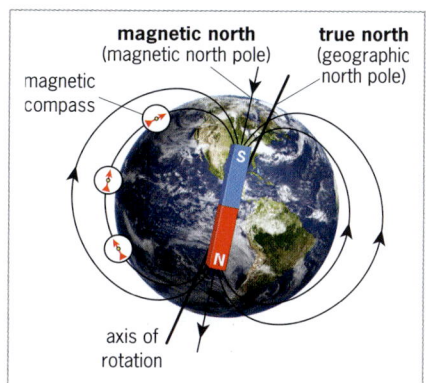

Figure 15.14 *Earth as a magnet*

Figure 15.15 *Gyrocompass*

Sound navigation and ranging (sonar)

The depth of water can be calculated using ultrasound by a technique known as **'depth sounding'**. A pulse emitted by a **transmitter, T**, reflects from the seabed to a **receiver, R**, and a timer measures the time taken for the round trip. Example 2 shows how the water's depth is calculated.

Sonar can also be used to **avoid collisions** with reefs and between submarines, to perform **search and rescue operations**, to **map the seabed**, and to **locate schools of fish**.

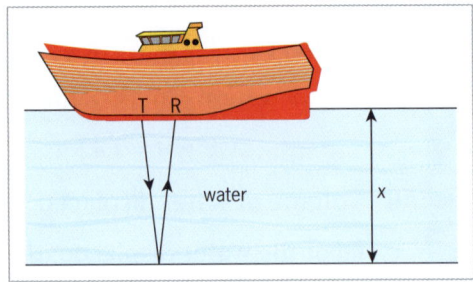

Figure 15.16 *Depth sounding*

Example 2

A ship emits an ultrasonic pulse to the seabed which returns in a time of 0.16 s (refer to Figure 15.16). Given that the speed of sound in water is 1500 m/s, determine the depth, x, of the water.

Solution:

$$\text{speed} = \frac{\text{distance}}{\text{time}} \qquad \therefore \text{distance} = \text{speed} \times \text{time}$$

$$2x = 1500 \, \frac{m}{s} \times 0.16 \, s \qquad 2x = 240 \, m \qquad x = \frac{240 \, m}{2} = 120 \, m$$

Radio Detection and Ranging (Radar)

A **rotating transmitter** sweeps a **narrow beam of radio pulses** across the water surface. On meeting an object, the **pulses are reflected** back to a **receiver** and displayed on a screen as a series of **'blips'**. Radar can be used to locate objects by measuring their bearings, speeds and distances from a vessel.

Global Positioning System (GPS)

This is a system of satellites (see page 171). Knowing the **speed of radio waves** through air and by measuring the **times** for radio pulses to be received by a GPS receiver from three of these satellites, the distances travelled to the receiver can be determined. From this, the location of the receiver can be calculated using a method of **trilateration**. A fourth satellite is required to correct the clock bias of the receiver since the satellites use expensive atomic clocks, whereas the receivers use less accurate quartz crystal clocks.

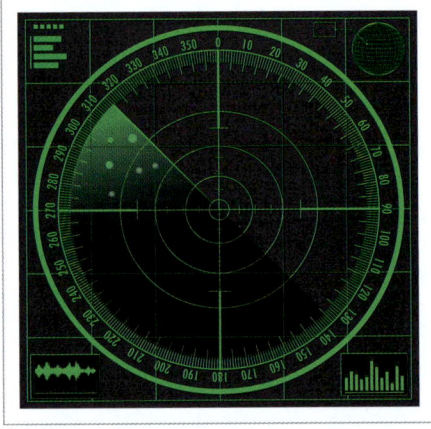

Figure 15.17 *'Blips' on radar screen*

Maritime safety standards

Maritime safety standards are a set of regulations and guidelines designed to ensure the safety of ships, crew members, passengers and cargo at sea.

Water safety devices

Water safety devices are designed to ensure the **safety** of persons taking part in water-related sporting or recreational activities. They can be **worn** by persons to help keep them **afloat** during the activity, e.g. swimming, or in the event of an accident, e.g. during sailing. They are also **carried** onboard larger watercraft and sea-going vessels for use in an emergency.

- **Life jackets** or **personal flotation devices (PFDs)** keep a person **afloat** even when unconscious. They are **worn** by persons taking part in water-related activities and are **carried** on sea-going vessels as part of their safety equipment. They contain **foam**, which is waterproof and buoyant, or are **inflated** by a cartridge of carbon dioxide when needed. The person is kept afloat because the jacket weighs less than the water displaced by the person.
- **Life rafts** are inflatable boats that are **carried** by larger watercraft and sea-going vessels as part of their safety equipment. They can be **manually inflated** using a **carbon dioxide cylinder** or they can **inflate automatically** on contact with the water. Persons in need of rescue then climb on board.
- **Inflatable tubes** or **rings** are buoyant flotation devices that are **worn** around a person's waist and can be used to keep non-swimmers and those learning to swim afloat.

a life jacket   b inflatable life raft   c inflatable tube

Figure 15.18 *Water safety devices*

Effects of diving on the human body

Humans can enter the underwater world by **free diving** or **scuba diving**. When **free diving**, divers hold their breath while underwater. When **scuba diving**, divers breathe underwater using self-contained underwater breathing apparatus (scuba). This usually consists of a **scuba tank** filled with **compressed air** or a **compressed breathing gas mixture**, e.g. nitrox, which contains a higher percentage of oxygen and lower percentage of nitrogen than air.

All divers experience **increasing pressure** from the surrounding water as they **descend**, and this affects the body and can lead to a number of potential **hazards**. Most affect scuba divers, not free divers, and include **decompression sickness, nitrogen narcosis, barotrauma** and **air embolism**.

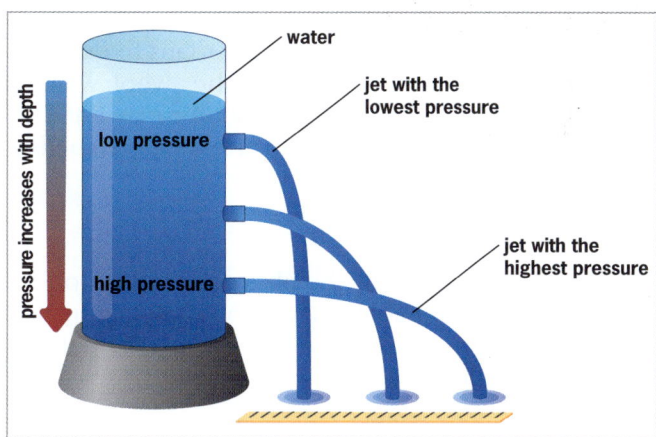

Figure 15.19 *Demonstrating that pressure increases with depth*

Decompression sickness or 'the bends'

The **solubility** of gases in liquids increases as pressure increases. As a scuba diver breathes compressed air from his or her tank during a dive, gases, mainly nitrogen, dissolve from the inhaled air into his or her blood and tissue fluids. The **deeper** the diver goes and the **longer** the diver's time at depth, the **greater** the amount of nitrogen dissolving.

As the diver **ascends**, pressure **decreases** causing the dissolved nitrogen to come out of solution. If the diver ascends **too quickly** and does not make any decompression stops to slowly eliminate the nitrogen, it forms **nitrogen bubbles** in the blood and tissue fluids during or soon after the ascent. These bubbles cause symptoms of **decompression sickness**, including joint pain, pressure bruising of the skin, tingling or numbness in the extremities, dizziness, shortness of breath and paralysis, and can be fatal.

Decompression sickness is **treated** by **recompression** in a **decompression** or **recompression chamber** as soon as possible to relieve symptoms, reduce the size of the bubbles, and help the nitrogen to dissolve back into the blood and tissue fluids so it can then be eliminated slowly.

Nitrogen narcosis

At depths greater than **30 m**, the nitrogen dissolved into a diver's blood and tissue fluids has a **narcotic effect** similar to drinking alcohol. This impairs the diver's judgment, sense of perception and decision making, reduces his or her coordination and creates a sense of detachment from the environment. The effects increase as the diver goes deeper and can be reversed by ascending to a **shallower depth**.

Barotrauma

Barotrauma refers to an injury that occurs when there is a **pressure difference** between the air-filled cavities in the body and the surrounding environment.

- **Ear barotrauma** can occur when there is a pressure difference between the middle ear and the surroundings, and can affect free divers and scuba divers. As a diver **descends**, the **ear drums** are pushed **inwards** by the increasing pressure, which can damage and even **rupture (burst)** them, causing vertigo. To prevent this, the diver must pinch the nose during descent and force air up the Eustachian tubes into the middle ears to **equalise** the pressure on each side of the ear drums.
- **Lung barotrauma** can occur when there is a pressure difference between the air inside the lungs and the surroundings. If a scuba diver inhales and starts to **ascend**, the air in the lungs **expands**. If the diver holds his or her breath or fails to exhale sufficiently, the air cannot escape and it stretches the lung tissue, which can cause the alveoli walls to **rupture**. This causes chest pain, difficulty breathing and coughing up blood. Air may enter the pleural cavity and cause a **collapsed lung**, or the bloodstream and cause an **arterial gas embolism**.

Embolism

An **embolism** is caused by the **obstruction** of a blood vessel. In a scuba diver, this can be caused by **air bubbles** or **nitrogen bubbles**. If a bubble blocks an artery, known as an **arterial gas embolism**, it can lead to a stroke, heart attack and even death. It is treated by immediate recompression in a decompression chamber to reduce the size of the bubbles.

Revision questions

13. Define density and provide a general equation which can be used to calculate density.

14. State: **a** the principle of Archimedes   **b** the principle of flotation.

15. A block of density 0.8 g/cm^3 has a volume of 5 cm^3.

    a Calculate its mass: **i** in grams   **ii** in kilograms.

    b Calculate its weight in newtons.

    c State with reason, whether the block will sink or float in water of density 1 g/cm^3.

16 Use Archimedes' principle to explain why a hot air balloon rises through the air.

17 Explain why a submarine can float at the surface of the ocean.

18 Explain why the water line (water level) on a ship rises as the ship travels from the ocean into the mouth of a river.

19 **a** What is a plimsoll line?

b Explain the importance of following the regulations set out by plimsoll line markings.

20 Briefly describe how the following devices work:

a magnetic compass **b** RADAR **c** GPS.

21 Why are gyrocompasses preferred to magnetic compasses for navigation?

22 **a** A ship emits an ultrasonic pulse to the seabed which returns in a time of 0.24 s. Assuming the speed of sound in water is 1500 m/s, determine the depth, x, of the water. Use the equation: $\text{speed} = \dfrac{\text{distance}}{\text{time}}$

b List TWO other ways that SONAR can be used by maritime vessels.

23 What are *maritime safety standards*?

24 Discuss the use of EACH of the following water safety devices:

a life rafts **b** life jackets **c** inflatable tubes.

25 Distinguish between free diving and scuba diving.

26 Outline what happens when a scuba diver suffers from EACH of the following:

a 'the bends' **b** lung barotrauma **c** nitrogen narcosis.

16 Forces

Forces play a key role in understanding how bodies interact in static and dynamic situations.

Types of forces and their effects

Principles of forces

A *force* is a push or pull which changes, or tends to change, the size, shape or motion of a body.

The **newton, N**, is the SI unit of force.

To find the **resultant force** of several forces acting along a line we assign **one direction** as **positive** and the **opposite** as **negative**. In Figure 16.1, forces directed to the right have been taken as positive.

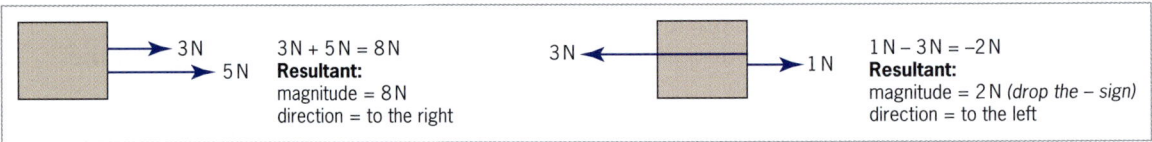

Figure 16.1 *Finding the resultant force of several forces*

Field forces (non-contact forces)

Forces due to **gravitational, electrical, magnetic** or **nuclear** fields.

Contact forces

Forces due to **physical contact with the environment**.
- **Friction** or **drag** forces prevent bodies from sliding over each other.
- **Tension** or **compression** forces extend or shorten materials.
- **Upthrust** due to Archimedes' principle acts upward on an object immersed in fluid.
- **Normal reaction** acts perpendicularly from a surface as a reaction to a force acting onto the surface.
- **Aerodynamic lift** acts upward on a body due to its shape and motion through the air.

Free-body force diagrams

A free-body diagram shows all the forces acting **ON** the selected body. For this course:
- **Field force** – weight of a body acting from its **centre of gravity** (page 211).
- **Contact forces** – forces from the environment acting on the body at contact points.

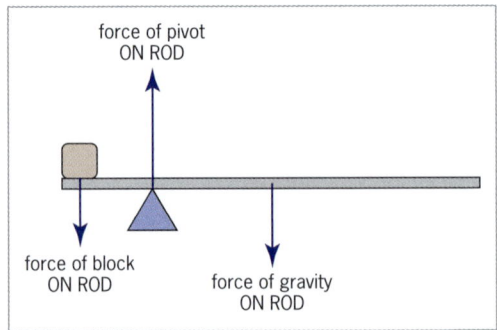

Figure 16.2 *Free-body diagram of a balanced rod*

Newton's laws of motion

Table 16.1 *Newton's laws of motion*

1st law	If there is no resultant force on a body, it continues in its state of rest or at constant velocity (no acceleration) in a straight line.
	resultant force = 0 ∴ **acceleration = 0**
2nd law	If there is a resultant force (net force) on a body, it produces a proportional acceleration.
	resultant force = mass × acceleration $F_R = ma$
3rd law	If body **A** exerts a force on body **B**, then body **B** exerts a force on body **A**, equal in *magnitude* but opposite in *direction*. (Every action has an equal but oppositely directed reaction.)

Note: A body can never have **both** forces of a **'Newton's 3rd law pair'** acting on it. For example, Figure 16.2 shows a free body diagram of the rod; we are only interested in the forces **ON THE ROD**. We are not concerned with the downward force of the rod **ON THE PIVOT** or the upward force of the rod **ON THE BLOCK**.

Note: 'Newton's 3rd law pairs' are easily identified by simply reversing the phrases. For example: force of *foot on floor* – force of *floor on foot*; force of *child on trampoline* – force of *trampoline on child*.

Applications of Newton's laws

- The force of brakes on a moving vehicle will cause the vehicle to decelerate (negatively accelerate). (**2nd law:** $F_R = ma$)
- Occupants of the vehicle not held by seatbelts will continue forward at constant velocity since the force is not on them. (**1st law:** $F_R = 0$ if $a = 0$)
- On colliding with objects in their paths, the *action force* exerted by the occupants on those objects results in a reaction force from those objects onto the occupants. (**3rd law: action = reaction**)

Example 1

Akeel pulls a block of weight 20 N across the floor using a force of 13 N against a frictional force of 7 N. A free-body diagram of the block is shown in Figure 16.3. Calculate:

a the mass of the block   b the acceleration of the block.

Solution:

a From ch15 page 203: $W = mg$ ∴ $m = \dfrac{W}{g} = \dfrac{20\ N}{10\ m/s^2} = 2\ kg$

b **Vertically:** acceleration = 0, therefore resultant force = 0

(Newton's 1st law: $F_R = 0$ if $a = 0$)

The normal reaction is therefore 20 N upward, cancelling the weight.

Horizontally: There is a resultant force (13 N – 7 N = 6 N).

(Newton's 2nd law: $F_R = ma$)

$F_R = ma$ ∴ $6\ N = 2\ kg \times a$ ∴ $a = \dfrac{6\ N}{2\ kg} = 3\ m/s^2$

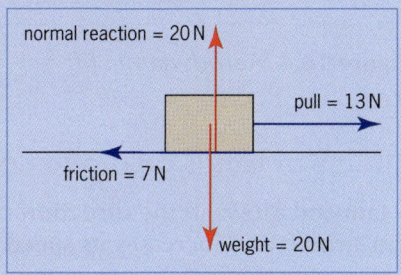

Figure 16.3 *Example 1*

Figure 16.3 *Example 1*

Friction

Friction is a force which opposes the motion of a body sliding over another body.

Advantages

Friction due to brakes **reduces the speed** of vehicles; friction between tyres and the road allows vehicles to **accelerate** and to **go around corners**; friction between our feet and the floor allows us to **walk and run**; friction between our fingers and an object helps us to **grip the object**.

Disadvantages

Friction can: **destroy surfaces**; **produce heat**; **reduce efficiency** and the **life** of machinery.

Reducing friction

To reduce friction: lubricate with oil, grease or a fine powder; use **wheels** or place **rollers** under objects; **sand** or **polish** surfaces; use **streamlined** shapes for objects sliding through liquids or gases.

Aerodynamic lift on an aircraft in flight

Figure 16.4 shows that as the wing of a plane (or bird) moves through the air, its smooth streamlined surface creates a pressure difference which results in an **aerodynamic 'lift' force**.

Figure 16.5 shows the forces acting on an aircraft as it moves horizontally through the air:
i at constant velocity and **ii** whilst accelerating.

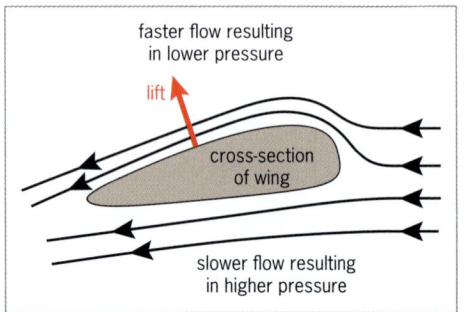

Figure 16.4 *Aerodynamic lift*

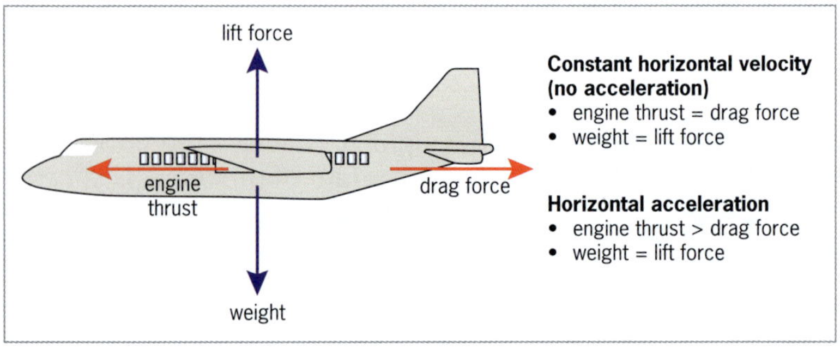

Figure 16.5 *Forces on an aircraft in horizontal flight*

Effects of winds on motion of aircraft

A **tailwind** blows in the direction of travel of the aircraft and **increases its speed**.

A **headwind** blows opposite to the direction of travel of the aircraft and **decreases its speed**.

Figure 16.6 shows how an aircraft can travel east when travelling through a **crosswind** headed north.

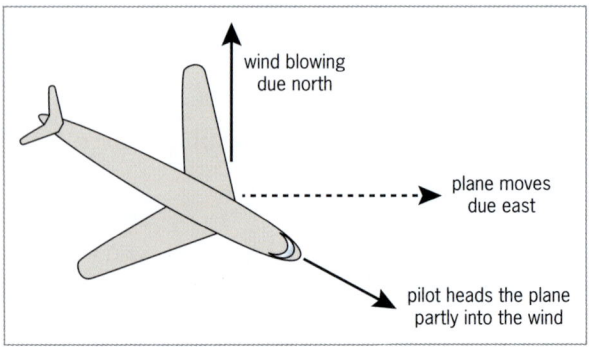

Figure 16.6 *Flying through a crosswind*

Gravitational force

Gravitational force is the attractive force between bodies due to their masses.

The **force is greater** if the **masses are greater** and if the **distance between the masses is smaller**.

> **Example 2**
>
> Determine the weight of a body of mass 4.0 kg on Earth if the acceleration due to gravity is 10 m/s².
>
> **Solution:**
>
> weight = mass × acceleration due to gravity $W = 4 \text{ kg} \times 10 \dfrac{m}{s^2}$ $W = 40 \text{ N}$

Centre of gravity

*The **centre of gravity** (C of G) of a body is the point through which the resultant gravitational force on the body may be **considered** to act.*

Locating the centre of gravity of regular shapes

- **Rectangle:** intersection of its diagonals.
- **Triangle:** intersection of its medians (lines from its vertices bisecting the opposite sides).
- **Circle:** midpoint of its diameter.

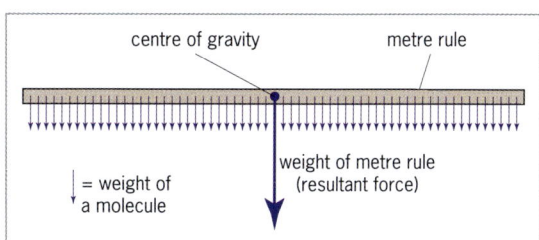

Figure 16.7 *Centre of gravity*

Locating the centre of gravity of irregular shapes

Figure 16.8 shows a **lamina** hung from a pin so that it swings freely. The position of a plumbline suspended from the pin is marked on the lamina. The process is repeated by suspending the lamina from other points near its edge. Where the lines intersect is the centre of gravity of the lamina.

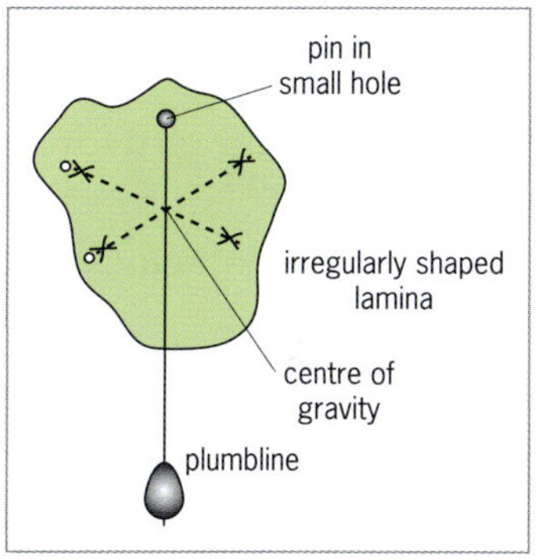

Centripetal force and circular motion

*A **centripetal force** is the force on an object required to keep it in circular motion.*

A centripetal force is **directed to the centre** of curvature of the path. A **greater force** is required for:

Figure 16.8 *Centre of gravity of an irregular lamina*

i a **greater mass**;   **ii** a **higher speed**;   **iii** a **smaller radius** of curvature.

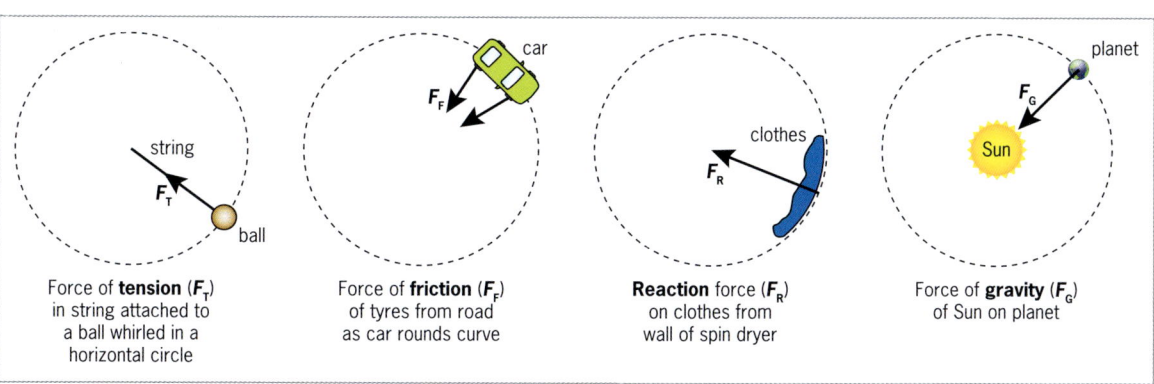

Figure 16.9 *Various types of forces acting as centripetal forces*

16 Forces

If the string in Figure 16.9 breaks, the centripetal force ceases and the object will **move along a tangent**. **Satellites depend on gravity** to keep them in orbit (refer to page 171, Figure 13.1).

Moments, equilibrium and stability

*The **moment of a force** about a point is the product of the force and the perpendicular distance of its line of action from the point.*

moment = force × perpendicular distance $M = F \times d_\perp$

Conditions for equilibrium under the action of parallel forces

1. The **sum of the forces** in any direction is equal to the **sum of the forces** in the opposite direction.
2. The **sum of the clockwise moments** about any point is equal to the **sum of the anti-clockwise moments** about that same point. (This 2nd rule is known as the **principle of moments**.)

Example 3

Demonstrating the conditions required of a body in equilibrium under the action of parallel forces.

A uniform metre rule of weight 2 N is suspended by a spring balance and weights are hung from it as shown in Figure 16.10 to keep it in balance. The spring balance reads 9 N.

The free-body diagram shows the **weight** (force of gravity) **ON THE RULE** and the three **contact forces** of the strings **ON THE RULE**. The following observations are noted:

1. sum of upward forces = sum of downward forces (9 N = 4 N + 2 N + 3 N)

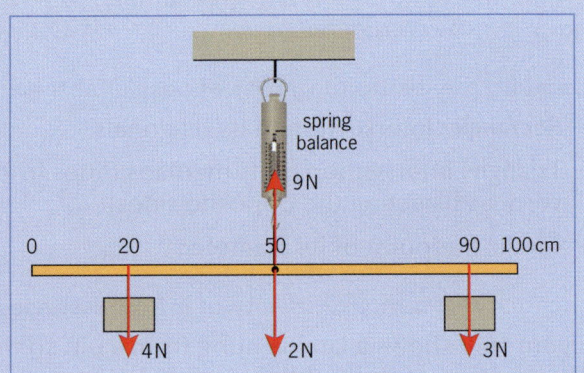

Figure 16.10 *Example 3*

2. sum of anti-clockwise moments = sum of clockwise moments (about centre of rule)

$$4 \text{ N} \times (50 - 20) \text{ cm} = 3 \text{ N} \times (90 - 50) \text{ cm}$$

$$4 \text{ N} \times 30 \text{ cm} = 3 \text{ N} \times 40 \text{ cm}$$

$$120 \text{ N cm} = 120 \text{ N cm}$$

VERY IMPORTANT: Moments have been taken about the pivot. The 2 N force and the 9 N force do not appear in the equation on moments – they have **no moment** about the pivot since they **act at the pivot** and so have **no distance** to the pivot.

Stable, unstable and neutral equilibrium

*A body is in **stable equilibrium** if, when slightly displaced, its centre of gravity **rises** and a **restoring moment** is created which returns it to its original position (its base).*

*A body is in **unstable equilibrium** if, when slightly displaced, its centre of gravity **falls** and a **toppling moment** is created which removes it from its original position.*

*A body is in **neutral equilibrium** if, when slightly displaced, its centre of gravity remains at the **same level** and **no moment** is created, leaving the body in the displaced position.*

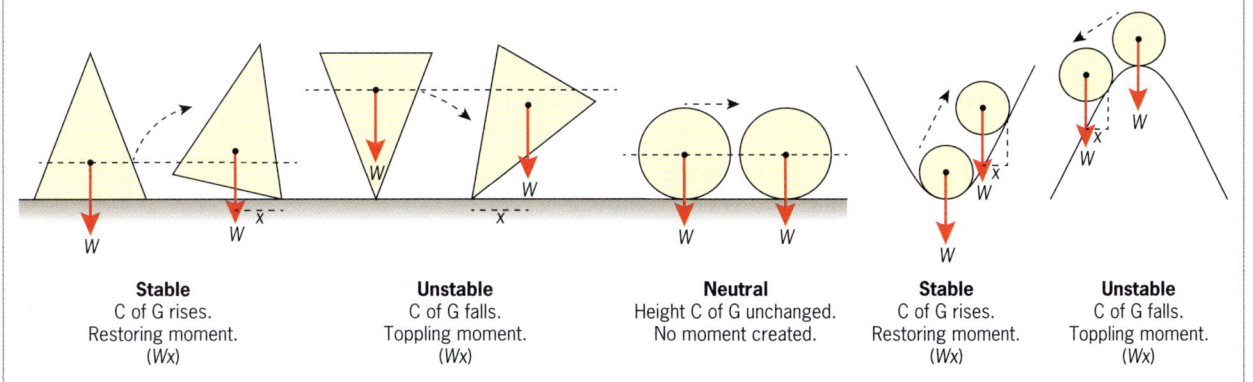

Figure 16.11 *Stable, unstable and neutral equilibrium*

Factors affecting the stability of an object

Figure 16.12 shows that the **stability** of an object **increases** if:

1. The **height** of its **centre of gravity decreases**.
2. The **width** of its base **increases**.

Go-karts have low centres of gravity and wide wheelbases to become more stable. Large buses have baggage compartments beneath the floor to lower their centre of gravity and enhance stability.

Tare weight and maximum loaded weight

Loading the top of a vehicle (e.g. a van or bus) raises its centre of gravity, making it less stable. Commercial vehicles therefore cannot exceed a maximum loading capacity. Their **tare weight** (unloaded weight) and their **maximum loaded weight** must be clearly marked on their sides.

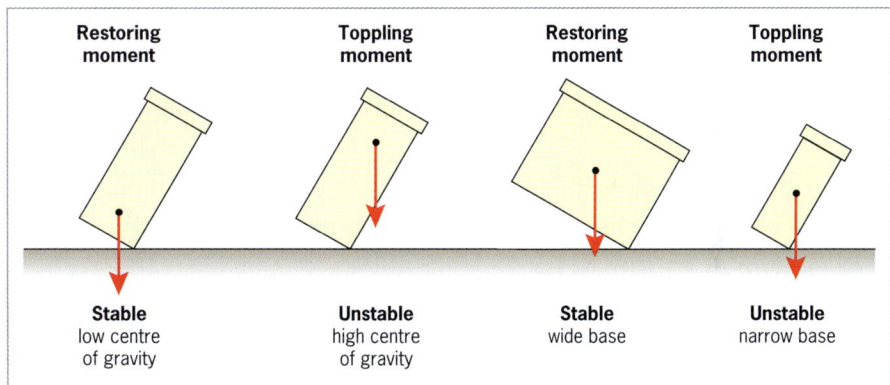

Figure 16.12 *Factors affecting the stability of an object*

Other applications of equilibrium

Homeostasis and biological equilibrium

*Homeostasis is the self-regulating biological process which maintains stability by monitoring and maintaining the various **physiological** and **behavioural** processes necessary for **optimum function**.*

Table 16.2 *Key applications of equilibrium in the body*

Osmoregulation	The **hypothalamus** detects **changes in blood osmolarity** and signals the release of **antidiuretic hormone** (ADH) to regulate **water balance in the kidneys**.
pH regulation	**Chemoreceptors** detect blood **pH changes**, triggering adjustments in **breathing rate** to **balance CO_2** levels and **maintain pH**.
Hormonal regulation	The **hypothalamus** senses the body's needs and directs the **pituitary gland** to release hormones that **regulate metabolism**, **growth** and other functions.

Blood pressure regulation	**Baroreceptors** in arteries **monitor blood pressure** and signal the **nervous system** to adjust **heart rate** and **vessel dilation** for stabilisation.
Thermo-regulation	**Temperature receptors** detect **fluctuations**, prompting the **hypothalamus** to activate responses like **sweating** or **shivering** to maintain optimal temperature.
Respiratory regulation	The **medulla oblongata detects CO_2** levels and controls **breathing rate** and depth to **optimise O_2** intake and **waste removal**.

Reversible chemical reactions and chemical equilibrium

A **reversible chemical reaction** is one in which the conversions of reactants to products and products to reactants occur simultaneously.

When the reaction begins, the forward reaction is faster, but as time goes on, the forward reaction decreases and the backward reaction increases until they reach and maintain **chemical equilibrium**.

Example: ammonium chloride ⇌ ammonia + hydrogen chloride

Linear momentum

The **linear momentum** of a body is the product of its mass and velocity.

momentum = mass × velocity $p = m \times v$

The **SI unit of momentum is kg m/s**.

The law of **conservation of linear momentum** states that for a system of colliding objects, their TOTAL momentum before the collision is equal to their TOTAL momentum after the collision.

Example 4

Car A of mass 1500 kg travels east at 15 m/s and car B of mass 2500 kg travels west at the same speed (Figure 16.13). Determine their total momentum.

Solution:

Designating **east as positive** and **west as negative**:

$p = (m_A \times v_A) + (m_B \times v_B)$

$p = (1500 \text{ kg} \times 15 \frac{m}{s}) + (2500 \text{ kg} \times -15 \frac{m}{s})$

$p = (22\,500 \text{ kg} \frac{m}{s}) + (-37\,500 \text{ kg} \frac{m}{s})$

$p = -15\,000 \text{ kg} \frac{m}{s}$

Total momentum: magnitude = 15 000 kg $\frac{m}{s}$ **direction** = west

Note: We drop the negative sign since it simply means 'west'.

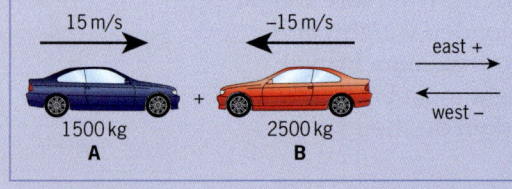

Figure 16.13 *Example 4*

Example 5

A toy car with a mass of 3 kg and speed 4 m/s to the right, collides head-on with another of the same mass which is initially stationary. Determine the following.

a The total momentum: **i** before the collision   **ii** after the collision.
b The magnitude (size) of their common velocity after the collision if they stick together.

Solution:

a i Taking to the **right as positive**, the total momentum **before** the collision is given by:

$p = (m_A \times v_A) + (m_B \times v_B)$ $p = (3 \text{ kg} \times 4 \frac{m}{s}) + (3 \text{ kg} \times 0)$ $p = 12 \text{ kg} \frac{m}{s}$

Momentum: magnitude = 12 kg $\frac{m}{s}$ direction: to the right (since the result is positive)

  ii From the principle of conservation of linear momentum, the total momentum **after** the collision must also be 12 kg m/s.

b After colliding, the masses stick together and have a **common velocity**, *v* (Figure 16.14).

$$p = (m_A + m_B)v$$
$$12 \text{ kg} \frac{m}{s} = (3 \text{ kg} + 3 \text{ kg})v$$
$$\frac{12 \text{ kg} \frac{m}{s}}{6 \text{ kg}} = v$$
$$v = 2 \frac{m}{s}$$

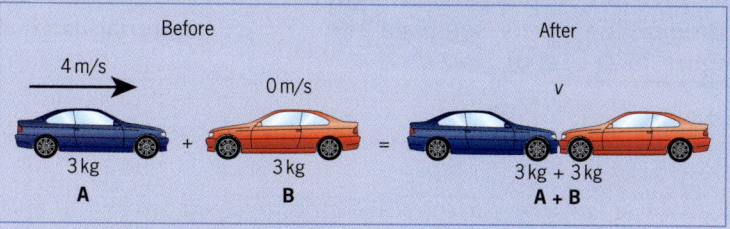

Figure 16.14 *Example 5*

Common velocity: magnitude = 2 m/s   direction = to the right (since the result is positive)

Understanding machines

*A **machine** is a device that makes doing work (using energy) easier.*

A machine **does not make doing work less**; it simply **alters** the **force** and **distance** components of work.
Recall Chapter 8, page 89:   **work (or energy) = force × distance (moved in direction of force).**

***Load** is the force to be overcome (usually the weight) without the use of the machine.*

***Effort** is the force required using the machine.*

***Useful work (energy) output** is the work (energy) needed to move **the load** without the machine.*

  useful work (energy) **output** = load × distance moved by load

***Work (energy) input** is the work (energy) needed by **the effort** in using the machine.*

  useful work (energy) **input** = effort × distance moved by effort

*The **law of conservation of energy** states that energy cannot be created or destroyed but can be transformed from one type to another.*

∴ **total energy output = total energy input**

However, unless the machine is IDEAL (100% efficient), only part of the output is **useful** because some is **wasted** as **heat** due to **friction** or **resistive forces** other than the load.

Efficiency

$$\text{efficiency} = \frac{\text{useful work (energy) output}}{\text{work (energy) input}} \times 100\%$$

$$\text{efficiency} = \frac{\text{load} \times \text{distance moved by load}}{\text{effort} \times \text{distance moved by effort}} \times 100\% \quad \text{efficiency} = \frac{L \times d_L}{E \times d_g} \times 100\%$$

An **IDEAL** machine has an **efficiency of 100%** (i.e. no energy is wasted).

Note: Only the **useful** energy output (due to moving the **load**) is used. The **wasted** energy output (usually due to **friction**) is not used in calculating efficiency.

Table 16.3 *Inefficiencies and ways to overcome them*

Factors causing inefficiency	Reducing inefficiencies
Friction between moving parts.	Reduce **friction** by using grease, oil or a fine powder.
Corrosion leads to instability and friction.	Reduce **corrosion** by rustproofing, greasing or oiling.
Resistive forces such as the weight of rising pulley blocks (see page 219, Figure 16.21 b, c, d).	Pulley systems should be made of strong, light materials such as aluminium to reduce **resistive forces**.

Mechanical advantage, force multipliers and distance multipliers

Mechanical advantage is the force amplification benefit of using a machine. $\left(MA = \dfrac{load}{effort}\right)$

Force multiplier – A machine in which **effort** < **load** ... (MA > 1).

Distance multiplier – A machine in which **distance** moved by **effort** < **distance** moved by **load** ... (MA < 1).

Consider a job where **work output** = $L \times d_L$ = 20 N × 4 m = 80 J

Using IDEAL machine A: **work input** = $E \times d_E$ = 10 N × 8 m = 80 J ...force multiplier

Using IDEAL machine B: **work input** = $E \times d_E$ = 40 N × 2 m = 80 J ...distance multiplier

MA of **A** = $\dfrac{20\ N}{10\ N}$ = 2   MA of **B** = $\dfrac{20\ N}{40\ N}$ = 0.5   ... but they have the same efficiency (100%)

NOTE: MA of **A** > MA of **B** but they have the **same efficiency**, since they use the **SAME ENERGY of 80 J**.

A is preferred if the user **prefers the smaller force of 10 N**.

B is preferred if the user **prefers to push through the smaller distance of 2 m**.

If a machine becomes rusted, its MA decreases since a greater **effort** is required to overcome friction.

Types of machines

Inclined plane

Example 6

Figure 16.15 shows a mass of 4 kg being pushed along an inclined plane through a distance of 5 m.

a Calculate the following.
  i The weight of the load.
  ii The work done by the effort (work or energy input).
  iii The work done against the load (useful work or energy output).
  iv The efficiency.

b Account for the difference between the work output and work input.

c Why is the ramp used even though more energy is needed to raise the object?

d Is the machine a force or distance multiplier?

e Calculate the mechanical advantage.

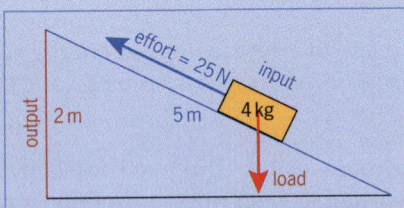

Figure 16.15 *Example 6*

Solution:

a i   weight = mg ∴ weight = 4 kg × 10 $\frac{m}{s^2}$ weight = 40 N
   ii work (energy) input = effort × distance moved by effort = 25 N × 5 m = 125 J
   iii useful work (energy) output = load × distance moved by load = 40 N × 2 m = 80 J
   iv efficiency = $\frac{\text{useful work (energy) output}}{\text{work (energy) input}}$ × 100   efficiency = $\frac{80\text{ J}}{125\text{ J}}$ × 100 = 64%

b More work is done (energy is used) using the ramp since **friction** has to be overcome. The energy used to overcome this friction is lost as **heat**.

c The user prefers to use the smaller force of 20 N; to lift it vertically would require 40 N.

d The machine is a force multiplier since the effort is less than the load.

e Mechanical advantage = $\frac{\text{load}}{\text{effort}}$ MA = $\frac{40\text{ N}}{25\text{ N}}$ MA = 1.6

Lever

*A **lever** is a pivoted rod on which an effort can be applied to overcome a load.*

In each of Figures 16.16 to 16.20, an effort (E) is balancing a load (L) and the lever is pivoted at P.

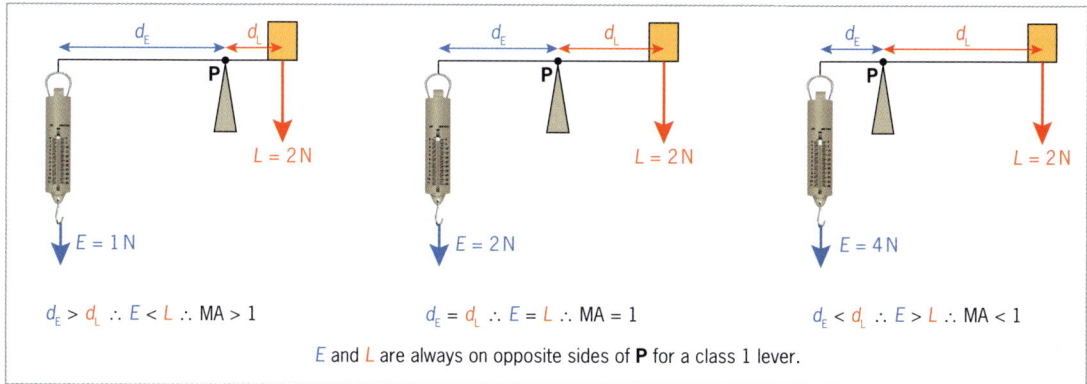

Figure 16.16 *Class 1 lever*

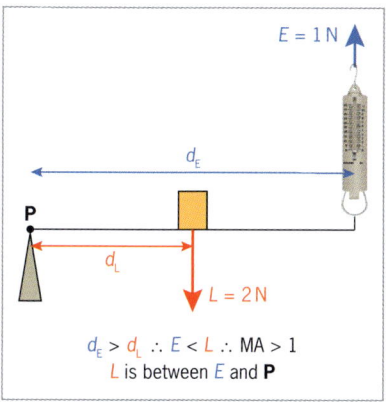

Figure 16.17 *Class 2 lever*

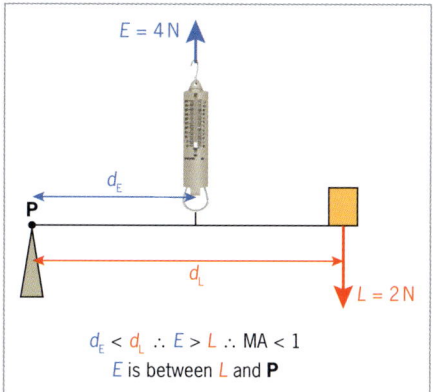

Figure 16.18 *Class 3 lever*

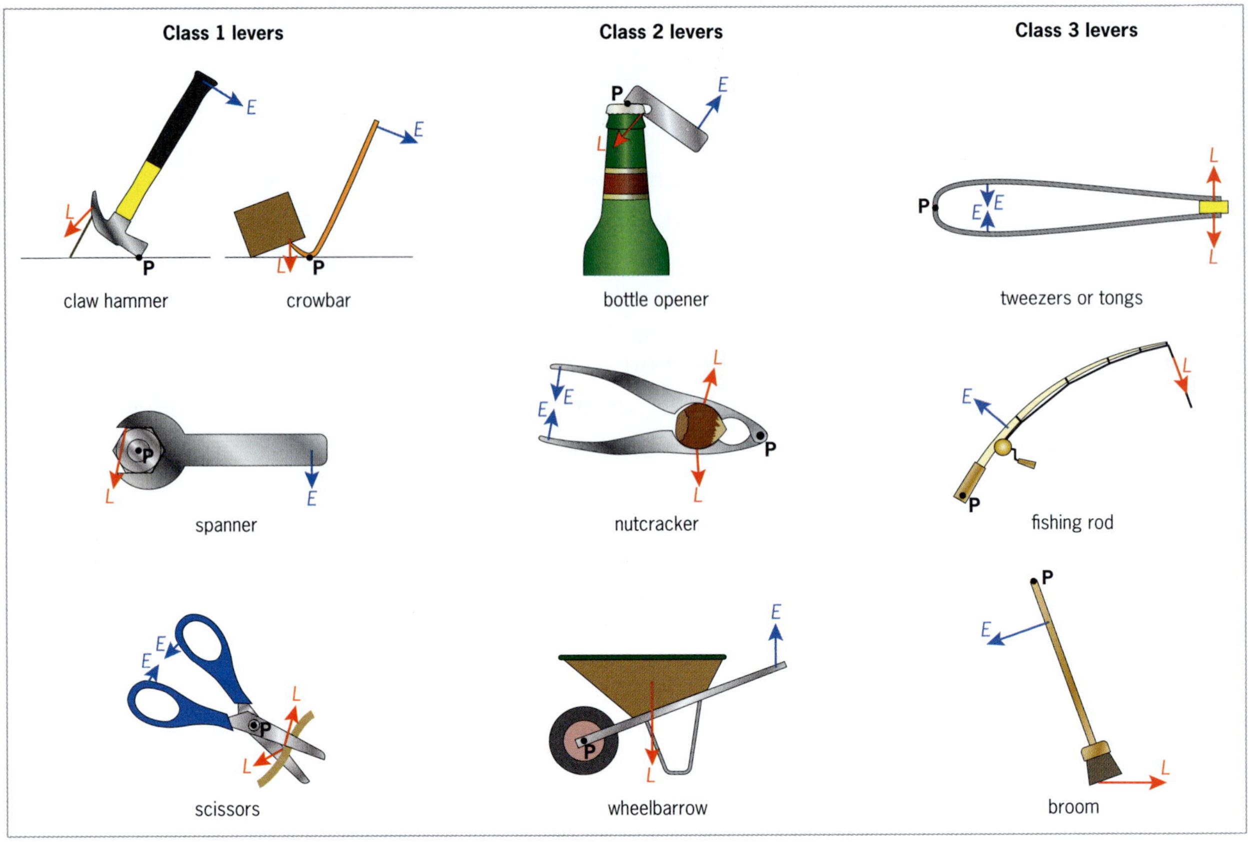

Figure 16.19 *Examples of practical levers*

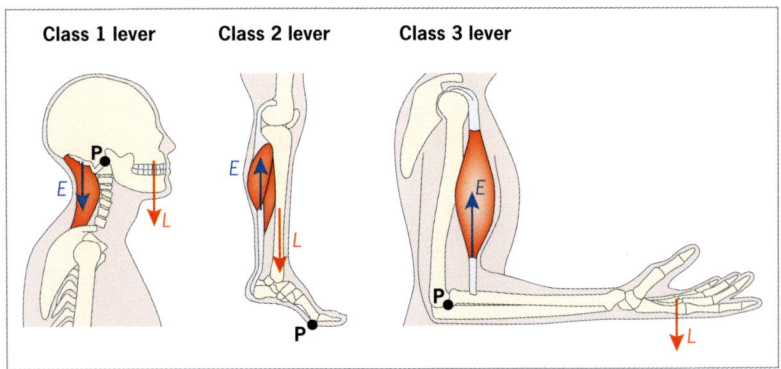

Figure 16.20 *Levers of the mammalian skeleton*

Pulley

- For simplicity, the following pulleys are assumed to be ideal (100% efficient) and therefore the work or **energy input** is **equal** to the work or **energy output**.
- An effort (E) applied to the string sets up an equal tension (T) within it, which **raises** the load (L).
- The sum of **these** tensions (T) is equal to the load (L) and so the greater the number of strings **raising** the load, the smaller is the required effort.
- Although the load is equal to the effort in Figure 16.21a, this machine is advantageous since:
  - the user may use his/her body weight to provide part of the effort
  - loads may be raised to great heights without the user having to move through those heights.
- In 16.21c and 16.21d, the user can also use his/her body weight to provide part of the effort.
- Draglines use pulley systems connected to their scoops to move very heavy loads (see Figure 16.22).

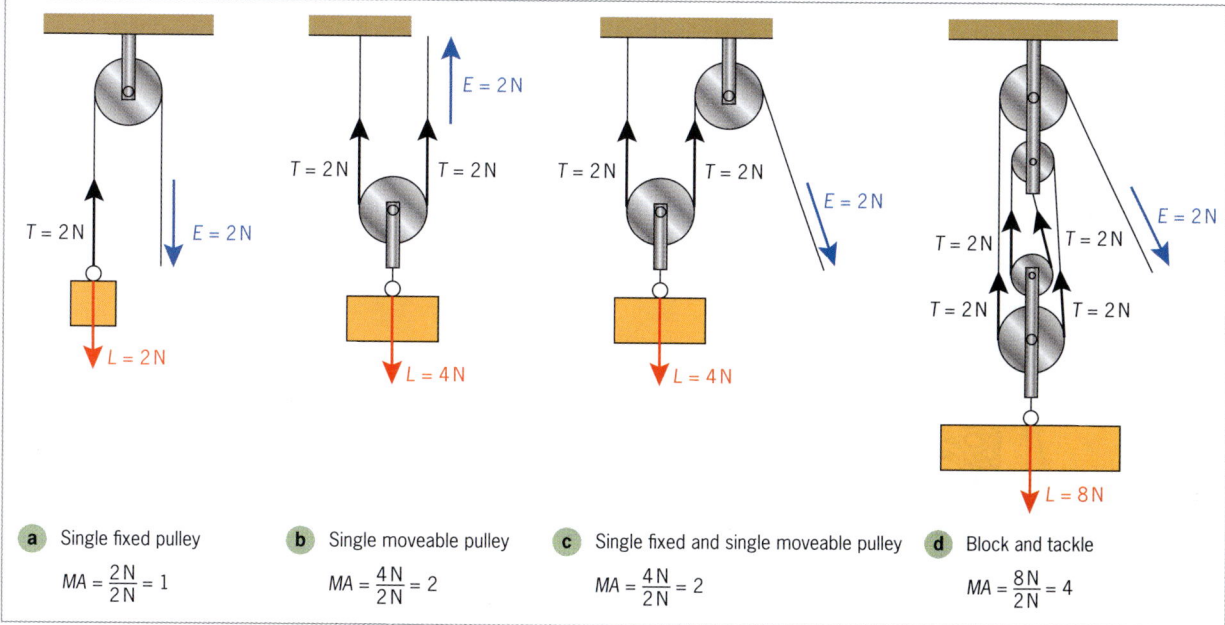

Figure 16.21 *Different types of pulley arrangements*

Figure 16.22 *Dragline excavator*

16 Forces **219**

Gears

One rotation of the pedals of a bicycle connected to the larger cogged wheel shown in Figure 16.23 will cause 1.5 rotations of the smaller cogged wheel, which is connected to the back wheel of the bicycle.

gear ratio = $\dfrac{\text{no. of teeth of driving wheel}}{\text{no. of teeth of driven wheel}} = \dfrac{24}{16} = 1.5$

Gears in a car's transmission work in the same way as the up and down motion of the engine's pistons transfers motion via a crankshaft to the driving cogged wheel.

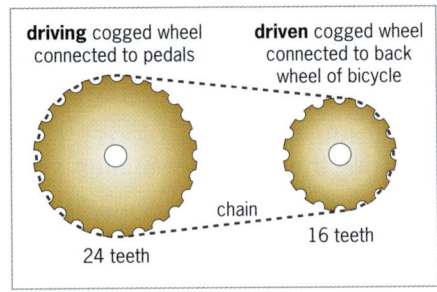

Figure 16.23 Cogged wheels

The human skeleton

Structure of the human skeleton

The **human skeleton** provides the **framework** for the body and is surrounded by skeletal muscles. An adult human skeleton is composed of 206 bones, which are held together at **joints** by tough, elastic **ligaments**. **Movement** is brought about by **skeletal muscles** working across these **joints**.

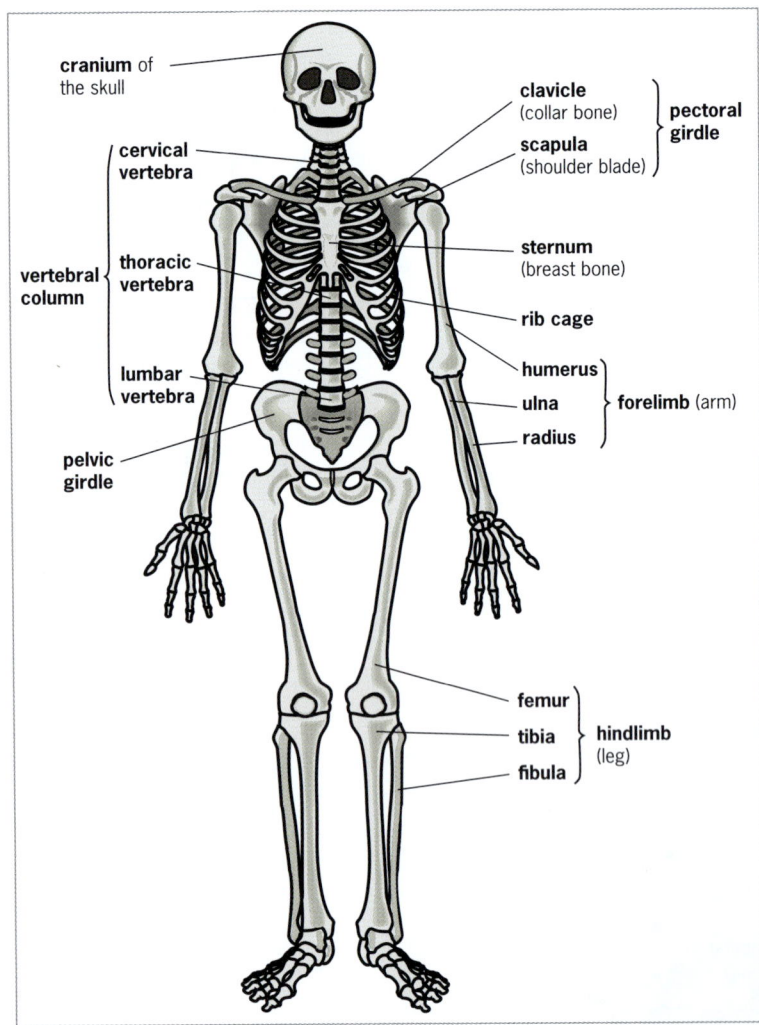

Figure 16.24 *The human skeleton*

The skeleton is made mainly from **two** types of tissues: **bone** and **cartilage**, and it can be divided into the **axial skeleton** and the **appendicular skeleton**.

The axial skeleton

The **axial skeleton** consists of the **skull**, **vertebral column**, **ribs** and **sternum**.

- The **skull** is made up of the **cranium** and **upper jaw**, which are fused, and the **lower jaw**, which articulates with the upper jaw. The skull encloses and protects the brain and sense organs of the head.
- The **vertebral column** or **spine** is composed of 33 bones called **vertebrae**, which have **intervertebral discs** of cartilage between. The **spinal cord** runs through a hole in the center of each vertebra. The column supports the head and the body, protects the spinal cord and allows some movement.
- The **ribs** are attached to the vertebral column, and together with the **sternum** or **breast bone**, they form the **rib cage**. The rib cage encloses and protects the heart, lungs and major blood vessels, and movements of the ribs and sternum are essential for **breathing**.

The appendicular skeleton

The **appendicular skeleton** consists of the **pectoral** and **pelvic girdles**, the **arms** (forelimbs) and the **legs** (hindlimbs).

- The **girdles** connect the limbs to the axial skeleton. They have broad, flat surfaces for the attachment of muscles that move the limbs. The **pelvic girdle** is fused to the bottom of the vertebral column to provide **support** for the lower body and to transmit the **thrust** from the legs to the vertebral column, which moves the body forwards.
- The **limbs** are composed of long bones which have **joints** between them to allow for easy **movement**. Being long, the bones provide a large surface area for the **attachment** of muscles, and those of the arms allow the arms to have a long **reach**, whilst those of the legs permit long **strides** to be taken.

Functions of the skeleton

The human skeleton has **five** main functions.

- **Movement** – The skeleton is jointed and muscles work across these joints to bring about **movement**. Most movement is brought about by the **legs** and **arms**, whilst the **vertebral column** allows some movement.
- **Protection** – The skeleton **protects** the internal organs. The **skull** protects the brain and sense organs of the head, i.e. the eyes, ears, nasal cavities and tongue. The **vertebral column** protects the spinal cord. The **rib cage** and **sternum** protect the lungs, heart and major blood vessels. The **pelvic girdle** protects the internal reproductive organs, bladder and lower part of the digestive system.
- **Support** – The skeleton **supports** the body's soft parts and internal organs, and provides the framework that gives **shape** to the body. The **vertebral column, pelvic girdle** and **legs** are responsible for providing support.
- **Production of blood cells** – Red blood cells, most white blood cells and platelets are produced in the **red bone marrow**, found inside **flat bones**, mainly the pelvis, scapula, ribs, sternum, cranium and vertebrae, and in the **ends** of the **long bones** of the limbs, mainly the humerus and femur.
- **Breathing** – Alternate contractions of the external and internal intercostal muscles between the ribs bring about movements of the **rib cage**, which draw air into the lungs and expel air from the lungs (see page 116).

Joints

*A **joint** is formed where two bones meet.*

Most joints allow the rigid skeleton to **move**. There are **three** types of joints in the human body.

- **Fixed joints** or **fibrous joints** – The bones are joined together firmly by **fibrous connective tissue** that allows **no** movement, e.g. the cranium of the skull is made of eight bones joined by fixed joints.

- **Partially movable joints** or **cartilaginous joints** – The bones are separated by **cartilage pads** which allow **slight** movement, e.g. the vertebrae are separated by intervertebral discs of cartilage.
- **Movable joints** or **synovial joints** – The articulating surfaces of the bones are covered with slippery **articular cartilage** and **synovial fluid** fills the joint cavity between the bones to provide **friction-free movement**. The bones are held together by tough, elastic **ligaments**. There are **three** types.
    - **Hinge joints** are formed where the **ends** of bones meet. They allow movement in **one plane** (direction) only. This limited movement provides **strength** and the joints can bear heavy loads, e.g. the elbow and knee joints, and the joints in the fingers and toes.
    - **Ball and socket joints** are formed where a **ball** at the end of one bone fits into a **socket** in the other bone. They allow **rotational** movement in **all planes**. This free range of movement provides less support and makes the joints more susceptible to dislocation than a hinge joint, e.g. the shoulder and hip joints.
    - **Gliding joints** are formed where the articulating surfaces of bones are **flat** or **nearly flat**. They allow the bones to **glide** or **slide** past each other in various directions along the **same plane** as the articulating surfaces, e.g. between the bones of the wrists and the ankles, where the joints provide **flexibility** and **stability**.

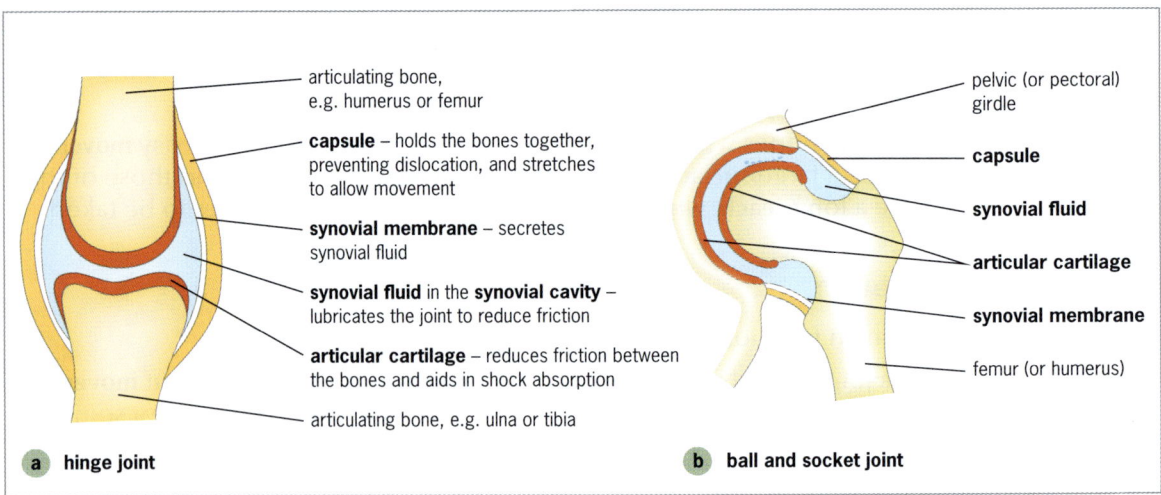

Figure 16.25 *Structure of a hinge joint and a ball and socket joint, and functions of the different parts*

Movement of a limb

Two muscles, known as an **antagonistic pair**, are always needed to produce movement at a movable joint. This is because, when a muscle contracts it **shortens** and exerts a **pull**, but it cannot exert a push when it relaxes and lengthens.

- The **flexor muscle** brings about the **bending** of a hinge joint when it contracts, and when moving a ball and socket joint, it moves the arm or leg **forwards** at the shoulder or hip.
- The **extensor muscle** brings about the **straightening** of a hinge joint when it contracts, and when moving a ball and socket joint, it moves the arm or leg **backwards** at the shoulder or hip.

Both muscles are attached, by **tendons**, to the bone that does not move at one of their ends, known as the **origin**, and to the bone that does move at the other end, known as the **insertion**. The **origin** is usually as **far** away from the joint as possible and the **insertion** usually very **close** to the joint. This arrangement **maximises** the **efficiency** and **effectiveness** of muscles to bring about movement.

Movement of the elbow joint

The **biceps muscle** and **triceps muscles** bend and straighten the **elbow joint.**
- The **biceps** is the **flexor** muscle.
- The **triceps** is the **extensor** muscle.

To **bend** the elbow joint, the **biceps contracts** and the triceps relaxes. To **straighten** the elbow joint, the **triceps contracts** and the biceps relaxes.

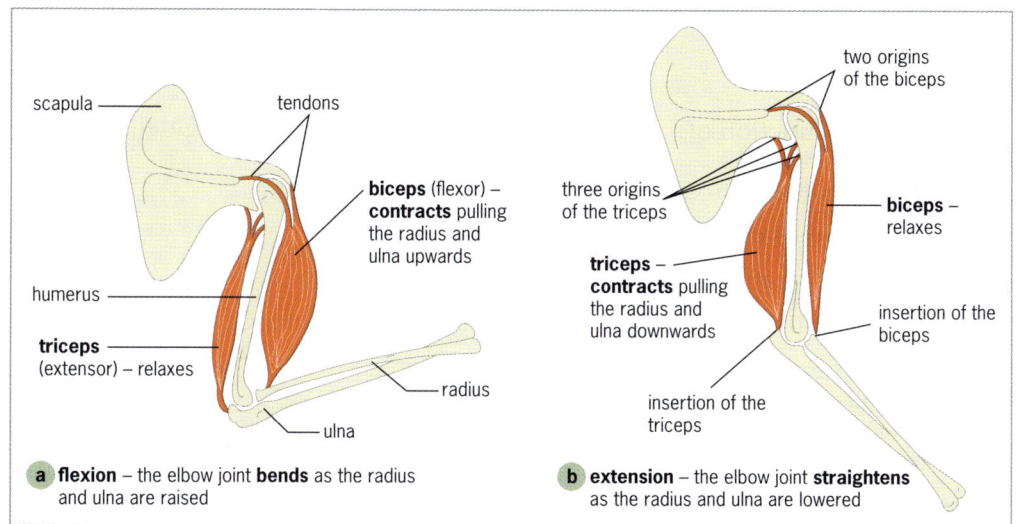

Figure 16.26 *Movement of the elbow joint*

Revision questions

1. **a** Define force.
   b Give TWO examples of EACH of the following: **i** contact forces   **ii** field forces.
   c Why is force considered a vector quantity?

2. State Newton's THREE laws of motion.

3. **a** Define friction.
   b State TWO of EACH of the following: **i** benefits of friction   **ii** problems of friction.
   c State THREE ways of reducing friction.

4. Figure 16.27 shows the cross-section of an aircraft wing. Explain how a lift force is created as the wing moves through the air.

   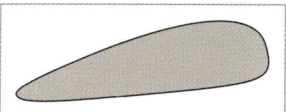

   Figure 16.27 *Revision question 4*

5. **a** Define: **i** the centre of gravity of a body
      ii stable equilibrium.
   b State the effect on the stability of an object if:
      i its centre of gravity is raised
      ii its base is widened.

16 Forces **223**

6 **a** Define centripetal force.
    b Sketch a diagram showing the centripetal force acting on a body in motion.
    c How is the required centripetal force affected if the body has:
      i greater mass?   **ii** greater speed?   **iii** greater orbit radius?

7 **a** Define the 'moment of a force about a point' and state an appropriate SI unit for a moment.
    b State the conditions required for equilibrium under the action of coplanar parallel forces.

8 **a** Define homeostasis.
    b Briefly describe the function of: **i** osmoregulation   **ii** respiratory regulation.

9 **a** Define the linear momentum of a body and state its SI unit.
    b State why momentum is considered a VECTOR quantity.
    c State the law of conservation of linear momentum.
    d Calculate the momentum of a body of mass 4 kg and velocity 20 m/s.
    e The body in **d** collides with a body of THREE times its mass that is at rest. Calculate the combined velocity of the two if they move on together.

10 **a** Define the following: **i** a machine   **ii** a force multiplier   **iii** a distance multiplier.
    b What class of lever is a force multiplier?
    c Distinguish between a class 2 and a class 3 lever.
    d Which of the following is a class 1 lever? tweezers, nutcracker, spanner, broom
    e What class of lever is utilised when the biceps muscle raises an object held in the hand?

11 Distinguish between the axial skeleton and the appendicular skeleton and make a list of the bones that make up EACH.

12 By referring to the different parts of the human skeleton, discuss FOUR of its functions.

13 **a** What is a joint?
    b Identify the different types of joints found in the human body and name ONE place where EACH is found.

14 **a** Why are two muscles needed to bring about movement of the knee joint?
    b Explain how the muscles in Joe's arm bring about bending and straightening of his elbow joint.

17 Metals and non-metals

Chemical elements can be classified as **metals** or **non-metals**, and **materials** used in everyday life can be classified as **metallic** or **non-metallic materials** based on their properties. Each class of materials has its own unique properties which make the materials suitable for different **uses**. **Metals** differ in their reactivity and the more reactive ones are affected by chemicals in the environment, which cause them to **corrode**.

Properties and uses of metals

Most **metals** have the following **common physical properties**.

- They have **high melting points**. Therefore, they are suitable for use in high temperatures, e.g. in furnaces, engines and many industrial processes.
- They are **good electrical conductors**, i.e. **electricity** passes through them easily. Therefore, they are used to make electrical wires, circuits and electronic components.
- They are **good thermal conductors**, i.e. **heat** passes through them easily. Therefore, they are used in industrial processes where heat transfer is essential and to make cooking utensils.
- They are **malleable**, i.e. they can be hammered, rolled or pressed into different shapes, and **ductile**, i.e. they can be drawn out into wires. Therefore, they are used to make metal tools, sculptures, decorative metalwork, jewellery, metal sheets, foil, electrical wires and cables.
- They have **high tensile strengths**, i.e. they are **strong** and can withstand tension, heavy loads and stress without deforming or breaking. Therefore, they are used in the construction industry.
- They have **high densities**, i.e. they have a relatively high mass per unit volume, which makes them heavy. Therefore, they are used as ballast in ships, counterweights in elevators and diving weights.
- They are **shiny** in appearance or can be **polished** to make them shiny, i.e. they have a **high lustre**. Therefore, they are used to make jewellery and for decorative purposes.
- They are **sonorous**, i.e. they make a **sound** when hit. Therefore, they are used to make musical instruments.

These **common properties vary** from metal to metal, and this makes certain metals **more suitable** than others for specific jobs.

Table 17.1 *Properties and uses of some important metals*

Metals	Specific properties making the metal suitable for its uses	Uses
Aluminium	• Good conductor of electricity and heat. • Relatively low density. • Very malleable and ductile. • Very shiny and reflective. • Resistant to corrosion.	To make overhead electrical cables, cooking utensils, cans to store drinks (see page 229), foil for cooking and window frames. **Alloys** made mainly of aluminium are used to construct aircraft.
Copper	• Good conductor of electricity and heat. • Very malleable and ductile. • Very resistant to corrosion.	To make electrical wires, bases of saucepans and water pipes.

Metals	Specific properties making the metal suitable for its uses	Uses
Iron	• High tensile strength. • Very malleable and ductile. • Easily welded.	To make ornamental iron work. **Steel**, an **alloy** made mainly of iron, is used to construct buildings, bridges, oil rigs, ships, trains and motor vehicles, and to make wire, nails, cutting tools, drill bits and many household items (see pages 229 to 230).
Zinc and **tin**	• Resistant to corrosion.	To coat iron and steel items to prevent rusting, e.g. zinc is used to coat nails and roofing sheets, and tin is used to coat 'tin cans'.
Silver and **gold**	• Very shiny. • Very malleable and ductile. • Very resistant to corrosion.	To make jewellery and medals. Silver is also used to make cutlery and ornaments. Gold is also used in dentistry to make crowns and fillings.

Properties and uses of non-metallic materials

Non-metallic materials include **natural** materials, e.g. wood, rubber and textiles made from natural fibres, and **manufactured** materials, e.g. plastics, glass, ceramics, carbon fibre and textiles made from synthetic (man-made) fibres. These materials have very **variable physical properties**:

- They are all **poor electrical** and **thermal conductors**, i.e. they are **insulators**.
- Some are **flexible**, others are **brittle**.
- Some are **strong**, others have **low tensile strengths**.
- Most are **dull** in appearance.
- Some have **high densities**, others have **low densities**.

Non-metallic materials have a great many **uses** because of their variable properties.

Wood

Wood is a hard, fibrous material obtained from the trunks and branches of trees. It is fairly easy to work with, shape, carve and join, it is a good insulator and is fairly flexible, allowing it to bend without breaking. Most **hardwoods** are hard, strong and durable, whereas most **softwoods** are softer, easier to work with and not as durable.

Wood is **used** to build houses and boats, and to make furniture, flooring, cupboards, window frames, doors, decorative items, musical instruments, toys, tool handles and handles for saucepans.

Ceramics

Ceramics are made from soft, non-metallic materials, mainly **clay**, which become **hard** and **brittle** when heated. Other materials, e.g. **feldspar**, **silica**, **alumina** (aluminium oxide) and **limestone** (calcium carbonate) can be added to **modify** and **improve** the properties of the final products. Ceramic items are produced by **shaping** mixtures of the raw materials, **firing** them in a **kiln**, cooling them and usually **glazing** them to give a protective, non-porous, attractive coating. Ceramics include stoneware, earthenware, porcelain and bone china. Most are hard, strong, durable, resistant to heat, chemicals and scratching, and are good insulators; however, they are **brittle**.

Figure 17.1 *Shaping a ceramic item before firing*

Ceramics are **used** to make bricks for building, floor tiles, wash basins, toilets, cookware, tableware, decorative items, components for electrical devices and for car, aircraft and spacecraft engines, artificial hips and knees, and dental implants.

Plastics

Most **plastics** are made using chemicals obtained from petroleum (crude oil) and are composed of very large organic molecules called **polymers**, e.g. polyethylene, polystyrene, polyvinyl chloride (PVC), polyesters and nylon. Plastics are durable, strong, light in weight, easy to mould, join, spin into fibres and colour, they are good insulators and they can be rigid or flexible.

Plastics are **used** to make bottles for drinks and cleaning products, food containers, shopping and garbage bags, toys, handles on saucepans, packaging materials, insulation for electrical wires, water pipes, guttering, window frames, clothing, boat sails, carpets, ropes, fishing lines and furniture.

Textiles

Textiles are materials composed of **natural fibres**, e.g. cotton, wool, silk, linen and jute, or **synthetic fibres**, e.g. nylon, polyester and acrylic, or a mixture of both. The fibres are **woven**, **knitted**, **crocheted** or otherwise **intertwined** to create textiles. After production, they can undergo various finishing processes to improve their appearance, texture and performance. Most textiles are strong, flexible, durable, good thermal insulators, hard to rip or tear, can be dyed easily and some are elastic.

Textiles are **used** to make clothing and accessories, bed sheets, towels, curtains, covers for furniture and car seats, carpeting, camping gear, boat sails, bandages, surgical gowns and many other items.

Figure 17.2 *Textiles can be produced by weaving fibres*

Materials used in sporting equipment

Materials used in making **sporting equipment** have evolved from mainly natural materials, e.g. wood, leather, gut, rubber and steel, to a variety of **high-technology materials** aimed at enhancing **performance**, increasing the equipment's **durability**, and improving **safety** and **comfort**. These include **carbon fibre** and **graphite composites**, made by embedding fibres made from carbon atoms in a resin; **glass fibre composites** made by embedding glass fibres in a resin; **Kevlar** made from long, rigid, parallel, synthetic fibres; **aluminium** and **aluminium alloys**; and **titanium** and **titanium alloys**.

Table 17.2 *Properties of materials used to make sporting equipment*

Material	Properties	Sporting equipment made
Carbon fibre and **graphite composites**	Strong, lightweight, stiff, durable, good impact and fatigue resistance, allow freedom of design.	Tennis and badminton rackets, bicycle frames, golf club shafts, hockey sticks, fishing rods, sailboat masts, skis, racing car parts.
Glass fibre composites	Strong, relatively lightweight, durable, can be made rigid or flexible, good impact resistance, easy to make and to mould into complex shapes.	Surfboards, skis, snowboards, boat hulls, kayaks, pole vaulting poles, fishing rods.
Kevlar	Extremely strong, tough, lightweight, durable, resistant to stretching, cutting and abrasion.	Racing boat sails, protective clothing used in cycling, motor sports, fencing, speed skating and rock climbing.

Material	Properties	Sporting equipment made
Aluminium and **aluminium alloys**	Strong, lightweight, durable, can be formed into complex shapes.	Baseball bats, tennis and badminton rackets, hockey sticks, bicycle frames, javelins.
Titanium and **titanium alloys**	Strong, lightweight, durable, extremely resistant to corrosion, fatigue and stress.	Bicycle frames and components, tennis rackets, golf clubs, archery equipment, racing car components.
Wood	Strong, flexible, good shock absorber, easy to shape and customise to the individual, has a classic look and good feel.	Cricket and baseball bats, hockey sticks.
Rubber	Elastic, good shock absorber, durable, resistant to wear and tear, non-slip.	Baseballs, tennis balls, soccer balls, rugby balls, grips on tennis rackets and golf clubs.
Leather	Durable, flexible, provides a good grip.	Baseball, cricket and boxing gloves.

The reactivity of metals

Some metals react **vigorously**, even violently, with other chemicals, e.g. **acids**, **oxygen** and **water**, whilst others are relatively **unreactive**. Potassium, sodium, calcium and magnesium are the most reactive, whilst aluminium, zinc, iron and tin are less reactive, and copper, silver and gold are relatively unreactive.

Reactions of metals with dilute acids

When a metal reacts with dilute hydrochloric or sulfuric acid, it forms a **salt** and **hydrogen**. Salts formed with **hydrochloric acid** are called **chlorides** and salts formed with **sulfuric acid** are called **sulfates**.

$$\text{metal} + \text{acid} \longrightarrow \text{salt} + \text{hydrogen}$$

E.g.:
$$\text{zinc} + \text{hydrochloric acid} \longrightarrow \text{zinc chloride} + \text{hydrogen}$$
$$\text{aluminium} + \text{sulfuric acid} \longrightarrow \text{aluminium sulfate} + \text{hydrogen}$$

Reactions of metals with oxygen

When a metal reacts with oxygen, it forms a **metal oxide**.

$$\text{metal} + \text{oxygen} \longrightarrow \text{metal oxide}$$

E.g.:
$$\text{tin} + \text{oxygen} \longrightarrow \text{tin oxide}$$

Reactions of metals with water as steam

When a metal reacts with **water** in the form of **steam**, it forms a **metal oxide** and **hydrogen**.

$$\text{metal} + \text{steam} \longrightarrow \text{metal oxide} + \text{hydrogen}$$

E.g.:
$$\text{iron} + \text{steam} \longrightarrow \text{iron oxide} + \text{hydrogen}$$

Table 17.3 *Summary of the reactions of some specific metals with dilute acids, oxygen in air and water*

Metal	Description of the reaction with dilute acids	Description of the reaction when the metal is heated in air	Description of the reaction with water
Aluminium (Al)	Reacts vigorously.	Burn when heated strongly, especially if powdered.	Do not react with cold or hot water. React with steam.
Zinc (Zn)	Reacts fairly vigorously.		
Iron (Fe)	Reacts slowly.	Burn when powdered and heated strongly.	
Tin (Sn)	Reacts very slowly.		
Copper (Cu)	Do not react with dilute acids.	Does not burn when heated, but forms an oxide coating if heated very strongly.	Do not react with water or steam.
Silver (Ag)		Does not react, even when heated very strongly.	

Aluminium cooking and canning utensils

The surface of any aluminium item is coated in a thin layer of **aluminium oxide** which is relatively **unreactive**. This layer **sticks** to the metal surface and protects it against **corrosion** (see page 230). **Aluminium** is **used** extensively to make **cooking utensils**, e.g. pots, pans, pressure cookers and baking trays, and to make **cans** to store beverages and some foods.

Table 17.4 *Advantages and disadvantages of using cooking and canning utensils made of aluminium*

Advantages	Disadvantages
• The utensils are **resistant to corrosion** due to the aluminium oxide coating. • The utensils are **unreactive** due to the aluminium oxide coating. • The utensils are very **good conductors** of heat. • The utensils are **light in weight** because aluminium has a low density. • The utensils can be polished to have a **shiny**, attractive appearance.	• The utensils can be **scratched** or **dented** easily, and they **warp** easily because aluminium is a soft metal. • The utensils can be **stained** easily, especially if cooking very acidic foods, e.g. tomatoes or citrus fruits. • If the utensils are used to cook or store very **acidic foods**, the acid may **react** with the aluminium oxide coating, reducing its effectiveness. This also causes **aluminium ions** to enter the food and alter the taste, and aluminium has been implicated in increasing a person's risk of developing **Alzheimer's disease**, a type of dementia. • If the utensils are **scratched** or **damaged**, aluminium ions may **leach** into food during cooking or storage, possibly altering its taste and increasing the risk of Alzheimer's disease.

Note: Some of these disadvantages can be overcome or reduced by increasing the **thickness** of the aluminium oxide layer on cooking utensils by a process called **anodising**.

Alloys in the home and workplace

Alloys are **mixtures** of two or more metals. Sometimes a non-metal may be added to the mixture. Alloys are produced to **improve** or **modify** the properties of metals. Alloys are usually **harder**, **stronger**, more **malleable** and more **resistant to corrosion** than the pure metals.

Table 17.5 *Alloys commonly found in the home and workplace*

Alloy	Composition	Properties	Uses
Steel	**Iron** alloyed with up to 1.5% **carbon**.	Harder and stronger than iron, and more malleable and ductile.	In the construction of buildings, and motor vehicles. To make 'tin cans' to store food, tools, nails, door hinges, gates, fences and cookware.
Stainless steel	Usually about 70% **iron** alloyed with 20% **chromium** and 10% **nickel**.	Harder, stronger and much more resistant to corrosion than iron or steel, malleable and ductile. Has a very shiny, attractive appearance.	To make cutlery, cooking utensils, kitchen equipment and appliances, sinks and surgical equipment.
Brass	**Copper** alloyed with up to 45% **zinc**.	Harder, stronger and denser than copper, malleable, ductile and resistant to corrosion. Has an attractive golden-yellow colour.	To make door and window fittings, taps, lamp fittings, nuts and bolts, ornaments and musical instruments.
Soft solder	About 60% **tin** alloyed with 40% **lead**.	Has a relatively low melting point, so melts easily when being used. More malleable and ductile than tin and lead.	To join metal items together at relatively low temperatures, e.g. electrical wires and water pipes.

Electroplating

Electroplating is also used to **modify** the properties of metal objects. An **electric current** is used to **coat** the object with a thin layer of an unreactive or less reactive metal. It is used to **protect** metal objects from corrosion and wear, to make them look more **attractive** or to make inexpensive objects appear more **valuable**. Silver, nickel, chromium and tin are often used to electroplate objects made of **steel**, which corrodes easily, but is relatively inexpensive. 'Tin cans' used to store food and beverages are made of steel electroplated with a thin layer of **tin**.

Tarnishing and rusting

Tarnishing and **rusting** are **two** forms of **corrosion**. Corrosion occurs when the **surface** of a metal **reacts** with chemicals in the environment, mainly oxygen and moisture, and it is speeded up by the presence of certain pollutants.

Tarnishing

A metal **tarnishes** when its freshly polished surface reacts with **oxygen**, or sometimes **sulfur dioxide**, in the air to form a thin layer of the **metal oxide** or **metal sulfide**, known as **tarnish**. This causes the metal to become **dull** and sometimes **discoloured**, and it adheres (sticks) to the surface and **protects** the metal below from reacting. **Aluminium**, **copper** and **nickel**, in particular, tarnish. **Silver** also tarnishes by its surface reacting with hydrogen sulfide or sulfur dioxide to form black **silver sulfide**. Tarnish can be removed by polishing or using chemical metal cleaners.

Figure 17.3 *Tarnished (left) and polished (right) silver salt and pepper shakers*

Rusting

Rusting occurs when **iron** and **steel** objects are exposed to both **oxygen** and **water (moisture)** in the air. The iron, oxygen and water react and form **hydrated iron oxide**, also known as **rust**.

iron + oxygen + water ⟶ hydrated iron oxide (rust)

Rust does not stick to the metal and protect it as tarnish does, instead it **flakes off**. This exposes fresh iron to the environment, which rusts and the rust flakes off. This continues, causing the iron to gradually wear away.

Figure 17.4 *Rust damages objects made of iron and steel*

Factors affecting the rate of rusting

Both **oxygen** and **water** must be present for **rusting** to occur, and the **rate** at which iron and steel rust is affected by several **environmental** factors.

- **Temperature** – The higher the atmospheric temperature, the faster rusting occurs.
- **Humidity** – The more water vapour in the air, the faster rusting occurs.
- **Salts** – Any salts, especially sodium chloride, dissolved in moisture in the air speed up rusting. Iron and steel fixtures in homes near the **sea** rust much faster than normal because they are exposed to **sea spray** composed of tiny droplets of seawater.
- **Pollutants** – Certain pollutants in the air speed up rusting, e.g. sulfur dioxide, nitrogen oxides and carbon dioxide speed it up because they are acidic. Iron and steel fixtures in homes near to **industrial plants**, especially those burning fossil fuels, rust faster than normal.

The **climate** of the **Caribbean**, characterised by high temperatures and high humidity, significantly increases the rate of rusting compared to cooler and dryer climates. The high **salt** content of the air, especially in coastal regions, also speeds it up.

Methods to reduce or prevent tarnishing and rusting

Metal objects and structures that are prone to **tarnishing** or **rusting** can be **protected** in various ways. Most prevent chemicals in the air from coming into **contact** with the object or structure.

Proper storage and using a drying agent

Metal objects can be stored in **dry, moisture-free** environments, e.g. sealed in airtight containers or bags. **Desiccants** that absorb moisture, e.g. packets of silica gel, can be added to the containers or bags.

Non-metallic protective coatings

Metal objects and structures can be **coated** with **paint**, **grease**, **oil** or **plastic** to create **physical barriers** between the objects and their surroundings, which prevent oxygen, moisture and other pollutants from reacting with the metal. **Clear lacquer** or **varnish** can be used to protect metal objects against tarnishing, whilst still preserving the original shiny appearance.

Electroplating

Electroplating is mainly used to prevent iron and steel objects from **rusting**. However, if the metal coating is scratched or damaged in any way, the iron or steel will rust because iron is more reactive than any of the metals used for the plating, e.g. **tin**.

Galvanising

Galvanising is used to prevent iron and steel objects from **rusting**. The object is **coated** with a thin layer of **zinc**. The surface of the zinc oxidises and a thin, adherent layer of unreactive **zinc oxide** forms, which protects the surface. If the zinc coating is scratched or damaged, the zinc is oxidised instead of the iron or steel because zinc is more reactive, and this protects the iron or steel from rusting. The zinc is said to provide **sacrificial protection**.

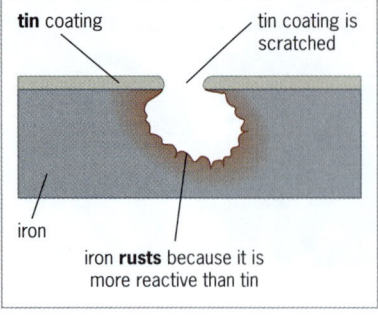

Figure 17.5 *Iron rusts if the metal it is plated with is scratched*

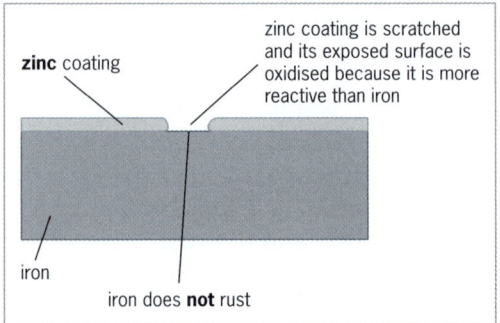

Figure 17.6 *Iron does not rust if the zinc coating is scratched*

Alloying

Tarnishing can be reduced by alloying metals prone to tarnishing with other metals. Iron can be prevented from **rusting** by alloying it with **chromium** and certain other metals, e.g. nickel and manganese, to form **stainless steel**. On contact with the air, the chromium at the surface is oxidised to form a layer of unreactive **chromium oxide**, which prevents the iron in the stainless steel from rusting.

Revision questions

1. Distinguish between metals and non-metals in terms of their:
   a conductivity   b tensile strength   c density.

2. Suggest TWO reasons for EACH of the following.
   a Copper is used to make electrical wires.   b Gold is used to make jewellery.
   c Aluminium is used to make foil for cooking.

3. Distinguish among plastics, ceramics and textiles.

4. Discuss the use of EACH of the following in making sporting equipment:
   a wood   b graphite composite   c glass fibre composite.

5. Write a word equation to summarise EACH of the following reactions.
   a Iron reacting with hydrochloric acid.   b Zinc reacting with oxygen.
   c The reaction between aluminium and steam.   d Tin reacting with sulfuric acid.

6. Suggest THREE reasons why aluminium is used to make cooking utensils and THREE disadvantages associated with its use for this purpose.

7. What is an alloy and why are alloys often used in place of the pure metals?

8. Explain what happens when an iron gate rusts and write a word equation to summarise the process.

9. Identify FOUR environmental factors that affect the rate at which the iron gate in Question **8** rusts and explain why an iron gate to a house on the coast of Dominica in the Caribbean rusts at a faster rate than a similar iron gate to a house in the countryside in Canada.

10. Discuss FOUR different ways to protect metallic objects against tarnishing and rusting.

18 Household chemicals

Household chemicals are non-food chemicals that are commonly found and used in and around the home. They play an essential role in everyday life, from maintaining cleanliness and hygiene to enhancing comfort and improving the household environment. Understanding their uses, what they are composed of and how to handle them properly is key to using them effectively, safely and responsibly.

The uses of common household chemicals

Household chemicals include **water**, **cleaning** products, **hygiene** products, **healthcare** products and **pest control** products.

Water

Water is the **most common** household chemical because it **dissolves** a very large number of substances. It is used to do laundry, wash dishes and clean floors, and for bathing, flushing toilets, watering gardens, cooking and drinking. Most household chemicals in the liquid state contain water as a **solvent**, e.g. window cleaners and liquid bleaches. Water is also relatively inexpensive, readily available in most households and non-toxic.

Other household chemicals

Table 18.1 *Common household chemicals and their uses*

Category	Use	Examples and their specific uses
Hard surface cleaners	To clean and sanitise non-porous surfaces, e.g. countertops, floors, appliances and bathroom fixtures.	• Household **ammonia**, a dilute solution of **ammonium hydroxide**, and other **degreasers**, used to remove oils, fats, grease and grime (greasy dirt). • Household **bleaches**, especially chlorine bleaches, used to remove stains, mould and mildew. • **Disinfectants**, used to destroy or inhibit the growth of microorganisms in or on non-living objects, e.g. **pure alcohols** such as ethanol and propanol, **chlorine-releasing** compounds such as **sodium hypochlorite** found in household bleaches, and **quaternary ammonium salts**.
Personal hygiene products	To maintain personal cleanliness, good health and well-being.	• **Hand soaps**, **bath soaps** and **body washes**, used to clean the hands and body. • **Shampoos** and **conditioners**, used to clean and condition hair. • **Deodorants** and **antiperspirants**, used to control body odours. • **Toothpastes** and **mouthwashes**, used to maintain oral hygiene.
Laundry products	To clean clothing and other textiles, e.g. towels and bedding.	• **Soaps** (soapy detergents) and **soapless** or **synthetic detergents** (**detergents**), used to remove greasy dirt from fabrics. • **Fabric softeners**, usually containing **quaternary ammonium salts**, added as laundry is being rinsed to soften fibres in fabrics.

Category	Use	Examples and their specific uses
Kitchen cleaners	To maintain cleanliness in the kitchen.	• **Dishwashing liquids**, detergents used to remove grease, food residue and bacteria from dishes, utensils and cookware. • **Oven cleaners**, usually containing a strong alkali, e.g. **sodium hydroxide**, used to remove baked on food and grease from inside ovens. • **Vinegar**, a dilute solution of **ethanoic (acetic) acid**, added to water and used to clean countertops, sinks, floors and other kitchen surfaces. • **Baking soda (sodium hydrogencarbonate)** and **baking powder** (**sodium hydrogencarbonate** mixed with an **acidic** component); mildly abrasive and used to scrub surfaces and absorb odours.
Healthcare products	To maintain and improve health within households.	• **Antiseptics**, used to destroy or inhibit the growth of microorganisms on living tissue, e.g. **hydrogen peroxide**, **rubbing alcohol** (about 70% **propanol**) and **iodine solution**. • **Hand sanitisers**, used to reduce or eliminate microorganisms on hands. • **Alcohols**, used to reduce the spread of infections in homes. • **Antacids** containing **basic** chemicals, e.g. **magnesium hydroxide**, used to **neutralise** excess hydrochloric acid in the stomach. • **Painkillers**, used to relieve pain.

Figure 18.1 *Common household chemicals*

Eco-friendly household chemicals

Eco-friendly household chemicals, especially cleaning products, should be used if possible to help create healthier living spaces and minimise damage to the environment caused by many household chemicals. These are usually made from **plant-based ingredients** that are effective cleaning agents without having the harmful effects of harsh, synthetic chemicals. They are also **biodegradable** and **non-toxic**.

The safe use of household chemicals

Guidelines for the **safe** use of **household chemicals** include:
- **Read** all labels very carefully, always follow the instructions given and use only as directed.
- Use only the **amount** of the chemical needed to do the job.
- Use chemicals in **well-ventilated** areas and avoid inhaling the fumes.
- **Never mix** household chemicals, especially chlorine bleach and products containing ammonia.
- Wear the appropriate **protective clothing**, e.g. gloves and goggles, when using harmful chemicals and do not use chemicals near food.
- **Wash hands** immediately after using any household chemicals.
- **Store** chemicals in their original containers and ensure the containers are tightly sealed and out of reach of children.

The economic use of household chemicals

Using household chemicals **economically** involves trying to maximise their efficiency and minimise waste. This can be achieved in the following ways.
- Use the **recommended quantity** of household chemical for each job, following the instructions given.
- Buy frequently used household chemicals in **bulk** and store them properly, but only buy sufficient to ensure all will be used before the expiration date.
- Buy household chemicals that have **long shelf lives**.

Safety symbols on chemicals

A number of household chemicals are potentially harmful and carry **safety symbols** to warn about the **potential hazards** associated with their use. These symbols are crucial to ensure the chemicals are handled, stored, used and disposed of **safely**. They may appear on a yellow background inside a black triangle or a white background inside a red diamond.

Figure 18.2 *Some important safety symbols*

Figure 18.3 *Safety symbols on a can of insecticide*

Properties of acids, bases and salts

The **properties** of **acids** and **bases** are **opposite** to each other, and they have the ability to **neutralise** each other. When an acid reacts with a base, the reaction always forms a **salt** and **water**.

Acids and bases

- **Acids** are substances that form positive **hydrogen ions** (H^+ **ions**) when dissolved in water. Solutions of acids are described as **acidic**. **Hydrochloric acid**, **sulfuric acid** and **nitric acid** are common acids found in the laboratory. **Citric acid** and **ascorbic acid (vitamin C)** are found in fruits.
- **Bases** include **metal oxides**, e.g. calcium oxide; **metal hydroxides**, e.g. magnesium hydroxide; and **ammonia**. Some are **soluble** in water and are called **alkalis**, e.g. **sodium hydroxide**, **potassium hydroxide**, **calcium hydroxide** and **ammonium hydroxide**, formed when ammonia dissolves in water. Solutions of alkalis are described as **alkaline** and they contain negative **hydroxide ions** (OH^- **ions**).

Table 18.2 *Properties of acids and alkalis*

Property	Acids	Alkalis
Taste	Sour	Bitter
Nature	Corrosive	Corrosive and feel soapy
Effect on litmus	Change blue litmus to red	Change red litmus to blue
pH value	Less than 7	Greater than 7

The concept of pH

Solutions of **acids** and **alkalis** can be classified as **strong** or **weak**, and their **strengths** can be measured on the **pH scale**, which is a numbered scale ranging from **0** to **14**. A solution with a pH of **7** is **neutral**.

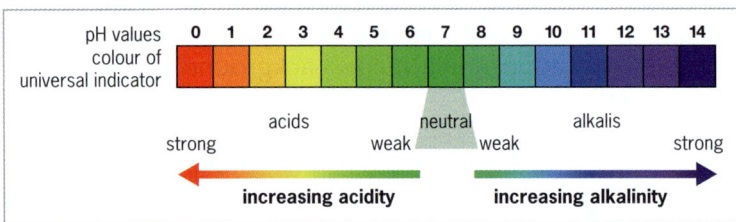

Figure 18.4 *The pH scale*

The pH of a **solution** can be **measured** using **universal indicator paper** or **solution**, which changes **colour** depending on the pH of the solution. When using the paper, a small piece is dipped into the test solution and its colour is compared with a pH colour chart.

Salts

Salts are formed when bases react with acids. Salts formed by hydrochloric acid are called **chlorides**, salts formed by sulfuric acid are called **sulfates**, and salts formed by nitric acid are called **nitrates**. Other salts include **carbonates**, **hydrogencarbonates** and **phosphates**.

$$\text{base} + \text{acid} \longrightarrow \text{salt} + \text{water}$$

Classification of household chemicals as acids, bases and salts

Household chemicals can be classified into **acids**, **bases** and **salts**. Most salts are **neutral**; however, some can be acidic and some can be basic.

Table 18.3 *Classification of some common household chemicals as acids, bases or salts*

Classification	Household chemical	Main chemical component
Acid	• Limescale remover	• Phosphoric, sulfamic or citric acid
	• Toilet bowl cleaner	• Hydrochloric or citric acid
	• Rust remover	• Phosphoric or oxalic acid
	• Battery acid	• Sulfuric acid
	• Vinegar	• Ethanoic (acetic) acid
	• Aspirin	• Acetylsalicylic acid
Base	• Drain cleaner	• Sodium or potassium hydroxide
	• Oven cleaner	• Sodium or potassium hydroxide
	• Chlorine bleach	• Sodium hypochlorite
	• Household ammonia	• Ammonia
	• Antacid	• Magnesium hydroxide or aluminium hydroxide
Salt	• Washing soda (also basic)	• Sodium carbonate
	• Toothpaste (also basic)	• Sodium fluoride and sodium hydrogencarbonate
	• Epsom salt	• Magnesium sulfate
	• Baking soda (also basic)	• Sodium hydrogencarbonate
	• Table salt	• Sodium chloride
	• Antacid (also basic)	• Sodium hydrogencarbonate or calcium carbonate

Neutralisation reactions

A **neutralisation reaction** is a reaction between a base and an acid to form a salt and water. Neutralisation reactions play important roles in **agriculture**, **health**, **nutrition**, **cooking** and **sanitation**.

- Basic **calcium hydroxide (lime)** or finely ground **calcium carbonate (limestone)** can be added to acidic soils to neutralise them, making them suitable for most plants to grow.
- **Antacids** are used to treat **indigestion** and **acid reflux**. One or more active ingredient is basic, and these neutralise excess hydrochloric acid in the stomach, helping relieve symptoms.
- **Toothpaste** contains certain basic ingredients which neutralise any acids produced by bacteria in the mouth, and this helps to prevent **tooth decay**.
- Basic **baking soda** is used to treat **bee stings**, which contain methanoic (formic) acid, and acidic **vinegar** is used to treat **wasp stings**, which are basic. Both neutralise the venom in the stings.
- Basic **baking soda** reacts with acidic ingredients, e.g. buttermilk, yoghurt or lemon juice, when producing baked goods. This produces **carbon dioxide** gas, which causes the mixture to rise and become light and fluffy.
- Many **household cleaning products** clean because of neutralisation reactions, e.g. rust removers and limescale removers (see page 238).
- The **pH** of the **treated wastewater** in **wastewater** and **sewage treatment plants** and of the **treated water** in **water treatment plants** is adjusted to neutral before the water is discharged by adding the appropriate acid or base to the water.

The effects of cleaning agents on household appliances

Cleaning agents used to clean **household appliances** work in different ways.

- By acting as **abrasives** to **physically** remove dirt and stains when rubbed on hard surfaces, e.g. **scouring powders**.

- By acting as **surfactants**, e.g. **detergents**.
- By their **chemical action**, e.g. **rust removers**, **limescale removers** and **oxidising agents**.

Common cleaning agents for household appliances

- **Scouring powders** contain **fine particles** of an insoluble mineral, e.g. limestone, quartz or silica, mixed with other powders and chemical cleaners, e.g. a detergent or bleaching agent. They are mixed with water to form a thick paste, and the mineral particles act as the **abrasive**.
- **Soapy** and **soapless detergents** remove grease and dirt. They **lower** the surface tension of the water, allowing it to **spread out** and **wet** surfaces more efficiently, i.e. they act as **surfactants** or **surface-active agents**, and they **break up** and **disperse** grease and dirt.

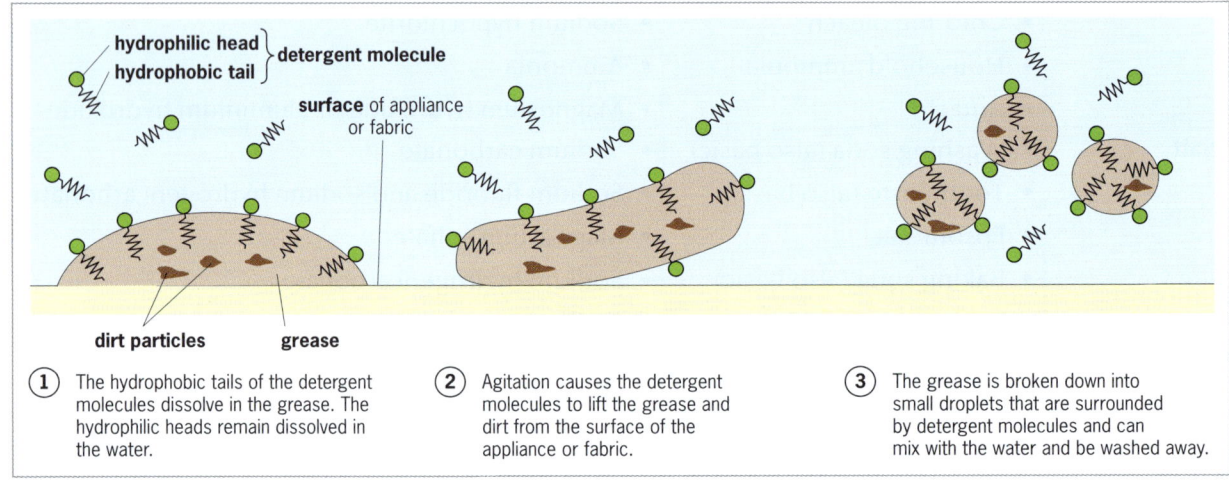

Figure 18.5 *How detergents remove grease and dirt*

- **Rust removers** usually contain **phosphoric** or **oxalic acid**. The acid reacts with the basic **rust (iron oxide)** to form a soluble salt that can be washed away, leaving bare iron or steel.
- **Limescale removers** usually contain **phosphoric**, **sulfamic** or **citric acid**. The acid reacts with the basic **limescale (calcium carbonate)** deposits that build up inside kettles and shower heads, and around taps. This forms a soluble salt that can be washed away.
- **Oxidising agents**, e.g. **chlorine bleach**, **hydrogen peroxide** and other **oxygen bleaches** remove coloured stains by oxidising them to their colourless form.

Cleaning metal appliances

Household appliances can be made of various **metals**, e.g. aluminium, copper, iron, tin, zinc and silver, and their alloys. These need to be cleaned regularly to remove grease, dirt, tarnish, corrosion, stains and discolouration, and to keep them shiny. The correct cleaning agents must be used to avoid scratching, corroding, discolouring or damaging surfaces.

Use a **mild detergent** to remove grease and dirt, diluted **vinegar** to remove mineral deposits, e.g. limescale, and a paste of **baking soda** and water to remove stains. After cleaning, **rinse** with water and **dry** using a clean, dry, microfibre cloth. **Cleaners** and **polishes** designed specifically for each metal can also be used.

Avoid using **abrasive cleaners**, which can scratch metal surfaces, and **chlorine bleaches**, which can discolour metals. **Harsh cleaners** containing strong alkalis, e.g. sodium hydroxide, or strong acids, e.g. hydrochloric acid, should **not** be used because they cause corrosion or pitting. However, rust removers can be used to remove rust from iron or steel appliances.

Soapy and soapless detergents

Detergents can be classified as **soapy** and **soapless**.
- **Soapy detergents** or **soaps** are made by boiling **animal fats** or **vegetable oils** with concentrated potassium or sodium hydroxide solution. An example is **sodium octadecanoate (sodium stearate)**.
- **Soapless detergents** or **detergents** are made from chemicals obtained from **petroleum**. An example is **sodium dodecyl sulfate**.

Scum formation

Soapy detergents do not lather easily in **hard water** (see page 193). The dissolved calcium and magnesium ions in hard water react with sodium octadecanoate (soap) and form insoluble calcium or magnesium octadecanoate, also known as **soap scum** or **scum**.

hard water + soap ⟶ scum

Scum is an unpleasant, greasy substance that can build up in clothes and discolour them, and forms an unpleasant grey, greasy layer on hard surfaces, e.g. around sinks, baths and showers. **Soap** will only **lather** and remove greasy dirt when all the calcium and magnesium ions have been removed as **scum**.

Soapless detergents do not form scum when added to hard water; therefore, they begin to remove dirt as soon as they are added to the water.

Advantages and disadvantages of soapy and soapless detergents

Table 18.4 Advantages and disadvantages of soapy and soapless detergents

Type of detergent	Advantages	Disadvantages
Soapy	• They are manufactured from fats and oils which are **renewable** and will not run out. • They are **biodegradable**, i.e. they are broken down by microorganisms in the environment, therefore they do not cause foam to form on waterways. • They do not contain phosphates, therefore they **do not** cause **pollution** of aquatic environments by causing eutrophication. • They contain **fewer irritants** than soapless detergents.	• They **do not** lather easily in hard water, therefore larger amounts are need than soapless detergents, making them **less cost-effective**. • They form unpleasant **scum** in hard water, which makes them **less efficient cleaning agents** than soapless detergents. • They tend to be **more expensive** to produce than soapless detergents.
Soapless	• They **lather** very easily in hard water, therefore smaller amounts are needed than soapy detergents to achieve the same result, making them **more cost-effective**. • They **do not** cause scum to form in hard water, therefore they are **more efficient cleaning agents** than soapy detergents. • They are generally **less expensive** to produce than soapy detergents.	• They are manufactured from petroleum, a **non-renewable** resource which will run out. • Some are **non-biodegradable**. These cause **foam** to build up on waterways, which prevents oxygen dissolving for aquatic organisms, resulting in their death. Most modern ones are biodegradable. • Some contain **phosphates**, added to improve their cleaning ability. These **pollute** aquatic environments by causing **eutrophication**. • They contain **more irritants** than soapy detergents.

Figure 18.6 *Foam build-up on a river caused by using soapless detergents*

Revision questions

1. Identify the MOST common chemical used in the home and state why it is used so extensively.

2. Give the use of EACH of the following household chemicals and list THREE examples of EACH:
   a laundry products
   b kitchen cleaners
   c personal hygiene products.

3. Suggest THREE guidelines that Pedro should follow in EACH case to ensure that he uses the household chemicals kept in one of his cupboards:
   a safely
   b economically.

4. You see a skull and cross-bones on a white background inside a red diamond on a container of pesticide. What message does the symbol convey to you?

5. State THREE properties of an acid and THREE properties of an alkali.

6. a What is the pH scale?
   b Give the pH value of EACH of the following: a strong alkali, a weak acid and a neutral substance.

7. a Provide a suitable definition for a salt.
   b Name THREE household chemicals that are acidic, THREE that are basic and THREE that are salts.

8. Explain how neutralisation reactions can be used:
   a to treat insect stings
   b in baking
   c to treat certain soils.

9. Identify THREE different types of cleaning agents that can be used to clean household appliances by their chemical action and explain the action of EACH.

10. Describe the steps that Nola would take to clean the outside of her stainless steel microwave.

11. a What is scum and how is it formed?
    b Give THREE advantages of using soapy detergents and THREE advantages of using soapless detergents in the home.

The states of matter

Matter is anything that occupies space and has mass.

All **matter** is made of **particles**. On Earth, matter can exist in **three** common **states**: the **solid state**, the **liquid state** and the **gaseous state**. **Three** different **types** of particles make up matter: **atoms**, **molecules** and **ions**, and the differences between the states lie in the **energy** and **arrangement** of their particles.

Table 18.5 *The properties of the three common states of matter compared*

Property	Solid	Liquid	Gas
Shape	Fixed.	Takes the shape of the part of the container it is in. The surface is always horizontal.	Takes the shape of the entire container it is in.
Volume	Fixed.	Fixed.	Variable – it expands to fill the container it is in.
Arrangement of particles	Packed closely together, usually in a regular way:	Have very small spaces between and are arranged randomly:	Have large spaces between and are randomly arranged:
Forces of attraction between the particles	Strong.	Weaker than those between the particles in a solid.	Very weak.
Energy possessed by the particles	Possess very small amounts of kinetic energy.	Possess more kinetic energy than the particles in a solid.	Possess large amounts of kinetic energy.
Movement of the particles	Vibrate about their mean position.	Move slowly past each other.	Move around freely and rapidly.

Changes of state

Matter can exist in any of the three states, depending on its **temperature**. It can be **changed** from one state to another by **adding** or **removing heat** because this changes the **energy** and **arrangement** of the particles.

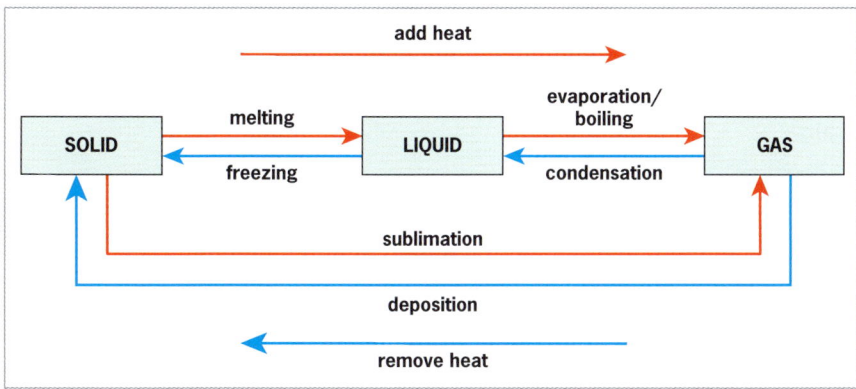

Figure 18.7 *Summary of the changes of state*

Melting

When a **solid** is **heated**, its particles **gain** kinetic energy and vibrate more vigorously. Eventually, the particles are able to overcome the strong forces of attraction between them and move freely past each other, forming a **liquid**. The solid is said to **melt**. The temperature remains constant whilst the solid is melting; this temperature is called the **melting point**.

Melting point is the constant temperature at which a solid changes into a liquid.

Evaporation

When a **liquid** is **heated**, its particles **gain** kinetic energy and move faster. Some of the particles near the surface have enough kinetic energy to overcome the forces of attraction between them, and they leave the liquid and become a **vapour**. These particles are said to **evaporate**.

Boiling

When the temperature of a **liquid** being **heated** reaches a certain point, the liquid particles have gained enough kinetic energy and move fast enough to change into a **gas**, both within the liquid and at its surface. The liquid is said to **boil**. The temperature remains constant whilst the liquid is boiling; this temperature is called the **boiling point**.

Boling point is the constant temperature at which a liquid changes into a gas.

Condensation

When a **gas** is **cooled**, its particles **lose** kinetic energy and move more slowly. The forces of attraction between the particles become stronger, causing the particles to move closer together and form a **liquid**. The gas is said to **condense**.

Freezing

When a **liquid** is **cooled**, its particles **lose** kinetic energy and move more slowly. The forces of attraction between the particles become stronger, causing the particles to move even closer together and form a **solid**. The liquid is said to **freeze** and the temperature at which this occurs is called the **freezing point**.

Freezing point is the constant temperature at which a liquid changes into a solid.

The freezing point of a **pure** substance has the same value as the melting point, e.g. pure water has a freezing point and a melting point of 0 °C.

Sublimation and deposition

When the forces of attraction between the particles in a **solid** are **weak**, the addition of a small amount of heat can cause the **solid** to change directly into a **gas**. The solid is said to **sublimate** or **sublime**, and the change in state is called **sublimation**. If the gas is cooled it will change directly back to the solid. The gas is said to **deposit**, and the change of state is called **deposition**. Iodine, carbon dioxide (dry ice), ammonium chloride and naphthalene undergo sublimation and deposition. Snowflakes also form by deposition of water vapour in the air.

Plasma

A plasma is an electrically charged gas.

Plasma is the fourth state of matter. It is rare on Earth, but the Sun and other stars are composed mainly of plasma and it is thought to be the most common state in the universe. Plasmas are **ionised gases** consisting of negatively charged **electrons** and positively charged **ions** known as **cations**, which possess very large amounts of kinetic energy. A plasma is usually formed when a gas is heated strongly enough that the atoms in it lose electrons, i.e. they **ionise**. Plasmas can emit **light** of various colours.

Like gases, plasmas do not have fixed shapes or volumes. **Unlike gases**, plasmas conduct an electric current and are attracted to magnetic fields. On Earth, plasmas are found in neon signs, fluorescent light bulbs, lightning and auroras, also known as the northern or southern lights.

Figure 18.8 *An aurora in Iceland*

The properties of mixtures

Mixtures are composed of two or more substances **physically combined**. Each component retains its own individual properties and has not reacted chemically with any other component. The components can be **separated** by physical means. **Solutions**, **suspensions** and **colloids** are all mixtures.

Solutions

*A **solution** is a homogeneous (uniform) mixture of two or more substances; one substance is usually a liquid.*

A **solution** is made by **dissolving** one substance, known as the **solute**, in another, known as the **solvent**.

*The **solvent** is a chemical substance that dissolves another chemical substance to form a solution.*

*The **solute** is a chemical substance that can be dissolved by another chemical substance to form a solution.*

The **solvent** is the substance that is present in the higher or highest concentration and is usually a liquid. **Solutes** can be solids, liquids or gases. A **solution** usually appears **transparent**. Based on the **nature** of the **solvent**, solutions can be classified into **two** types: **aqueous solutions** and **non-aqueous solutions**.

- **Aqueous solutions** have **water** as the solvent, e.g. sea water, acids, iced tea and carbonated drinks.
- **Non-aqueous solutions** have substances other than water as the solvent. Common **non-aqueous solvents** include ethanol, kerosene, gasoline, turpentine, acetone and methylated spirits.

Solvents play an important role in **removing stains** and certain other substances, such as oil and grease, in the home and places of work. They do this by **dissolving** the stains, oil or grease.

Suspensions

*A **suspension** is a heterogeneous (non-uniform) mixture in which minute, visible particles of one substance are dispersed in another substance, which is usually a liquid.*

A **suspension** appears **opaque**, e.g. muddy water, chalk dust or flour stirred in water, and oil shaken vigorously in water.

Colloids

*A **colloid** is a heterogeneous mixture in which minute particles of one substance are dispersed in another substance, which is usually a liquid. The dispersed particles are larger than those of a solution, but smaller than those of a suspension.*

A colloid can appear **translucent** or **opaque**, e.g. smoke, fog, gelatin and whipped cream. Mayonnaise, milk and emulsion paint are colloids known as **emulsions**, which contain minute liquid droplets, e.g. oil droplets, dispersed in a water-based mixture.

Comparing solutions, colloids and suspensions

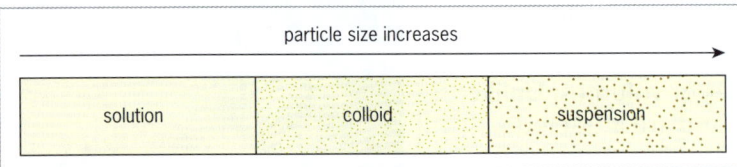

Figure 18.9 *Comparing the particle sizes in a solution, a colloid and a suspension*

Table 18.6 *Comparing the properties of solutions, colloids and suspensions*

Property	Solutions	Colloids	Suspensions
Size and visibility of dispersed particles	Extremely small. Not visible, even with a microscope.	Between those in a solution and those in a suspension. Not visible, even with a microscope.	Larger than those in a colloid. Visible to the naked eye.
Separation of components on standing	Dispersed particles do not separate out if left undisturbed.	Dispersed particles do not separate out if left undisturbed.	Dispersed particles separate out if left undisturbed.
Passage of light and appearance	Light usually passes through, making them appear **transparent**.	Most will scatter light, making them appear **translucent**. Some are **opaque**.	Light does not pass through, so they appear **opaque**.
Separation of components by filtration	Dispersed particles cannot be separated by filtration.	Dispersed particles cannot be separated by filtration.	Dispersed particles can be separated by filtration.

Classification of household chemicals

Nearly all **household chemicals** are **mixtures** and can be classified as **solutions**, **colloids** or **suspensions**.

Table 18.7 *Classification of some common household chemicals as solutions, colloids or suspensions*

Solutions	Colloids	Suspensions
• Chlorine bleach • Limescale remover • Window cleaner • Household ammonia • Vinegar • Hydrogen peroxide • Rubbing alcohol • Iodine solution	• Aerosol sprays, e.g. insecticides • Liquid detergents • Fabric softeners • Shaving cream • Hand cream • Conditioner	• Liquid scouring (abrasive) cleaners • Metal polish • Calamine lotion

Separation techniques

Separating mixtures into their component parts can be important and the **technique** used depends on the **physical properties** of the components being separated.

Distillation

Simple distillation separates the components of a **solution** based on the **difference** in their **boiling points**. The difference must be significant and the solvent must have a lower boiling point than the solute.

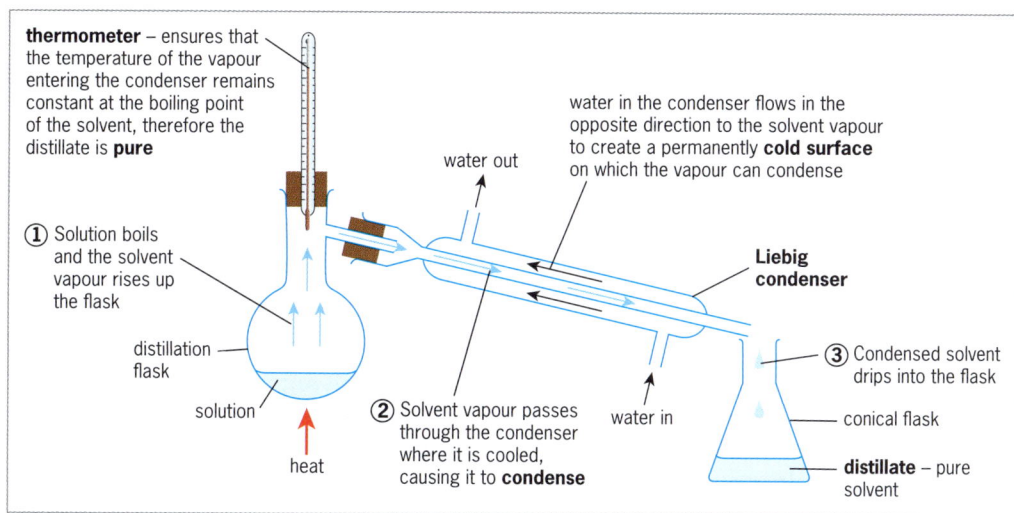

Figure 18.10 *Simple distillation to separate the components of a solution*

Distillation can be **used** to produce **distilled water** from tap water, to **desalinate** seawater or brackish water to produce **fresh water** in a **desalination plant** (see page 198) and to produce a variety of **spirits** and **fortified wines** in the **wine-making industry**.

Filtration

Filtration separates the components of a mixture of **solid** particles suspended in a **liquid** or **gas**. The components are separated due to their different **particle sizes**, using the appropriate **porous** material that allows the smaller liquid or gas particles to pass through and retains the larger solid particles.

Filtration is **used** in **cooking** to separate items such as potatoes or pasta from the cooking water, or to remove the coffee grounds or tea leaves when making coffee or tea. It is also used in **water filtration systems** to remove impurities from tap water (see Table 15.6, page 199), **air purifiers** to remove dust and other airborne particles from indoor air, and **fuel filters** to remove impurities from the fuel in car engines.

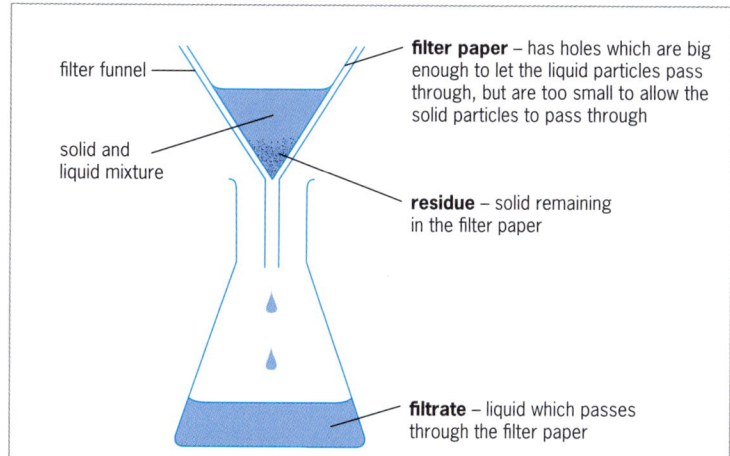

Figure 18.11 *Separating the components of a suspension by filtration*

18 Household chemicals

Chromatography

Chromatography separates several **solutes** present in a **solution**. The solutes are usually coloured and can move through some kind of stationary material, e.g. absorbent paper, when dissolved by a solvent. The solutes are separated based on **two** factors.

- How **soluble** each solute is in the solvent used, which is usually water or ethanol.
- How strongly each solute is **attracted** to the stationary material used.

The **bleeding** of **colours** from one garment to another during **washing** occurs if the garments are made of **different fabrics** and the fabric of one garment attracts one or more of the dyes more readily than the fabric of the original, coloured garment.

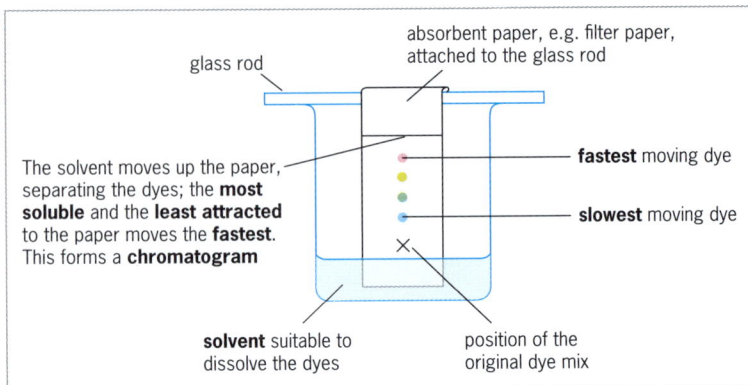

Figure 18.12 *Separating the coloured dyes in a mixture by paper chromatography*

Crystallisation

Crystallisation separates and retains the solid solute from a **solution**. The solvent is allowed to **slowly evaporate** from the solution, usually at room temperature, leaving the solid behind as **crystals**. Crystallisation is **used** to obtain **sugar crystals** from the concentrated syrup formed during the production of sugar from **sugar cane**. It can also occur in **honey**, which is a supersaturated solution of sugars, mainly glucose and fructose. Over time, the **glucose** can crystallise in the honey, making it gritty.

Evaporation

Evaporation separates and retains the solid solute from a **solution** by **heating** the solution to evaporate the solvent. Evaporation is **used** to produce **table salt** from seawater or brine (concentrated salt solution). In hot climates, the **heat** from the Sun evaporates the water from seawater in shallow ponds or pans. Alternatively, brine can be heated to a temperature of about 80 to 90 °C in large containers to evaporate the water and leave salt crystals.

Revision questions

12. Distinguish among a solid, a liquid and a gas in terms of their shape, the arrangement of their particles, the energy possessed by their particles and the forces of attraction between their particles.

13. By referring to particles, explain what happens when EACH of following changes of state occurs and, in EACH case, name the change of state.

    a A liquid is heated until it becomes a gas.   b A liquid is cooled until it becomes a solid.

14. Explain why heating a crystal of iodine causes it to change directly into a gas. Name the change of state occurring.

15. What is a plasma and where would plasmas be found?

16. What defines a mixture?

17 a Using particle size, passage of light and separation of the components on standing, distinguish among a solution, a colloid and a suspension.

  b Identify TWO household chemicals that are solutions, TWO that are colloids and TWO that are suspensions.

18 Provide a suitable definition for EACH of the following:

  a solution **b** solvent **c** solute.

19 Solutions can be classified into TWO types, name these and distinguish between them.

20 Explain how you would:

  a remove the tea leaves when making tea
  b separate the dyes in a sample of black ink
  c obtain pure water from tap water.

  EACH answer must include an explanation of the principles involved.

19 Pollutants and the environment

Human activities produce waste and harmful substances which, if released into the environment, can **pollute** the air, the land and water, and cause harm in many other ways. In order to reduce the harmful effects of any waste to a minimum or to prevent them totally, and to maintain a clean and healthy environment, all waste must be properly treated and properly disposed of.

Air pollution and its effects

Air pollution is mainly caused by the release of unpleasant and harmful substances into the **atmosphere** by **human activities**, e.g. the combustion of fossil fuels, mining and quarrying, manufacturing, cigarette smoking, livestock farming and rice agriculture. Certain **natural activities** also contribute, e.g. wildfires, volcanic eruptions, lightning strikes and dust storms.

The major air pollutants include **sulfur dioxide**, **carbon monoxide**, **carbon dioxide**, **nitrogen oxides**, **methane**, **volatile organic compounds**, **carbon particles** or **soot** in smoke and other fine **particulate matter**.

Effects of air pollution on human health

Air pollution can cause a variety of health issues, including **allergies**, **asthma**, **lung cancer**, **emphysema** and other **respiratory disorders**, and **cardiovascular disease**.

- Certain air pollutants can trigger or worsen **allergies** in susceptible individuals, especially asthma and hay fever (see page 71). **Carbon particles** in smoke and other **particulate matter** can irritate the eyes, nasal passages and airways, make individuals more susceptible to other allergens and carry other allergens into the respiratory system. **Nitrogen oxides** can weaken the body's defences against allergens.
- **Volatile organic compounds** are known to be **carcinogenic**, i.e. have the potential to cause cancer. When inhaled, they increase a person's risk of developing **lung cancer** in particular.
- **Sulfur dioxide, nitrogen oxides, carbon particles** and other **particulate matter** can irritate and inflame the airways and cause, or worsen, certain **respiratory disorders**, e.g. chronic bronchitis, asthma and emphysema (see page 120).

Effects of air pollution on the environment

Air pollution has adverse effects on ecosystems and natural processes worldwide.

- **Sulfur dioxide** and **nitrogen oxides** react with water vapour and oxygen in the air to form sulfuric acid and nitric acid, respectively, also known as **acid rain**. Acid rain damages **crops**, **trees** and other **plants**. This reduces crop yields and causes the decline or destruction of forests in industrialised regions. It turns **soils acidic** and causes **bodies** of **water**, e.g. ponds, lakes, streams and rivers to become **acidic**, which harms aquatic organisms. It also **corrodes** buildings and statues.

Figure 19.1 *Trees damaged by acid rain*

- **Particulate matter** and **carbon particles** can settle on the leaves of plants. The particles **block light**, reducing photosynthesis, and can **block** the **stomata**, which prevents carbon dioxide from entering leaves, also reducing photosynthesis.
- **Carbon dioxide, nitrogen oxides** and **methane** are known as **greenhouse gases** because they are **enhancing** the **greenhouse effect**. This enhancement is causing a gradual increase in the Earth's temperature, known as **global warming**, which is leading to **global climate change** (see pages 125 to 126).

Community hygiene

Community hygiene refers to the practices that promote and maintain health and cleanliness within communities. It involves individuals, households, local authorities and community organisations working together to ensure that public spaces and shared environments are **clean** and **safe**, and promote **good health**.

Types of waste

Waste produced by humans can be divided into **five** categories.

- **Domestic waste** is produced by **households**. It includes **sewage** and **refuse**. **Sewage** is wastewater from toilets, showers, baths, washing machines, dishwashers and sinks. It can contain human faeces and urine, detergents and food particles. **Refuse** consists of **solid waste**, including plastics, paper, cardboard, glass, metal, and food and garden waste.
- **Industrial waste** is generated by **industrial activity**, including manufacturing, mining, food and construction industries. It includes **solid waste**, **sewage** and other **liquid waste**, and **gaseous waste**. The components depend on the industry producing it.
- **Biological waste** or **biohazardous waste** contains, or has been contaminated by, potentially **hazardous biological materials**, e.g. bacteria, viruses and human cells. It comes from hospitals and other healthcare facilities, laboratories and research institutions. It can contain human blood, blood products, tissues and organs, surgical dressings and gloves, syringes, needles and blood vials.
- **Chemical waste** contains potentially **harmful chemicals** produced mainly by industry, laboratories and agriculture. It can contain mineral oils, cyanides, acids and alkalis, solvents, heavy metals, e.g. mercury and lead, radioactive waste, expired drugs and pesticides.
- **Electronic waste** or **e-waste** consists of discarded **electrical** or **electronic devices**, e.g. computers, tablets, mobile phones and televisions, whose components contain potentially harmful chemicals, e.g. lead, cadmium and mercury, plastics, glass and certain valuable metals, e.g. gold and silver.

Figure 19.2 *Biohazardous waste*

Biodegradable and non-biodegradable waste

Waste can also be classified as **biodegradable** and **non-biodegradable**.

- **Biodegradable waste** can be broken down by living organisms, mainly bacteria and fungi, into harmless materials that can be recycled into the environment. It includes food waste, most paper, farmyard and garden waste, natural fabrics, e.g. cotton and wool, and some plastics.
- **Non-biodegradable waste** cannot be broken down by living organisms, therefore it remains in the environment. It includes metal, glass, rubber, construction waste, synthetic fabrics, e.g. nylon and polyester, and most plastics.

The impact of improper waste disposal

Improper disposal of waste poses a significant **threat** to the environment, public health and the well-being of communities, including **pollution**, an increase in **pest populations** and the spread of **pathogens** and **parasites**.

Figure 19.3 *Improper waste disposal poses a threat to human health and the environment*

Pollution

Improper waste disposal **pollutes** the **land**, the **air** and **water**, including surface water, groundwater, sources of potable (drinkable) water and bodies of water, e.g. lakes, rivers and oceans.

- **Toxic chemicals** can seep out and contaminate the soil and water.
- **Plastics** can persist in the environment for hundreds of years and pollute the land and water, and they gradually break down into **microplastics**, which contaminate the soil, affect plant growth and harm animals (see page 253).
- **Greenhouse gases**, e.g. methane and carbon dioxide, can be released into the atmosphere from the breakdown of organic material in the waste and contribute to the **greenhouse effect**.
- **Hydrogen sulfide gas** can be released into the air and create foul odours, and irritate the eyes and respiratory system.
- **Plant nutrients**, e.g. nitrates and phosphates, in improperly treated or untreated sewage can pollute water, leading to **eutrophication** (see Table 15.5, page 197).

Increased pest populations

Waste that has not been disposed of properly provides a **breeding ground** for **pests** and **vectors** of disease (see page 77), causing their population numbers to increase. This increases the **damage** caused by the pests and the spread of **infectious diseases** transmitted by vectors.

Spread of pathogens and parasites

Pathogenic microorganisms, mainly bacteria, some viruses and some fungi, and **intestinal parasites** or their **eggs**, which are present in human faeces, raw or partially treated sewage and solid waste, can enter water supplies and contaminate potable water. This leads to the spread of **infectious diseases**, e.g. cholera, typhoid fever and dysentery, and **parasites** such as worms.

Recommended practices for waste management

Proper **waste management practices** help minimise the impact posed by improper waste disposal.

Refuse, reduce, reuse, repair, refill, repurpose and recycle

The amount of **solid waste** to be disposed of can be decreased by practising the **7Rs** of waste management.

Table 19.1 *The 7Rs of waste management*

7Rs	Description
Refuse	Actively **avoid** or **decline** buying or using products that create unnecessary waste.
Reduce	**Cut down** on what is produced, purchased and used.
Reuse	**Use** the same item **again**, preferably many times, for its original purpose or function.
Repair	**Mend** items that are broken or damaged, if possible, instead of discarding them.
Refill	**Replenish** a container with its original contents or an alternative product.
Repurpose	Find a **new** use for an item that is **different** to its original use.
Recycle	**Separate**, **recover** and **reprocess** materials into new raw materials that can be used to make new products.

Composting

Waste **organic matter**, e.g. fruit and vegetable peelings, tea leaves and grass cuttings, can be converted into nutrient-rich **compost** by **composting**. This can be carried out in compost heaps or composters at home, or in a commercial composting facility using agricultural waste. Bacteria and fungi in the waste break it down **aerobically** to produce **compost**, which can be used as a **soil conditioner** or **mulch**.

Biogas production

Organic waste, e.g. manure, crop residues and food waste, can be used to produce **biogas**, a renewable energy source. The waste is placed into an **anaerobic digester** and certain **bacteria** in the waste break it down **anaerobically** to produce **biogas**, a mixture of about **60% methane**, **40% carbon dioxide** and traces of other gases. Biogas can be used as a **fuel** for cooking, heating and to generate electricity, and the mineral-rich **residue** that remains can be used as **fertiliser**.

Figure 19.4 *Producing compost in a home composter*

Proper sewage disposal practices

Sewage can be disposed of in an **on-site sewage system**, e.g. a **septic tank**, **cesspool** or **suckwell**, which treats the sewage close to a building and releases the treated water into the same area. A **soakaway** can be used to collect and disperse surface runoff. Alternatively, sewage can be treated in a **sewage treatment plant**. This produces treated wastewater or **effluent**, which is released into the environment and semi-solid **sludge**, which can be used as **fertiliser**. Households, schools, offices, hospitals and factories must also have adequate **toilet facilities** linked to sewage systems to dispose of faeces and urine.

Proper collection and disposal of refuse

Refuse should be **separated** into different types before collection and disposal, e.g. organic waste, which should be **composted**, and recyclable materials, which should be transported to a **recycling facility**. The remaining refuse should be **stored** in durable bins with tight-fitting lids, **collected** at least once per week and transported to a **disposal facility**, e.g. a landfill.

Benefits of community hygiene

Good **community hygiene** has many **benefits**.

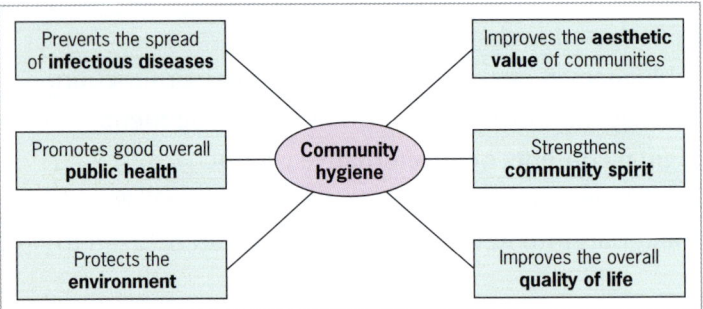

Figure 19.5 *The benefits of good community hygiene*

Plastics

Plastics are synthetic (manufactured) materials made from very large organic molecules called **polymers**. The **properties** of plastics make them **superior** to many other materials; however, their extensive use poses serious **threats** to living organisms and the environment.

Advantages of using plastics

Two of the main **advantages** of using plastics are their **ease of production** and their **durability**.

Ease of production

Plastics are relatively **easy to produce** compared with other materials.

- Most plastics are made from petrochemicals obtained from **petroleum (crude oil)** which is **widely available** and **inexpensive**.
- Plastic production is a **straightforward process** which can be done on a **large scale**.
- Plastics are more **cost-effective** to produce than alternatives, e.g. metals, glass and ceramics.
- A wide variety of manufacturing processes can be used to produce plastics, which offer manufacturers a high degree of **design flexibility**.
- Plastics are easy to **customise** by manufacturers for **specific uses**.

Durability

Plastics are extremely **durable**, which gives them a **long life**.

- Plastics are resistant to **wear** and **physical damage**.
- Plastics are resistant to **corrosion** and attack by other **chemicals**, e.g. acids, alkalis and solvents.
- Most plastics are resistant to **biological decay**.

Use of plastics in the medical and construction industries

In addition to being relatively **easy to produce** and **durable**, other **unique properties** of plastics make them useful in the **medical** and **construction industries**.

- Plastics are used in the **medical industry** to make medical devices, implants, surgical instruments, items of medical equipment and pharmaceutical packaging. This is because they are **biocompatible**, **lightweight**, can be **precisely moulded** into complex shapes, can be **sterilised** by various methods and are **waterproof**.
- Plastics are used in the **construction industry** to make fixtures and fittings, and building materials, e.g. window frames, doors, guttering, roofing, flooring, plumbing and insulation materials. This is because they are easily **moulded**, **lightweight**, **strong**, **waterproof**, and good **electrical** and **thermal insulators**.

Figure 19.6 *An artificial hand made from plastic*

Negative effects of plastics on the environment

Plastics can have significant **harmful effects** on ecosystems and the environment.

- **Toxic chemicals** are released into the environment during the manufacture of plastics and some continue to be released during their use and when disposed of.
- Most plastics are **non-biodegradable**. When disposed of, they build up and remain in the environment for hundreds of years, causing land and water pollution.
- Plastics are directly harmful to **aquatic organisms**, e.g. sea turtles, due to ingestion, entanglement and suffocation.
- Many plastics break down over time into **microplastics**, which are less than 5 mm in size. The tiniest of these can enter the bodies of organisms, where they harm tissues and organs.
- Many plastics are **flammable**, therefore they pose fire hazards.
- Plastics produce **dense smoke** and **poisonous gases** when burnt, causing air pollution.
- Plastics are made from petroleum, a **non-renewable resource**. Their manufacture is contributing to the depletion of petroleum worldwide.

Figure 19.7 *Microplastics*

Benefits of recycling plastics

Recycling plastics has environmental and economic **benefits**.

- It **conserves natural resources**, mainly petroleum.
- It **reduces energy consumption** associated with obtaining and processing raw materials.
- It **reduces** the **volume** of solid waste that must be disposed of in landfills.
- It **lowers greenhouse gas emissions** and the emissions of other air pollutants produced when manufacturing plastics from new raw materials.
- It **reduces** the **loss** of potentially useful materials and it keeps materials in use for as long as possible.

- It **reduces pollution** of land and water that can result from the disposal of plastics.
- It **creates jobs** to collect, sort, process and manufacture recycled materials.

Revision questions

1 Outline the effects of the following air pollutants on human health and/or the environment:

　a nitrogen oxides　　**b** carbon dioxide　　**c** particulate matter.

2 List FIVE different categories of waste produced by humans and outline what EACH might contain.

3 Distinguish between biodegradable waste and non-biodegradable waste.

4 **a** Explain the term 'community hygiene'.

　b Discuss the impact of the improper disposal of waste on the hygiene of communities.

　c Suggest FOUR benefits of practising good community hygiene.

5 Explain EACH of the following terms as it relates to waste management:

　a reduce　　**b** recycle　　**c** refuse　　**d** repair.

6 Discuss EACH of the following waste disposal practices:

　a biogas production　　**b** composting　　**c** sewage disposal.

7 What are plastics?

8 Explain the reason for EACH of the following.

　a Plastics are durable.　　**b** Plastics are considered easy to produce.

9 Outline the reasons why plastics are used so extensively in:

　a the construction industry　　**b** the medical industry.

10 Provide FIVE reasons to support EACH of the following statements.

　a Plastics are harmful to the environment.　　**b** Plastics should be recycled.

Exam-style questions – Chapters 13 to 19

1 **a)** **i)** How long does the Earth take to make one complete rotation on its axis? **(1 mark)**

  ii) Explain the occurrence of day and night on Earth. **(1 mark)**

 b) **i)** At what angle is the axis of rotation of the Earth tilted? **(1 mark)**

  ii) Draw a diagram to illustrate how this results in the northern hemisphere receiving more sunlight than the southern hemisphere for 6 months each year, followed by it receiving less sunlight than the southern hemisphere for the next 6 months. **(5 marks)**

  iii) Label the northern hemispheres to show the occurrence of summer and the occurrence of winter. **(1 mark)**

 c) Draw a well-labelled diagram to illustrate the formation of a solar eclipse. Mark on the diagram the points T and P showing where the observed eclipse is total and where it is partial. **(6 marks)**

Total 15 marks

2 **a)** **i)** Explain how a tropical cyclone is created and how it develops into a hurricane. **(6 marks)**

  ii) How does the weather change as the eye of a hurricane passes through a region? **(3 marks)**

  iii) Describe how a storm surge is formed. **(1 mark)**

 b) List:

  i) THREE ways you can prepare for the threat of a hurricane. **(3 marks)**

  ii) TWO precautions you can take after the hurricane passes. **(2 marks)**

Total 15 marks

3 **a)** Salmon hatch in freshwater rivers, where they remain for several months before migrating to the ocean to grow and mature. Once fully mature, they return to fresh water to spawn.

  i) Explain the challenges faced by the salmon when they are living in

   - fresh water

   - seawater. **(4 marks)**

  ii) What changes occur in the gills of the mature salmon as they return from the ocean to fresh water to spawn? **(2 marks)**

 b) Harpoon fishing, long-lining and setting fish pots or traps are three methods used to catch fish in the Caribbean.

  i) Suggest, giving ONE reason, which method would be MOST suitable for commercial fishermen to use. **(2 marks)**

  ii) Which of the three methods has the potential to cause the MOST harm? Give ONE reason for your answer. **(2 marks)**

  iii) Identify ONE other method that fisherfolk could use to catch fish in the Caribbean. **(1 mark)**

 c) Davard recently moved to a new city and noticed that the soap he had always used to wash his hands and face did not lather as it always did when he used it in his old home. Instead it formed an unpleasant grey substance that floated on the surface of the water.

  i) Provide an explanation to help Davard understand why the grey substance formed. **(2 marks)**

ii) Suggest ONE advantage that Davard could gain by drinking the water in his new home and ONE other disadvantage of the water. (2 marks)

Total 15 marks

4 a) Kalinda observed that the trees growing in the mangrove swamp close to her home were beginning to look unhealthy and a few appeared to be dying.

   i) Define the term 'pollutant'. (1 mark)

   ii) Identify TWO different pollutants that could be responsible for the unhealthy condition of the mangrove trees and explain the effects of EACH. (4 marks)

b) In order to make water potable or safe to drink, it must be treated in large-scale water treatment plants before it is piped to homes.

   i) Identify ONE source of water used in homes that are located several hundred kilometres from the coast. (1 mark)

   ii) Suggest ONE method that households can use to further purify the water before it is drunk and explain how the method works. (2 marks)

   iii) How is it possible to convert seawater to potable water? (1 mark)

c) Naymar has a deep-sea fishing boat that he also uses for scuba diving. To ensure he can locate the reef when he wishes to dive, he has sonar onboard.

   i) Explain how Naymar uses his sonar when he wishes to dive. (2 marks)

   ii) After his dive, Naymar begins to develop tingling in his arms and legs, and pain in his elbow and shoulder joints. Explain the possible cause of Naymar's symptoms. (3 marks)

   iii) What treatment must Naymar undergo to relieve his symptoms? (1 mark)

Total 15 marks

5 a) i) The combined mass of Abi and her small boat is 200 kg. Determine the resultant force on the boat as they ACCELERATE together at 5 m/s^2. (2 marks)

   ii) Determine the resultant force on her boat when it moves at a CONSTANT SPEED of 4 m/s in a straight line. (1 mark)

   iii) At this constant speed, if the forward force of the engine is 200 N, what is the frictional drag force on the boat? (1 mark)

   iv) Draw a free body diagram showing the HORIZONTAL forces on the boat in a) iii). (2 marks)

b) Figure 1 shows a trapdoor.

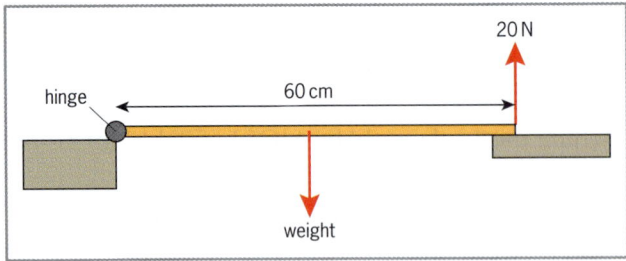

Figure 1 *Trapdoor*

   i) Determine the moment about the hinge created by the 20 N force. (2 marks)

   ii) If the trapdoor is uniform, calculate its weight. (2 marks)

c) A toy car with a mass of 5 kg and speed 8 m/s collides with another of the same mass which is initially at rest. Determine:
  i) The total momentum before the collision. **(2 marks)**
  ii) The total momentum after the collision. **(1 marks)**
  iii) The magnitude of the cars' common velocity after the collision if they stick together. **(2 marks)**

Total 15 marks

6 a) Figure 2 shows a graph of load against effort when Joshua used a machine.

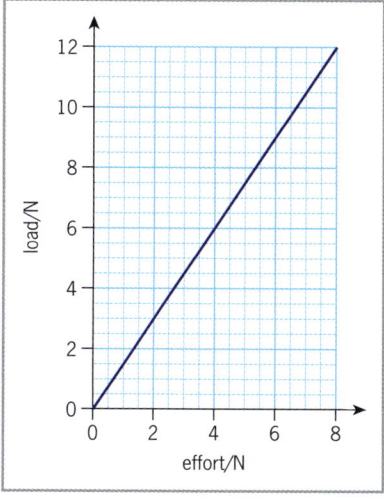

Figure 2 *Graph of load against effort*

  i) Determine the gradient (slope) of the graph and state what it represents. **(3 marks)**
  ii) What efforts are required to raise loads of 6 N and 15 N respectively? **(3 marks)**

b) Kailey must raise a block of weight 300 N (mass 30 kg) through a height of 2 m using the ramp shown in Figure 3.

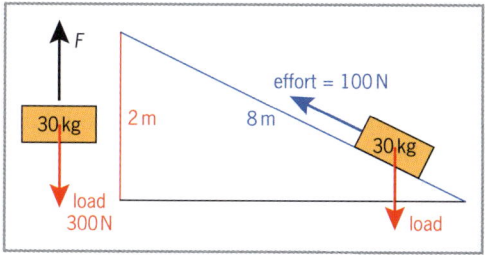

Figure 3 *Ramp to assist raising a block*

Determine the following.
  i) The work done by the effort (work input). **(2 marks)**
  ii) The work done against the load (work output). **(2 marks)**
  iii) The efficiency of the machine. **(3 marks)**

c) Kailey did more work by using the ramp than she would have done if she had simply lifted the block. Explain why she still preferred to use the ramp. **(2 marks)**

Total 15 marks

7 **a)** Metals are used extensively in today's world.
   i) Explain why aluminium is widely used to make cooking utensils even though it reacts readily with acidic foods. **(2 marks)**
   ii) Write a word equation to summarise the reaction between aluminium and hydrochloric acid. **(1 mark)**
   iii) Identify TWO metals that can be alloyed to make brass and explain why metals are often alloyed. **(2 marks)**

b) When bicycles were first made in the early 1800s, their frames and wheels were made out of wood. Today, however, it is usually only handcrafted bicycles that are constructed of wood.
   i) Identify ONE non-metallic material that is currently used to make bicycle frames and give ONE reason why it has replaced wood. **(2 marks)**
   ii) Wood is still used to make certain items of sporting equipment, especially cricket and baseball bats. Suggest TWO reasons for this. **(2 marks)**

c) Odette set up the experiment illustrated in Figure 4 to investigate the conditions needed for an iron nail to rust.

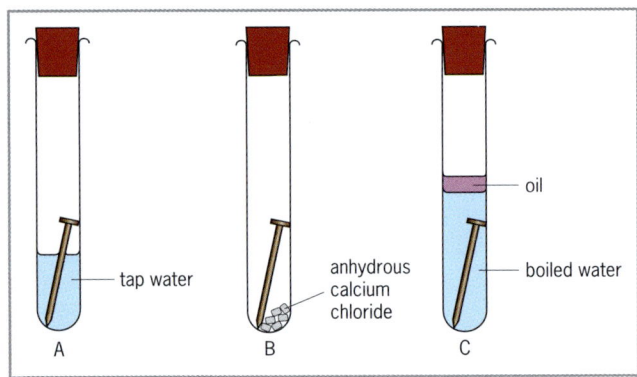

Figure 4 *Experiment to investigate the conditions needed for an iron nail to rust*

   i) Given that anhydrous calcium chloride absorbs moisture and when water is boiled, dissolved oxygen is removed, in which tube would Odette expect to see the nail rust? Provide an explanation for your answer. **(3 marks)**
   ii) Victor recently moved from his home in the countryside to live on the coast and he notices that the rust spots on his car seem to be quickly increasing in size. Suggest an explanation for Victor's observation. **(2 marks)**
   iii) Suggest the best way for Victor to prevent the chain of his bicycle from rusting. **(1 mark)**

Total 15 marks

8 **a)** Yolande, who likes to keep her home spotlessly clean, notices that a white deposit is slowly building up on her stainless-steel taps.
   i) Suggest a possible identity for the white deposit. **(1 mark)**
   ii) Name ONE household chemical that Yolande could use to remove the white deposit and explain how the chemical functions. Your answer must include reference to the type of reaction occurring as the white deposit is removed. **(4 marks)**

b) Both soapy and soapless detergents can be used to remove grease and dirt from household appliances.
   i) Distinguish between a soapy detergent and a soapless detergent. **(2 marks)**
   ii) Environmentalist, Selwyn, is extremely concerned about the environmental damage that certain household chemicals can cause and decides not to use any soapless detergents in his home. Suggest TWO reasons for Selwyn's decision. **(2 marks)**

Exam-style questions – Chapters 13 to 19

c) A student carries out an investigation to determine the pH value of three different household chemicals he has in his kitchen cupboard. His results are given in Table 1.

Table 1 *The pH values of three household chemicals*

Household chemical	pH
Drain cleaner	13
Vinegar	3
Household ammonia	11

i) Describe the pH scale and explain how the student determined the pH value of each of the chemicals in Table 1. **(3 marks)**

ii) Two of the chemicals in Table 1 are alkalis. Explain which one is the stronger. **(1 mark)**

iii) Whilst using the drain cleaner, the student accidentally spills some on the floor. Recommend TWO precautions he should take when cleaning the spill. **(2 marks)**

Total 15 marks

9 a) Figure 5 shows how the three states of matter can be changed from one to another. The letters A and B represent two of the processes that bring about the changes.

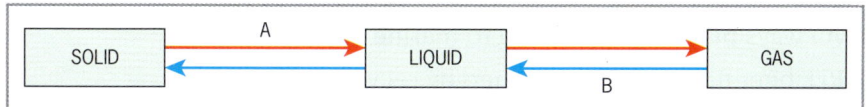

Figure 5 *Changing state*

i) Name the processes taking place at A and B. **(2 marks)**

ii) Using a small circle to represent ONE particle, draw nine particles to show how they would be arranged in a solid. **(2 marks)**

iii) Explain why solid carbon dioxide is often referred to as 'dry ice'. **(2 marks)**

b) Whilst playing in the garden, Marie fell and got grass stains on her new, pink dress. On seeing the green stains, her mother decided to try using chlorine bleach to remove them.

i) Would you classify chlorine bleach as a solution, a suspension or a colloid? **(1 mark)**

ii) Suggest why chlorine bleach would not be suitable to remove the grass stains from Marie's dress. **(1 mark)**

iii) Recommend, with a reason, a suitable method that Marie's mother could try instead of using bleach to remove the stains. **(2 marks)**

c) Mr. Joseph, a teacher of Integrated Science, wants to find out who spilt black ink in the Science Laboratory, so he asks Jo and Zak to each provide him with a sample of ink from their pens. He then prepares three chromatograms and his results are shown in Figure 6.

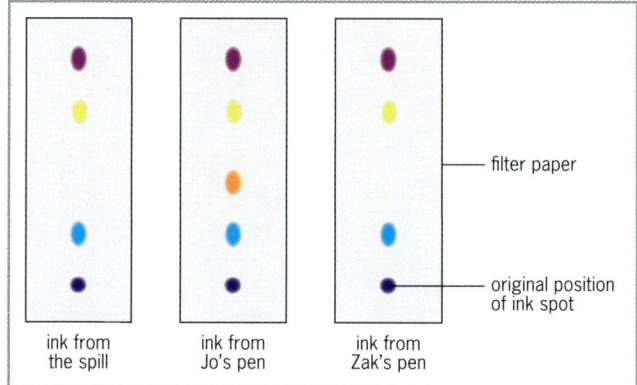

Figure 6 *Chromatograms produced by three samples of black ink*

       i) Which student spilt the black ink? (1 mark)

       ii) Explain why the dyes separated and moved up the filter paper. (4 marks)

Total 15 marks

10 a) The Dottin family lives in a community that is downwind from several factories and they accumulate their refuse in plastic bags outside their gate until it is collected once every two weeks. The children in the family experience frequent health challenges and they often find rodent droppings around their refuse.

       i) Suggest TWO different types of health challenges that the children are most likely to experience. (2 marks)

       ii) Explain the MAIN reason why the family should be concerned about the presence of rodent droppings. (2 marks)

       iii) Discuss TWO recommendations that you would give to the Dottin family to help them manage the disposal of their refuse. (4 marks)

b) Discarded plastic waste items and other waste have formed extremely large, floating 'plastic islands' in the world's oceans which threaten marine ecosystems and the health of the oceans.

       i) What are plastics and why are they used so extensively in today's world? (2 marks)

       ii) Discuss TWO ways plastics are harmful to marine life. (2 marks)

       iii) Explain THREE benefits that can be gained by recycling plastics. (3 marks)

Total 15 marks

Index

Page numbers along with 't' and 'f' refer to tables and figures, respectively.

A

abnormal gene, 70
ABO blood grouping system, 42
abrasive cleaners, 238
abrasives, 237
absorption, 111
abstinence, 29
accommodation, 55, 55f
acetic acid bacteria, 65
acid rain, 124–125, 248, 248f
acid reflux, 237
acids, 236
acquired immunity, 69
activation energy, 109
active transport, 4
    in living organisms, 4–5
adipose (fat) tissue, 50, 112
adrenal glands, 63t
adrenaline, 63
aerobic respiration, 3, 99, 113
aerodynamic 'lift' force, 210, 210f
afferent arteriole, 47
age, dietary needs and, 106
agents of pollination, 12
agglutination of red blood cells, 42
agricultural productivity, 21
air, thermal conduction of, 154
air masses, 180, 180t
air pollution, 248
    effects on environment, 248
    effects on human health, 248
air purifiers, 245
alcohol thermometer, 159–160
alkalis, 236
allergens, 69
allergic diseases, 69
allergic rhinitis, 71
allergies, 71
alloying, 232
alloys, 229, 230t
alternating current (AC), 138

alternative sources of energy, 127–131
    economic benefits, 132
    environmental benefits, 132
    geothermal energy, 130
    hydro-electric plants, 129–130, 129f
    hydrogen, 131
    nuclear plants, 130–131
    social benefits, 132
    solar energy, 127–128, 127f, 127t
    wave energy plants, 130
aluminium cooking and canning utensils, 229, 229t
alveoli (alveolus), 115, 115f
    of lungs, 118
    walls of, 118
Alzheimer's disease, 70
amino acids, 3–4, 37, 99, 104, 109, 112
ammeters, 136
ammonium compounds, 19–20
ammonium hydroxide, 236
amnion, 28
amniotic fluid, 28
amphetamines, 75
anaemia, 107
anaerobic respiration, 114
    in sports, 114
anaerobic training, efficiency of, 114
anal sphincter, 112
anaphylaxis, 71
animal cells, 1–2
    structure, 1
animals, as pollination agent, 12
annual plant, 14
antacids, 237
antagonistic pair, 222
anterior vena cava, 40
antibodies, 69, 112
antidiuretic hormone (ADH), 48, 63
antigen–antibody interaction, 69f
antigens, 69
anti-Rh antibodies, 42
antitoxins, 69
anus, 112
anxiety disorders, 70
aorta, 40

aquaculture, 195
aquatic organisms, 192–193, 195
aqueous solutions, 243
aquifers, 198
Archimedes principle, 200, 200f
arterial gas embolism, 206
arteries, 40
artificial colourings, 107
artificial immunity, 70
artificial lens, 57
artificial light, 143
artificial sources of light, 143, 144f, 144t
artificial sweeteners, 107
artificial vegetative propagation, 10–11
asexual reproduction
    advantages and disadvantages, 8, 9t
    in animals, 23–24
    in plants, 9–11, 10f
    role of mitosis in, 7
    runners in, 10f
ash, 5
assimilation, 112
asteroid belt, 170t
asteroids, 170t
asthma, 71
astigmatism, 56
athletic performance, 39
atria, 40
atrial systole, 41
audio frequency range, 59
autoimmune diseases, 69, 71

B

back-up electrical heating, 128
bacteria, 65–66, 73, 77, 80
baking soda, 237
balanced diet, 32, 103
barotrauma, 206
bases, 236
battery, 135
    emf of, 135f
bean seed, 13f
benzoates, 82
biceps, 223
biconcave discs, 38
bile
    duct, 111
    pigments, 46
    role in digestion, 111
bimetallic strip, 158, 158f
binary fission, 23, 23f
biodegradable waste, 249
biodiesel, 129
biofuels, 128
biogas, 128–129, 128f
    production, 251
biological control of pests, 78
biological equilibrium, 213
birth control (contraception), 30t–32t
    impact on human population growth, 35
    methods, 29–32, 32f
birthing process, 29
    contractions, 29
    dilation of cervix, 29
    expulsion of baby, 29
    labour, 29
birth rate, 34
blips, 204, 204f
blood, 38
blood capillaries, 111
blood cells, 38, 38f
    structure and functions of, 38–39, 38t–39t
blood doping, 39
blood groups, 41
    ABO system, 42, 42f, 42t
    O Rh-negative blood, 42
    Rhesus or Rh system, 42
blood transfusions, 42
blood vessels, 40t, 50
blue hydrogen, 131
body odour (BO), 73
body's metabolism, 46
boiling point, 242
bone, 220
Bowman's capsule, 47
brain, 61, 61f
breast milk, 33
breathing, 115, 147, 221
    mechanism of, 116–117, 116t
    movements, 118
    rescue, 117
    role of kinetic energy in, 117
bronchi and bronchioles, 120

bronchus, 115, 115f
budding, 10
    in animals and unicellular organisms, 23
    in hydra, 23f
buds, 10, 23
bulbs, 9
buoys, 130
bush fires, 148

C

caffeine, 75
calcium hydroxide, 236
calderas, 187
calluses, 11
cancer-causing chemicals, 121
candidiasis, 67
canines, 108, 109f, 109t
canning, 82
capillaries, 40
carbohydrates, 104, 104f, 107
    sources and functions of, 105t
carbonates, 236
carbon dioxide, 3, 37, 46, 117, 117t, 118–120
    emissions, 131
carbon particles, 248
carcinogens, 121
cardiac cycle, 41
cardiac muscle, 40
cardiopulmonary resuscitation (CPR), 117, 147
Caribbean climates, 162
Caribbean food groups, 103, 103f, 103t
Caribbean Volcanic Arc, 188f
Caribbean weather patterns, 183
carnivores, 99
carpels, 12
cartilage, 220
cataract, 56–57, 57f
causative agents of disease, 66
celestial bodies, 170t
cells
    animal and plant, 1–2, 1f–2f
    definition, 1
    disintegration, 42
    division, 7
    embryo, 28
    fragments, 39
    functions of, 2t

    hair, 58
    light-sensitive, 54
    membrane, 1–2
    movement of substances into and out of, 2–5
cellulose, 99
cell wall, 1–2
central nervous system (CNS), 59
centre of gravity, 211f
    of irregular shapes, 211, 211f
    of regular shapes, 211
centripetal forces, 211–212, 211f
ceramics, 226–227, 226f
cerebral palsy, 62
cerebrum, 62
certified inspection, 146
cervical cap, 30
charcoal, 122
chemical contaminants, 80
chemical control of pests, 78
chemical energy, 122
chemical equilibrium, 214
chemical fires, 148
chemical messengers, 62
chemical potential energy, 90
chemical reactions, 161
chest cavity, 115, 115f
chest compressions, 117
chlamydia, 67
chlorides, 228, 236
chlorine bleaches, 238
chlorophyll, 98
chloroplasts, 1–2
chromatography, 246, 246f
chromosomes, 7–8
chronic bronchitis, 120
chronic obstructive pulmonary disease (COPD), 120, 120f
cilia, 120
ciliary muscle, 55
Cinder volcano, 188f
circuit breaker, 140, 140f
    placement of, 141
circular motion, 211–212
circulatory system in humans, 38–42
    heartbeat and circulation, 41
citric acid, 238
clay soil, 18–19

cleanliness, 73
coastal erosion, 185
cocaine, 75, 75f
cochlea, 57
cogged wheel, 220, 220f
collecting ducts, 47
colloids, 243, 244t
colon, 111
colorectal (bowel) cancer, 106
colour blindness, 57
comets, 170t
communicable or infectious diseases, 66–68
community, 99
community hygiene, 249–251, 252f
    benefits of, 252f
compact fluorescent lamp (CFL), 143
companion cells, 44
compasses, 203, 203f
composite volcanoes or stratovolcanoes, 187, 187t, 188f
compost, 20
composting, 20, 251
concentration gradient, 3
condensation, 242
condensed oils, 122
condensing point, 122
condoms, 29–30
conduction
    kinetic theory and, 153
    metals and non-metals, 153
    thermal, through solids, 153, 154f
conductors, 134, 134t
connective tissue, 50
conserving energy, 143
constipation, 106
container gardening, 16t
contour farming, 21
contraceptive implant, 31
contraceptive injection, 31
contraceptive patch, 31
contraceptive pill, 31
convection
    applications of convection currents, 155–156, 156f
    kinetic theory and, 154, 155f
    in liquids and gases, 155
converging (convex) lenses, 56
coral reefs, 198

Coriolis force, 183–184
corms, 9
coronary arteries, 40
corrosion, 230
cotyledons, 13
cover crops, 21
COVID-18 vaccines, 70
cranial nerves, 59
crests, 92
crop growers, 9
crop production, 16, 16–17, 16t
crop rotation, 16t, 21
crop yields, 21
cross-pollination, 12
crosswind, 210, 210f
crowning, 29
CRT (cathode ray tube) TVs and screens, 147
crude oil, 122, 122t
crystallisation, 246
crystals, 246
curing, 81
curved reflectors, 92f
cyclones, 184
    tropical, 184t
cystic fibrosis, 70
cytoplasm, 1–2

D

daylight, 143
death rate, 34
decibels or dB, 58
decomposers, 19, 100–101
decompression sickness (bends), 205–206
deficiency diseases, 107
deforestation, 20, 20f
degenerative diseases, 70
dendrites, 60
denitrifying bacteria, 20
density, 200, 200f
deoxygenated blood, 40
deposition, 242
depressants, 75
depression, 70
depth sounding, 204, 204f
dermis, 50, 148
desalination of seawater, 199
desalination plant, 245

desertification, 21
desiccants, 231
desirable traits, 9
diabetes
    causes of, 71
    complications of, 71
    effects of, 71
    risk of skin infections, 71
    symptoms of, 71
    type 1 or insulin-dependent, 71
    type 2 or non-insulin dependent, 71, 106
dialysis machine or dialyser, 49
diaphragm, 30, 115, 115f
diastole, 41
dietary fibre, 103, 106
differentially permeable membrane, 3
diffusion, 3f
    definition, 3
    limitations of, 37
    in natural and artificial environments, 5
    role in living organisms, 3
digestion, 4, 195
    chemical, 110, 111t
    in humans, 107, 108f
    mechanical, 108
digestive juices, 110
digital thermometers, 161
dioxins, 80
direct current (DC), 138
direct solar dryers, 128
disaccharides, 104
distal convoluted tubule, 47
distance multiplier, 215
distillation, 194, 245, 245f
diverging (concave) lenses, 56
domestic refuse, 77
domestic wiring
    codes, 141t
    Earth wire and short circuits, 142, 142f
    for high powered circuits, 142
    live and neutral wires, 142, 142f
    three-core flexible cable, 141, 141f
    for three-pin plug with a fuse, 142, 142f
dragline excavator, 219f
drugs
    nervous system, effects on, 75
    non-prescription, 75
    over-the-counter, 75
    physiological effects on body, 75
    prescription, 75
    psychological effects on body, 75
    recreational, 75
    social and economic effects of use and abuse, 76
dry fruit, 13f
drying agent, 231

E

E. coli bacterium, 65f
ear, 57–59, 57f
    bones, 57
    control of balance and posture, 58
    inner, 57
    middle, 57
    ossicles, 57
    outer, 57
    regions, 57
    structure and functions, 57
ear barotrauma, 206
Earth, 170t, 185
    characteristic features of, 172t
    core, 187
    Coriolis force, 183–184
    day and night, 173
    as magnet, 203f
    seasons, 173
earthquakes, 189
    relationship between volcanoes and, 189t
Earth wire and short circuits, 142, 142f
earthworms, 19
eating disorders, 70
ecologically balanced ecosystem, 101
ecological pyramid, 101, 101f
ecosystem, 99
ecstasy, 75–76
eczema, 71, 71f
effectors, 53, 59–60
    connection between receptor and, 61f
efferent arteriole, 47
efficiency, 215
effort, 215
egestion, 46, 112
egg cells, 24
eggs, 79
elasticity of lens, 55

elastic potential energy, 89
elbow joint movement, 223, 223f
electrical burns, 148
electrical devices, safety features of, 140–141
electrical energy, 90
electrical energy consumption
    average power consumption, 140
    using kilowatt-hour, 138
electrical fires, 148
electrical impulses, 53, 59
electrical power grid, 138
electrical quantities and their SI units, 135t
electric circuits, 134
    symbols, 134t
electricity bills, 139
electricity meters, 139
    analogue, 139, 139f
    digital, 139, 139f
electrocution, 143, 146
electromagnetic wave energy, 90t–91t
    transporting energy between bodies, 92f
electromotive force (emf) of a cell or battery, 135
electronic cigarettes (e-cigarettes), 120
electron microscope, 1
electroplating, 230–231
embolism, 206
embryo, 28
embryonic root, 13
EMF (electromagnetic field) radiation, 147
emissions, 5
emphysema, 120
emulsions, 243
endocrine (hormonal) system, 62–63, 62f
    hormones secreted by, 63t
energy, 89, 112, 114
    alternative sources, 127–131
    chemical, 122
    conservation, 143
    law of conservation of, 92
    in life processes, 103–121
    non-renewable, 127
    renewable, 127
    transformation, transfer and transport of, 91–95, 91f
    types of, 89t–90t
energy-rich foods, 107
environment, 99

environmental sustainability, 101
environment degradation, 35
enzymes, 109–112
    digestive, 110, 111t
    temperature and pH on, effects of, 110
epicentre, 189
epidermis, 50, 148
epididymis, 25
equilibrium
    applications of, 213–214, 213t
    factors affecting, 213f
    neutral, 212, 213f
    stable and unstable, 212, 213f
erythropoietin (EPO), 39
ethanoic (acetic) acid, 81
eutrophication, 198, 250
evaporation, 184, 242, 246
e-waste, 249
excretion
    in flowering plants, 51
    purpose of, 46
    waste products during metabolism, 46
excretory organs in humans, 46t
exercise, 33, 72f
    circulatory system, effects on, 72
    muscle toning, effects on, 72
    physiological effects of, 72
exhalation, 117
exhausts of vehicles, 5
exoplanets, 175
explants, 11, 16
extensor muscle, 222
eye, 53–54, 54f
    cones, 55, 57
    control of light entering, 55, 55f
    detection of light intensity and colour by, 54–55
    image formation, 54
    physical injuries, 57
    retina damages, 57
    rods, 54
    sight defects and their corrections, 56–57

F

faeces, 112
false feet, 39
fat, 112
    cells, 50

fatty acids, 104, 112
female gametes, 12, 24
female reproductive systems, 24, 24f
fertile soil, 19
fertilisation, 28f
    in flowering plants, 12
    in humans, 27, 29
    membrane, 27
fertiliser, 251
filtration, 245, 245f
fire extinguishers, 149, 149f
fires
    classes, 149t–150t
    methods of extinguishing, 150, 150t
    types, 148
fire triangle, 148, 148f, 148t, 150
fishing methods, in Caribbean, 196, 196t
fleas, 77
flexor muscle, 222
flooding, 21
flotation, 200, 201f
flower structure, 11, 11f
fluorescent tube, 143, 144f
foliage leaves, 15
food additives, 80, 107
food allergies, 71
food chains, 99, 100f
    energy transfer in, 100–101, 101f
    examples, 99f
food contaminants, 80
food security and insecurity, 21
food poisoning, 80
food preservation
    methods used for, 81t–82t
    principles used in, 81
food webs, 100, 100f
force multiplier, 215
forces, 89, 89f
    aerodynamic 'lift,' 210, 210f
    contact, 208
    field (non-contact forces), 208
    gravitational, 211
    principles, 208
foreign objects, 80
fossil fuels and their formation, 96, 122, 123f, 132
    advantages, 124
    disadvantages, 124–132
    equation for reaction, 126
    examples of interconversion of energy using, 123f–124f
    harmful emissions from, 126t
    negative effects of, 126–127
    problems associated with, 127t
fovea, 55
fragmentation, 24
free-body diagrams, 208
free diving, 205
freezing, 82
    point, 242
friction
    advantages, 210
    disadvantages, 210
    methods to reduce, 210
fronts
    cold, 181, 182f, 182t
    occluded, 183, 183f
    stationary, 182, 182f
    warm, 181, 182f, 182t
fruit development, 13
fuel filters, 245
fumaroles, 187
fungi, 65, 77, 80
fuse, 140, 140f
    placement of, 141
    three-pin plug with, 142

G

galaxy, 170t
gall bladder, 111
galvanising, 231–232
gametes, 7
gaseous exchange, 117–118, 118f
    effects of smoking on, 120
    in fishes, 119
    in flowering plant, 119, 119f
    in humans, 118
gasohol, 129
gastrointestinal infections, 80
gears and gear ratio, 220
genetically identical offspring, 7
genital herpes, 67
genitalia, 73
    hygiene practices, 73

geostationary satellites, 171
geothermal energy, 130
germination, 14–15, 15f
    growth patterns of annual plants after, 15, 15f
    reproductive phase, 15
    sigmoid shaped or S-shaped growth curve, 15
gestation period (pregnancy), 29
gill filaments, 119
gills of fish, 119
glaucoma, 56
global climate change, 248
global positioning system (GPS), 204
    of satellites, 171, 171f
global warming, 125–126, 248
    effects of climate change due to, 126
glomerulus (glomeruli), 47
glucose, 3–5, 98, 112, 246
glycerol, 104, 112
glycogen, 112
goitre, 107
gonorrhoea, 67
Graafian follicle, 26
grafting, 10, 10f
gravitational attraction, 185
gravitational field strength, 200
gravitational force, 171, 211
gravitational potential energy, 89
greenhouse effect, 248, 250
greenhouse farming, 16t
greenhouse gases, 125, 248
    generation of, 126t
green hydrogen, 131
grey hydrogen, 131
gum disease, 109
gyrocompass, 203, 203f

H

habitat, 99
haemoglobin, 38
hair cells, 58
hair erector muscles, 51
hair follicles, 50
hallucinogens, 76
hard water, 193, 193t
    permanent, 194
    softening, 194
    temporary, 194
hardwoods, 226
hay fever, 71
hazardous biological materials, 249
hazards
    associated with electricity, 146–147
    from faulty electrical equipment, 147
    from faulty gas supplies, 147
    from heated cooking oil, 147
    overloading electrical circuits, 147, 147f
    treatment of victims, 147
    working electrical equipment in damp conditions, 146
    from X-ray machines and radioactive materials, 147
headwind, 210
healthy people, 65
hearing loss, 58
hearing mechanism, 58f
heart, 40–41, 41f
heart beats, 41
heat, 46
    energy, 37, 90
heating, 82
heat transfer processes, 153–159
    applications, 157
    emissive and absorptive properties of different surfaces, 157, 158f
hepatitis B, 67
    prevention and control of, 68
herbivores, 99, 101
heroin, 75
hibiscus plant, 10
high tides, 185
HIV/AIDS, 66
    prevention and control of, 68
homeostasis, 46, 213
hormones, 37, 62–63, 112
houseflies, 77
    control of, 78–79, 79t
    life cycle of, 79
household chemicals
    as acids, bases and salts, 236, 237t
    classification of, 244, 244t
    eco-friendly, 234
    economic use of, 235
    effects on household appliances, 237–238, 238f

safety symbols, 235, 235f
safe use of, 235
uses of common, 233–235, 233t–234t
household hygiene, 78, 78f
Hubble Space Telescope (HST), 175, 176f, 176t
human embryo/foetus, development of, 29, 29f, 30t
human population growth, 34–35
growth pattern, 33–34, 33f–34f
human skeleton
appendicular skeleton, 221
axial skeleton, 221
functions of, 221
structure of, 220–221, 220f
humidity
negative effects of high, 162
negative effects of low, 163
relative, 162
humus, 19
Huntington's disease, 70
hurricane
after, 185
cross-section through, 184f
preparedness, 185
satellite view of, 184f
hydrocarbons, 122
hydrochloric acid, 228, 236
hydro-electric plants, 129–130, 129f
hydro-electric power, 94f
hydrogen, 131
hydrogencarbonates, 236
hydrogen sulfide gas, 250
hydroponics, 16t
hypertension, 70
hypertension or high blood pressure
causes of, 72
complications of, 72
effects of, 72
symptoms, 72
hyphae (hypha), 65, 80
hypocentre, 189
hypothalamus, 48, 50
hysterectomy, 32

I

identical offspring, 9
imago, 79
immature ova, 25

immune system, 68
acquired immunity, 69
artificial immunity, 70
diseases, 70
impact of diseases on, 69
natural immunity, 69–70
immunity
acquired, 68
adaptive, 68
innate, 68
immunodeficiency diseases, 69
implantation, 27, 28f
incandescent (filament) lamp, 143, 144f
incisors, 108, 109f, 109t
inclined plane, 216–217
indirect solar dryers, 128
inefficiencies, 216t
infectious diseases, 33, 66, 250
inflammatory substances, 71
inflatable tubes or rings, 205, 205f
influenza virus, 65f
inhalation, 117
inherited disorders, 70
insects, as pollination agent, 12
insulators, 134, 134t
insulin, 63
internal combustion engine (ICE), 96–97
international public health, 65
International Space Station (ISS), 175, 175f
intra-uterine device (IUD or coil), 31
involuntary actions, 62

J

James Webb Space Telescope (JWST), 175, 176f, 176t
joints, 220–221
ball and socket, 222
cartilaginous, 222
fixed and fibrous, 221
gliding, 222
hinge, 222
movable, 222
partially movable, 222
synovial, 222
joule, 89, 138
Jupiter, 172t
exploration, 177

K

Kick'em Jenny, 188
kidneys, 47–49, 47f
    dialysis for malfunctioning, 49
    main functions of, 47
    osmoregulation and, 48–49
    regions, 47
    tubules, 47
    waster excretion, 106
kilowatt-hour, 138
kinetic energy, 90
Kuiper belt, 170t
kwashiorkor, 107

L

lacteals (lymph capillaries), 111
lactic acid, 114
landers, 177
larva (larvae), 79
larynx, 115, 115f
latent heat of vaporisation, 161
law of conservation
    of energy, 92, 215
    of linear momentum, 214
lead-lined absorbers, 147
lenticels, 51f, 119
Lesser Antilles Volcanic Arc, 188, 188f
lever, 217, 217f–218f
life jackets or personal flotation devices (PFDs), 205, 205f
life rafts, 205, 205f
lifestyle diseases, 70
light-emitting diode (LED), 143
light energy, 98
light microscope, 1
limb movement, 222
limbs, 221
limescale (calcium carbonate), 193t, 194f
limescale removers, 238
linear momentum of a body, 214–215
lipids, 99, 104, 104f
    sources and functions of, 105t
liquid hydrocarbons, 122
liquid-in-glass mercury thermometer, 159–160, 159f–160f
liquid natural gas (LNG), 123
liquid petroleum gas (LPG), 123

live and neutral wires, 142, 142f
living organisms
    active transport importance in, 4–5
    diffusion and, 5
    environmental conditions, importance of, 8–9
    osmosis and, 4
    variation in, 7–8
load, 215
loam soil, 18–19
lobed nucleus, 39
longitudinal waves, 92f
long-sightedness (hyperopia or hypermetropia), 56, 56f
loop of Henle, 47
loudness of a sound, 58
low tides, 185
lung barotrauma, 206
lung cancer, 120, 248
lupus, 69
lymph, 111
lymphocyte memory cells, 69
lymphocytes, 39t, 68

M

machines, 215
    types of, 216–220
macronutrients, 103
magnetic compass, 203, 203f
male gametes, 12, 25
male reproductive systems, 25, 25f
mangrove swamps, 198
marasmus, 107, 107f
mariculture, 195
marijuana, 75–76, 76f, 120
marine organisms, 195
maritime safety standards, 204
Mars, 172t
    atmosphere on, 176t
    exploration, 177
    rovers, 177
mass of object, 200
material resources, 35
matter, 241, 241t
    change of state, 241–242, 241f
mature ovum, 25
mechanical advantage, 215
mechanical control of pests, 78

meiosis, 8, 8f, 25
    role in sexual reproduction, 8
melting point, 242
menopause, 25, 27
menstrual cycle, 26
    events occurring in ovary and uterus during, 26t, 27f
    hormonal control of, 26
menstruation, 26
mental health problems, 70
Mercury, 172t
messenger RNA (mRNA), 70
metabolic rate, 161
metal appliances, cleaning, 238
metals
    properties and uses, 225, 225t–226t
    reactivity of, 228, 229t
meteorites, 170t
meteoroids, 170t
meteors, 170t
methamphetamine, 75
methamphetamine, 75
methane, 248
microbes or microorganisms, 65
    negative effects of, 65–66
    positive effects of, 65
microbial life, 177
micronutrients, 103
microorganisms, 19
    conditions for growth of, 80
    in food spoilage, production and processing, 80
    pH range for, 80
    slow downing or stopping the growth of, 82
micropyle, 15
microwave ovens, 147
Milky Way, 170t
mineral ions, 5
mineral nutrients, 19, 65, 100
minerals, 105, 105t, 106
mitochondria (mitochondrion), 1, 2
mitosis, 7, 7f, 13, 23–24
mixtures, 243, 244t
molars, 108, 109f, 109t
moment of a force, 212
monosaccharides, 104, 112
monosodium glutamate (MSG), 107

Moon
    eclipse of, 173, 173f
    orbital around Earth, 174, 174f
    viewing, 173, 173f
mosquitoes, 77
    control of, 78–79
    life cycle of, 79, 79f, 79t
    vector-borne diseases, 78
motor neurones, 60f, 62
moulds, 65, 66f, 80
mouth, 115, 115f
mouth-to-mouth resuscitation, 117
mulch, 251
multicellular organisms, 1
muscle toning, 72
mycelium, 65

N

narcotics, 76
nasal cavities, 115, 115f
natural contours, 21
natural immunity, 69–70, 70f
    of baby, 70
neap tides, 186, 186f
nematodes, 19
nephrons, 47
Neptune, 172t
nerve endings, 50
nerve fibres, 60
nerve impulses, 53
nervous system, 59–62, 60f
    malfunctioning of, 62
network of blood capillaries, 115, 115f
network of capillaries, 47
net zero-energy (NZE) buildings, 127
neurones, 59–60, 61
    cell body of, 60
    motor, 60, 61
    relay or intermediate, 60
    sensory, 60–61
neurosis, 70
neutral equilibrium, 212, 213f
neutralisation reactions, 237
Newton's laws of motion, 209
    applications, 209
nicotine, 75
night blindness, 107

nitrates, 19–20, 236
nitric acid, 236
nitrifying bacteria, 20
nitrogen-fixing bacteria, 20
nitrogen narcosis, 206
nitrogenous waste, 37, 46
nitrogen oxides, 248
nodes, 10
non-aqueous solutions, 243
non-biodegradable waste, 249
non-communicable diseases (NCDs), 70
non-metallic materials
    properties and uses, 226–227, 227t–228t
non-metallic protective coatings, 231
non-prescription drugs, 75
non-renewable resources, 34
northern or southern lights, 243, 243f
nuclear equations, 96
nuclear fission, 96, 96f
nuclear fuelback-up, 132
nuclear fusion, 96, 96f
nuclear plants, 130–131
nuclear potential energy, 89
nucleoid, 65
nucleus, 1–2, 38
nutritional deficiency diseases, 70

O

obesity, 70, 72, 106–107
occluded front, 183, 183f
occupation, dietary needs and, 106
oestrogen, 26–27, 63
Ohm's law, 135
omnivores, 99
on-site sewage system, 251
opioids, 76
orbiters, 177
organelles, 1
organic bile salts, 111
organic compounds, 98
organic farming, 16t
osmoregulation, 48–49, 49f
osmoregulation, 47
osmosis, 3f, 81, 192–193
    definition, 3
    in living cells, 4
    in living organisms, 4
    using Visking tubing, 4f
    water content in, 3
osteoporosis, 70, 106
outgrowths, 9
ovaries, 24, 63t
overcrowding, 35
over-the-counter drugs, 75
oviducts, 27
ovulation, 25–26
ovum (ova), 24–26, 25f–26f
oxalic acid, 238
oxidising agents, 238
oxygen, 3, 37, 81, 98–99, 117, 117t, 118–120
    content as condition to germinate, 14–15
    debt, 114
    removal for extinguishing a fire, 150
oxygenated blood, 40
oxytocin, 29

P

pacemaker, 41
pain, 76
pancreas, 63t
parallel circuit
    of appliances in domestic wiring, 138, 138f
    current in, 136
    voltage in, 136, 137f
paralysis, 62
parasites, 76–78, 250
parasitic protozoans, 80
Parkinson's disease, 62, 70
parthenogenesis, 24
particulate matter, 248
pasteurisation, 82
pathogenic diseases, 66
pathogenic microorganisms, 250
pathogens, 65–66, 68, 70, 77, 80
pectoral girdle, 221
pelvic girdle, 221
peptide links, 104
perennating organs, 9, 10
perimenopause, 27
peripheral nervous system (PNS), 59
personal hygiene, 73
    benefits of, 73
pesticides, 78

pests, 76, 250
    household, 76, 77f
    methods of control, 78, 78f
petals, 12–13
petroleum, 239
pH, 19, 110, 192t
    optimum, 110
    range for microorganisms, 80
    of solution, 236
    of treated wastewater, 237
phagocytes, 39t
phagocytosis, 39
pharynx, 115, 115f
phloem, 43
    sieve tubes, 44, 44f
    tissue, 44
phosphates, 236
phosphoric acid, 238
photochemical reaction, 98
photoreceptors, 54
photosynthesis, 3–4, 15, 98–99, 101, 119, 195, 248
    equations, 98
    products of, 98–99
    substrates and conditions for, 98
photovoltaic (PV) systems, 127, 127f
physical contaminants, 80
physical disabilities, 62
pickling, 81
pitch of sound, 59
pituitary gland, 48, 63t
placenta, 28–29
planetoids, 170t
planets, 170t
    characteristic features of, 172t
    gas giants, 172
    terrestrial, 172
plant cells, 1–2
    structure, 2
plantlet, 11
plant nutrients, 250
plasma, 38, 242–243
    membrane, 1
plasticisers, 80
plastics, 227, 250
    advantages of using, 252
    in medical and construction industries, 253
    negative effects of, 253
    recycling, 253–254
platelets (thrombocytes), 38, 39t
pleural fluid, 115, 115f
pleural membranes, 115, 115f
plimsoll line, 203, 203f
plumule (embryonic shoot), 13, 15
plutonium, 130
polar satellites, 171
pollen grains, 12
pollination, 11
pollutants, 180t
pollution, 250
polychlorinated biphenyls (PCBs), 80
polymers, 227
polysaccharides, 104
population, 99
posterior vena cava, 40
postnatal care, 33
potassium hydroxide, 236
potential difference (pd), 135
poverty, 35
power outlet sockets, 138, 138f
predators, 78
pregnancy, 28–29
premolars, 108, 109f, 109t
prenatal (antenatal) care, 32–33
prenatal check-ups, 32
prescription drugs, 75
preservatives, 81, 107
pressure filtration, 48
primary consumer, 99
primary follicle, 25
principle of moments, 212
progesterone, 26–28, 63
protective gear/wear in work and in sport, 150, 150t–151t
protein-energy malnutrition, 70
protein-energy malnutrition (PEM), 107
proteins, 99, 104, 104f
    sources and functions of, 105t
protozoans, 65–66, 77
pseudopodia, 39
puberty, 25–26, 33–34
pulleys, 219, 219f
pulmonary artery and aorta, 40–41, 115, 115f
pulmonary veins, 40–41, 115, 115f

pupa (pupae), 79
pyramid of biomass, 101
pyramid of energy, 101
pyramid of numbers, 101

R

radicle, 13, 15
radioactive materials, 147
radio detection and ranging (radar), 204
radius of curvature, 211
rainfall, 184
rainwater, 195
rats, 77
receptacle, 12
receptor, 53
receptor cells, 53
recompression, 206
recreational drugs, 75
red blood cells (erythrocytes), 38–39, 38f
red-green colour blindness, 57
reflex actions, 62
refrigeration, 82
refuse, 251
relative humidity, 162
renewable resources, 34
reproductive organs, 24
rescue breathing, 117
resistance of rocks or bushes, 135
resource-based conflict, 35
resource shortages, 35
respiration, 46, 72, 112–113, 120
    aerobic, 113
    anaerobic, 114
respiratory surfaces, 120t
    in fishes, 119
    in flowering plant, 119
    in humans, 118
respiratory system, 115, 115f
reversible chemical reaction, 214
Rhesus disease, 42
rheumatoid arthritis, 69
rhizomes, 9
rib cage, 115, 115f
ribosomes, 1–2
ribs and rib cage, 115, 115f, 221
Richter scale, 189, 189t
rickets, 107

rising, 200, 201f
    in relation to density, 202–203
rootstock or stock, 10
roughage, 103
rovers, 177
7Rs of waste management, 251t
rusting, 231, 233f
    factors affecting rate of, 231
    methods to reduce or prevent, 231
rust removers, 238

S

Saharan dust storms, 181
salting, 81
salts, 107, 236
sandy soil, 18–19
sanitary control of pests, 78
satellite navigational systems, 171
satellites, 171
Saturn, 172t
schizophrenia, 70
scion, 10
scouring powders, 238
scuba diving, 205
scurvy, 107
secondary consumer, 99
second convoluted (coiled) tubule, 47
second-hand smoke, 121
sediment, 21
sedimentation test, 18, 18f
seed development, 13
seismograph, 189, 189f
selectively permeable membrane, 3
selective reabsorption, 48
self-pollination, 12
semen, 27
semicircular canals, 57–58
semiconductors, 134, 134t
sense organs in humans, 53–63, 53t
sensory neurone, 60f
sepals, 12–13
separation techniques, 245–246
septum, 40
series circuit
    current in, 136
    voltage in, 136, 137f
sett, 10

severe combined immunodeficiency (SCID), 69
sewage, 77
    disposal practices, 251
    treatment plant, 251
sex cells, 7, 24
sex-linked characteristics, dietary needs and, 106
sexually transmitted infections (STIs), 29, 33, 66–68, 66t–68t
    causes, signs/symptoms and treatment of, 66–68
    prevention and control of, 68
sexual reproduction, 7–8
    in flowering plants, 11–13
    in humans, 24–33
    by implantation, 27, 28f
shield volcano, 188f
shock waves (seismic waves), 189
short-sightedness (myopia), 56, 56f
sickle cell anaemia, 70
simple pendulum, 95, 95f
sinking, 200, 201f
    in relation to density, 202–203
skin, 49–50
    excretion and, 50
    structure, 50
    temperature control and, 50
skull, 221
sludge, 251
smog, 5
smoke, 5
smoke-free environments, 121
soakaway, 251
soap scum or scum, 239
soapy and soapless detergents, 238–239, 239t, 240f
social inequality, 35
sodium dodecyl sulfate, 239
sodium hydroxide, 236
sodium octadecanoate (sodium stearate), 239
soft water, 193, 194t
softwoods, 226
soil, 16–21
    bacteria cycle, 20
    degradation, 21
    organisms, 19
    productivity, 21
    profile, 17, 18f
soil erosion, 21
    causes, 20
    effects of, 21
    effects of, 21
    impact on food production, 21
    natural and human-induced factors, 20
    prevention of, 21
soil fertility, 18–20
    composting and, 20
    contribution of bacteria and fungi, 65
    cycling of nitrogen and, 20
    humus and, 19
    organic matter and mineral nutrient content, 18
    types of soil and, 18, 19t
solar cookers, 128
solar dryers, 82f, 128, 128f
solar energy, 127–128, 127f, 127t
    variables affecting, 131
solar systems, 127t, 172, 172f
solar water heater
    heating cycle, 156
    usage cycle, 157
solid waste, 77
solute, 243
solutions, 243, 244t
solvents, 243
sorbates, 82
sound energy, 90
sound navigation and ranging (sonar), 204
sound waves, 58–59, 58f, 92f
space exploration, 174, 174t
    challenges, 175t
species, 99
sperm, 25
sperm cell, 26
spinal cord, 60–61, 61f, 62, 221
spinal nerves, 59, 61
spine, 221
spores, 80
sporting equipment, 227, 227t–228t
spring tides, 185, 185f
stable equilibrium, 212, 213f
    factors affecting, 213, 213f
stamens, 12–13
starch, 99
stem tubers, 9, 9f
stigma, 13
stimulants, 75

stimulus, 53
stolons, 9
stomata (stoma), 51f, 119
stored food, 9
storm surge, 184
stress, 70
stretch receptors, 29
strip planting, 16t
stroke, 62
sublimation, 242
submarines, 201, 201f
submarine volcano, 188
succulent fruit, 13f
sucrose, 37, 99
sugar cane stem, 10
sugar solution, 81
sulfamic acid, 238
sulfates, 228
sulfites, 82
sulfur dioxide, 82, 248
sulfuric acid, 228, 236
Sun, 170t
    eclipse of, 174, 174f
surface-active agents, 238
surface area to volume ratio of an organism, 37
surfactants, 238
survival in cross-pollination, 12
suspensions, 243, 244t
sweat and sweat glands, 50, 51f
    sweat formation, 50f
synthetic oxygen-carrying chemicals, 39
syphilis, 67
systole, 41

T
tailwind, 210
tarnishing, 230
    methods to reduce or prevent, 231
tectonic plates, 130, 187
teenage pregnancy, 35
teeth
    tooth care, 109
    types, 108, 109f, 109t
temperature of a body, 159
temperature regulation in humans, 161
temperature sensors, 161

tendons, 222
terracing, 21, 21f
terrestrial environment, 180
terrestrial planets, 172
tertiary consumer, 99
testa, 13
testes, 25, 63t
    seminiferous tubules of, 25
testosterone, 63
tetanus vaccine, 70
textiles, 227
thermal conduction, 153, 154f
    good and poor thermal conductors, 154
thermal energy, 90
thermal radiation, 156
    emitters and absorbers of, 156, 156f
thermometers, 159
    digital, 161
    liquid-in-glass, 159–160
thermostats
    of electric iron, 158, 158f
    in electric oven, 158, 158f
    of gas oven, 159, 159f
thorax, 115, 115f
three-core flexible cable, 141, 141f
throat, 115, 115f
thunderstorms, 184
thyroid gland, 63t
thyroxine, 63
ticks, 77
tidal barrage generators, 130
tidal power, 196
tidal stream generators, 130
tidal waves, 186, 186t
tides, 185
tissue culture, 11, 16t
tobacco, 120
tooth decay, 109
tooth-healthy foods, 109
topsoil, 20
torrential rains, 184
toxic chemicals, 65, 250
toxins, 70
trachea, 115, 115f
transesterification, 129
transfusing blood, 39

transpiration, 43
    environmental factors affecting, 45
    in photosynthesis, 45
    pull, 44
    role in plants, 44–45
transport system in multicellular organisms
    materials transported, 37
    need for, 37
transport system in plants, 37, 43–45
    active transport, 44
    soluble organic food, 44
transverse waves, 90, 92
    along rope, 92f
triceps contracts, 223
trilateration, 204
trophic level, 99
troughs, 92
tsunamis, 186, 186t
tubal ligation, 32
turgid, 4
type 1 diabetes, 69
type 2 diabetes, 70

U
ultra-filtration, 48
ultra-high temperature treatment (UHT), 82
ultrasound scans, 32
umbilical artery, 28
umbilical cord, 28–29
umbilical vein, 28
unbalanced diet, 107
underground structures, 9
unemployment, 35
unicellular organisms, 1
universe, 170t
unprescribed drugs, 33
unstable equilibrium, 212, 213f
upthrust, 200–201, 201f
uranium, 130–131
Uranus, 172t
urea, 37
urine, 48, 48f
useful work (energy) output, 215

V
vaccination and vaccine, 70
vacuole, 1, 2

vacuum flask, 157, 157f
valves, 40
variation, 7–8
    in cross-pollination, 12
vascular tissue in plant, 43f
vasectomy, 32
vasoconstriction, 51
vasodilation, 51
vectors, 77, 250
    of disease, 76
vegetative growth, 15
vegetative propagation
    artificial, 10–11
    natural, 9–10
veins, 40
ventilation, 161–162
    of enclosed spaces, 162
    mechanical, 162
    natural, 162f
ventricles, 40
ventricular systole, 41
venules, 40
Venus, 172t, 176
vertebrae, 61, 221
vertebral column, 221
vestibular nerve, 58
villi, 28
viral vectors, 70
viruses, 65, 66, 77, 80
vitamin and mineral deficiency diseases, 70
vitamins, 105, 105t
volatile organic compounds, 248
volcanic ash and gases, 181
volcanic eruptions, 5, 187
    types of, 187
volcanoes, 187
    ecological consequences of, 189t
    relationship between earthquakes and, 189t
    types of, 187t
voltage (energy per unit charge), 135
voltage and power rating of an appliances, 135

W
waste
    biodegradable, 249
    biohazardous, 249
    biological, 249

chemical, 249
disposal, 77, 77f
domestic, 249
electronic, 249
food, 77
impact of improper disposal of, 249–250, 250f
industrial, 249
non-biodegradable, 249
organic, 78
organic matter, 251
recommended practices for management, 250, 251t
solid, 77
substances, 37
wastage, 195
wasted energy, 215
waste energy, 143
water, 103
    chemical properties of, 192t
    conservation, 195
    content as condition to germinate, 14–15
    filtration systems, 245
    freezing point of, 159
    fresh, 192
    hard, 193–194, 193t
    heater systems, 128
    as household chemical, 233
    for hydration, 106
    importance to health, 106
    latent heat of vaporisation, 161
    physical properties of, 192t
    pollution, 197–198, 197t–198t
    production by respiration, 46
    reabsorption, 48
    reactions of metals, 228
    safety devices, 204–205, 205f
    seawater, 192
    soft, 193, 194t
    sources for domestic use, 198–199, 199f, 199t
    thermal conduction of, 154, 154f
    transportation in plants, 43
    uses of, 195–196
wave energy, 92
    plants, 130
wave power, 196
weakened or dead pathogens, 70
weight gain, 106
weight loss, 106
weight of object, 200
    of floating body, 200
well-being, 73
wetland ecosystems, 198
white blood cells (leucocytes), 38, 38f, 39t
white hydrogen, 131
wind, as pollination agent, 12
windbreaks, 21
wind energy, 131
wind farm, 129
wine-making industry, 245
wood, 226
work, 89, 215
World Health Organization (WHO), 65
worm casts, 19

X
xylem, 43
    tissue, 43
    vessels, 43, 44f

Y
yeast, 65

Z
zero energy buildings (ZEB), 127
zinc coating, 232–233, 233f
zygotes, 13, 27